Global Development in the Arctic

Viewing the Arctic as a key region for global development in the 21st century, this book offers a cross-disciplinary conceptual framework for understanding what international cooperation is, why it is difficult and what kind of alternative views can apply in the Arctic.

Written by Arctic experts, the book presents major trends and scenarios for international cooperation in the Arctic up to 2035 and future prospects for international cooperation in the Arctic in various sectors: energy, business and economy, transportation and logistics, climate change, diplomacy and security, culture, innovations, higher education and research. Implications of the scenarios for global development are discussed in the light of the United Nations Agenda for Global Development and Sustainable Development Goals. The book offers a cross-disciplinary conceptual framework of international cooperation in the Arctic and discusses implications of this framework for global development.

Filling the gap in analytical understanding of international cooperation, this book will be of interest to academics, students and professionals concerned with global development and the Arctic region.

Andrey Mineev, PhD, is a Postdoctoral researcher at Nord University, Norway.

Anatoli Bourmistrov, PhD, is Professor and Head of Section, Bodø Graduate School of Business, Nord University, Norway.

Frode Mellemvik, PhD, is Director of the High North Center, Nord University, Norway.

Routledge Explorations in Development Studies

This Development Studies series features innovative and original research at the regional and global scale. It promotes interdisciplinary scholarly works drawing on a wide spectrum of subject areas, in particular politics, health, economics, rural and urban studies, sociology, environment, anthropology, and conflict studies.

Topics of particular interest are globalization, emerging powers, children and youth, cities, education, media and communication, technology development, and climate change.

In terms of theory and method, rather than basing itself on any orthodoxy, the series draws broadly on the tool kit of the social sciences in general, emphasizing comparison, the analysis of the structure and processes, and the application of qualitative and quantitative methods.

Leadership and Organisational Culture in Development
Challenging Exceptionalism
Violeta Schubert

New Leadership of Civil Society Organisations
Community Development and Engagement
Edited by Ibrahim Natil

European Civil Society and International Development Aid
Organisational Incentives and NGO Advocacy
Balázs Szent-Iványi

Global Development in the Arctic
International Cooperation for the Future
Edited by Andrey Mineev, Anatoli Bourmistrov and Frode Mellemvik

For more information about this series, please visit: www.routledge.com/Routledge-Explorations-in-Development-Studies/book-series/REDS

Global Development in the Arctic

International Cooperation for the Future

Edited by
Andrey Mineev,
Anatoli Bourmistrov
and Frode Mellemvik

Routledge
Taylor & Francis Group
LONDON AND NEW YORK

First published 2023
by Routledge
4 Park Square, Milton Park, Abingdon, Oxon OX14 4RN

and by Routledge
605 Third Avenue, New York, NY 10158

Routledge is an imprint of the Taylor & Francis Group, an informa business

British Library Cataloguing-in-Publication Data
A catalogue record for this book is available from the British Library

Library of Congress Cataloging-in-Publication Data
Names: Mineev, Andrey, editor. | Bourmistrov, Anatoli, editor. | Mellemvik, Frode, editor.
Title: Global development in the Arctic : international cooperation for the future / edited by Andrey Mineev, Anatoli Bourmistrov and Frode Mellemvik.
Description: Abingdon, Oxon ; New York, NY : Routledge, 2023. | Includes bibliographical references and index. |
Identifiers: LCCN 2022032221 (print) | LCCN 2022032222 (ebook) | ISBN 9781032158624 (hardback) | ISBN 9781032158655 (paperback) | ISBN 9781003246015 (ebook)
Subjects: LCSH: Arctic regions—Economic development. | Arctic regions—International cooperation. | Economic development—Arctic regions.
Classification: LCC HC735 .G56 2023 (print) | LCC HC735 (ebook) | DDC 330.911/3—dc23/eng/20221025
LC record available at https://lccn.loc.gov/2022032221
LC ebook record available at https://lccn.loc.gov/2022032222

ISBN: 978-1-032-15862-4 (hbk)
ISBN: 978-1-032-15865-5 (pbk)
ISBN: 978-1-003-24601-5 (ebk)

DOI: 10.4324/9781003246015

Typeset in Goudy
by codeMantra

Contents

Figures

Tables

Contributors

Evgenii Aleksandrov is an Associate Professor at the High North Center for Business and Governance, Nord University Business School in Bodø, Norway. His research interests include public sector accounting and budgeting, performance measurement and management, dialogic accounting, public sector reforms, democratic governance, citizens' involvement, Smart Cities governance and sustainable development.

Natalia Andreassen is an Associate Professor of Organization and Management at Nord University Business School and NORDLAB, Nord University's innovation center for emergency preparedness. She is a leader of University of the Arctic Thematic Network on Arctic Safety and Security. Dr. Natalia Andreassen has been involved in the projects on maritime preparedness in the Arctic "MARPART", coordinated by the High North Center for Business and Governance. She holds a doctorate from Nord University and a Master's degree from St. Petersburg State University of Technology and Design. She is conducting research within the field of emergency management with special focus on maritime emergency preparedness organization, incident response management and risk in the Arctic.

Alexei Bambulyak is a Senior Adviser, Manager Russia & Eastern Europe at Akvaplan-niva company and a doctoral candidate at the Norwegian University of Science and Technology (NTNU). He has 25 years' experience in managing international research and capacity building projects in the Arctic and the Barents Region under international, bilateral and national programs and foundations, including the Arctic Council, the Barents Euro-Arctic Council and the Joint Norwegian-Russian Commission on Environmental Cooperation. Mr. Bambulyak is a Russian certified environmental auditor, expert in the international and regional working groups on environment. His primary research interests are in environmental risk assessment, Arctic shipping and international cooperation.

Odd Jarl Borch is Professor of Strategy and Management at the Nord University Business School, Bodo, Norway, and the leader of NORDLAB – the Nord University Innovation Center for Safety and Security. He holds a University of the Arctic chair in Societal Safety and Security. He received his MSc from the Norwegian School of Economics and Business Administration in Bergen, Norway, in 1979 and his PhD from Umea University in Sweden in 1990. He has a Master Mariner's education from Bodin Maritime Academy. Dr. Borch is conducting research in the field of safety, security and emergency preparedness, with special emphasis on the Arctic.

Anatoli Bourmistrov is a Professor and Leader of the MACE Research Group at Nord University Business School (Norway) and Research Coordinator at the High North Center for Business and Governance. He has a doctoral degree from the Norwegian School of Economics (NHH, Norway). He and his colleagues were awarded the Research and Innovation Prize from Nord University for 2020; the Outstanding Paper award in the 2018 Emerald Literati Awards in the Journal of Accounting & Organizational Change; and the David Solomons Prize for the best paper in the annual volume of Management Accounting Research for 2013. His research publications cover a wide range of topics, including but not limited to public sector accounting and performance management reforms, management control innovations in public and private sectors (including Beyond Budgeting), relationships between information and managerial attention, and issues of long-term planning via use of scenarios.

Ken Coates, raised in Canada's Yukon Territory, is Canada Research Chair in Regional Innovation, Johnson-Shoyama Graduate School of Public Policy and Munk Senior Fellow with the Macdonald-Laurier Institute. He has published extensively on northern and Indigenous issues and is active in Circumpolar comparative studies. Most recently, his has coauthored the book *From Treaty People to Treaty Nation: A Roadmap for all Canadians.* He was the cochair of the Thematic Network on Circumpolar Innovation for the University of the Arctic. He is currently working on a comprehensive study of innovation in the Circumpolar world, emphasizing the need for a practical vision for the application of new technologies in northern settings.

Salve Dahle is Project Director, Digital Solutions at Akvaplan-niva Ltd. He is educated as a marine biologist and has worked for more than 30 years studying marine ecosystems in Norwegian and Russian Arctic and human impact on the environment. This includes studies of marine biodiversity and distribution and pathways of pollutants and their accumulation in biota. This information has been made available to management and the private sector for advice on Environmental Impacts Assessments and monitoring in connection with development of offshore industries. Working across borders has made it clear to him that cooperation between scientists, businesses and countries is a must.

Jan Dietz is a scenario practitioner and strategy consultant working through Dietz Foresight (Norway). He holds the degree of Master of Management from the BI Norwegian Business School. His varied career includes being responsible for foresight methodology at the Research Council of Norway and working for the Ministry of Defence as an analyst. Dietz is an experienced facilitator of scenario processes and has worked extensively in Northern Norway, helping both business clusters and national and local authorities explore future opportunities and threats. He is also a frequent speaker at conferences and other events.

Elena Dybtsyna, PhD, is an Associate Professor at the Business School and High North Center for Business and Governance at Nord University (Norway). Her research and teaching focus on the Arctic and are in the areas of international governance and business, sustainable and smart cities, circular economics and management control in the context of the private and public sectors. She recently coauthored chapters in the Palgrave *Handbook of Arctic Policy and Politics* and the Routledge *Advances in Tourism: Asian Mobilities Consumption in a Changing Arctic.*

Bjarge Schwenke Fors is the Head of Department at the Barents Institute at UiT – The Arctic University of Norway. He holds a PhD in social anthropology with a specialization in border studies. His research relates to identity construction, cross-border interaction, and political and cultural performances in borderlands. Dr. Fors has conducted extensive ethnographic research on the Norwegian-Russian border. He is a member of the research group Russian Space: Concepts, Practices, Representations.

Ove Tobias Gudmestad has Master degree in Applied Mathematics and a DSc in Hydrodynamics from the University of Bergen in Norway. His research interests relate to marine and maritime engineering, as well as to cold climate engineering. Previously, he has been working in Statoil (now Equinor, the national oil company of Norway) and as Full Professor of Marine Technology at the University of Stavanger, where he now enjoys the status of Professor Emeritus. Mr. Gudmestad is Adjunct Professor Emeritus in Cold Climate Engineering at UiT – The Arctic University of Norway and Adjunct Professor of Maritime Engineering at Western Norway University of Applied Sciences, Haugesund. He has been involved in cooperation with Russian universities since the mid-1990s.

Björn Gunnarsson is an Associate Professor and senior researcher at the Center for High North Logistics (CHNL) at the Nord University Business School (Norway). He holds a PhD in geochemistry from Johns Hopkins University (USA). His research interests include Arctic maritime transportation and logistics, transport infrastructure development and socioeconomic development of the Arctic regions.

Larry Hinzman is the Assistant Director of Polar Sciences in the White House Office of Science and Technology Policy and the Executive Director of the Interagency Arctic Research Policy Committee. He recently served as the Vice Chancellor for Research and as a Professor of Civil and Environmental Engineering at the University of Alaska Fairbanks (UAF). Hinzman served as the Director of the UAF International Arctic Research Center from 2007 to 2015. Professor Hinzman's primary research interests involve permafrost hydrology.

Kim Holmén is a Senior Advisor at Norwegian Polar Institute. He received a PhD in Chemical Meteorology from Stockholm University in 1987 and has worked with human perturbations of the climate system and how they relate to natural climate variations. His main research focus is on change in the Arctic, but his work includes some projects in Antarctica and the Himalayas. Based in Tromsø and Svalbard since 2003, he has served as Research Director and International Director at the Norwegian Polar Institute. This work has kindled an interest in how scientific results and insights are used to shape societal progress and development. His present interests are in how pure science-based policy recommendations interact with normative and value-based expressions in different cultures and societies and how such a dialog can be respectful and meaningful.

Carin Holroyd is a Professor at the Department of Political Studies, University of Saskatchewan. She was involved in the establishment of the University of Northern British Columbia, where she negotiated agreements with several Arctic partners and has traveled extensively in the Canadian and Circumpolar North. In recent years, she has worked on aspects of innovation in the Circumpolar world, looking at the intersection of national innovation policies and practical northern developments, and on East Asia's involvement in the Arctic. She was a coeditor (with Ken Coates) and contributor to the *Palgrave Handbook on Arctic Policy and Politics*. Carin's work on Arctic innovation builds on her studies of the political economy in East Asia. Her major works include *Government, International Trade and Laissez Faire Capitalism*, *Green Japan: Environmental Technologies, Innovation Policy and the Pursuit of Green Growth* and the coauthored books *Japan and the Internet Revolution*, *Innovation Nation: Science and Technology in 21st Century Japan* and *Digital Media in East Asia: National Innovation and the Creation of a Region*.

Igor Khodachek is Associate Professor in Social Sciences at the High North Center for Business and Governance at Nord University Business School (Norway). He holds Master of Science in Geography from Saint Petersburg State University and PhD in Business from Nord University. His research interests include governance and management in public sector organizations, strategic planning in organizational and spatial dimensions, sustainable regional development, urban governance and management with a focus on smart city initiatives. Dr. Khodachek also has an extensive fieldwork and expedition experience in the Russian Arctic, Northern Norway and Svalbard.

Stefan Kirchner is Research Professor of Arctic Law and the Head of the Arctic Governance Research Group at the Arctic Centre of the University of Lapland in Rovaniemi, Finland. The focus of his work is the governance of the Arctic through international law, with particular attention to the intersection of human rights and the protection of the natural environment, especially through the international law of the sea. He has taught international law at universities in Finland, Greenland, Germany, Italy, Lithuania and Ukraine, and pursues a practice-oriented approach to governance research that is based on his work experience as a practitioner of international law in government, public administration and international law firms.

Timo Koivurova is a Research Professor at the Arctic Centre of the University of Lapland, Finland. He has a multidisciplinary specialization in Arctic law and governance, but has also conducted broader research on global law. He has been involved as an expert in several international processes globally and in the Arctic region.

Andrey Krivorotov is an Assistant Professor at Odintsovo branch, Moscow State Institute of International Relations (MGIMO University), Russia, and has participated in several cooperation efforts between MGIMO and Nord University. He holds degrees in Economics and Law, plus a PhD in International Economics. His main research interests include socioeconomic development of the Arctic (with a strong focus on Russia and Norway), the oil and gas industry in Russia and the Caspian region, and China in the Arctic. He is a member of the International Arctic Social Science Association (IASSA) and an expert of the Russian International Affairs Council (RIAC).

Lars-Henrik Larsen, PhD, is Section Manager in Akvaplan-niva, Tromsø, Norway. Larsen has been with Akvaplan-niva since 1990 and has worked with environmental impact assessments and monitoring of Arctic and marine ecosystems. Larsen has long experience in cooperation with Russian colleagues and institutions. He has field experience from the Barents Sea, the Kara Sea and the White Sea. In an era of climate change, cross-border and international understanding and cooperation are becoming of even greater importance.

Frode Mellemvik is the Director of the High North Center for Business and Governance and a Professor at Nord University Business School. Mellemvik holds a PhD and has written books and articles on management and High North-related topics. He has worked at academic institutions in Europe and the USA, and is Honorary Doctor and Honorary Professor at several universities. Mellemvik has been rector of Bodø University College, today Nord University, for ten years. He has participated in several boards and councils, for instance, in the Norwegian Government's Expert Committee for the High North, been Chair of the Norwegian Government's High North Council and is the representative of Norway in the Social, Economic and Cultural Expert Group in SDWG, Arctic Council.

Andrey Mineev is a researcher at the High North Center for Business and Governance at Nord University Business School (Norway). He holds a PhD in Business Economics from Nord University Business School and a Master's degree in IT from the Baltic State Technical University in Saint Petersburg (Russia). His research interests include socioeconomic development of the Arctic regions, business networks and strategic management control. Since 2005, he has been involved in cooperation projects in the sector of research and education between Norway and Russia.

Arild Moe is a research professor at the Fridtjof Nansen Institute (Norway). His degree in Political Science is from the University of Oslo, where he also studied Russian language and public law. Most of his research has been devoted to Russia, especially the energy sector, energy and climate politics, and Arctic shipping, with an emphasis on the political and legal conditions for navigation on the Northern Sea Route. His work also includes Russian-Norwegian relations in the North and Norwegian High North policies.

Denis Moiseev is a research director at Murmansk Marine Biological Institute of Russian Academy of Sciences – MMBI RAS (Russia). He holds a PhD in Oceanography from MMBI RAS and a Master's degree in Oceanography from Saint Petersburg State University (Russia). His research interests include ocean physics, climate geography, bio-oceanology, GIS, remote sensing, cartography, radioecology and nature management. Since 1997, he has been involved in cooperation projects in the sector of research and education between Norway and Russia.

Leif-Kristian Monsen is a docent in political science and has held a full-time position at the Faculty of Social Sciences, Nord University since 1982. Monsen has primarily taught leaders from various fields of practice, and his research has been related to organization and management. His publications mainly deal with leadership, change, innovation and development in the public sector.

Nadezda Nazarova holds a PhD in Business and is an Associate Professor at the Nord University Business School, Norway, where she has also contributed to many research and education cooperation projects at the High North Center for Business and Governance. With a background in accounting and management control, her research interests have covered the role of human and nonhuman actors in organizations in the context of logistics, supply chain management, risk management and smart societies in the High North.

Petter Nore is a Professor at the High North Center for Business and Governance at Nord University Business School (Norway). He also works as a senior consultant at the Norwegian Agency for Development Cooperation (NORAD). He has a broad background ranging from Senior Vice President and President of Hydro Russia to Director of Gas and Strategy in the Norwegian Ministry of Oil and Energy and Chief Energy Analyst in the Norwegian Ministry of Foreign Affairs. Dr. Nore has also worked as Team Leader for Natural Gas in the World Bank. He serves on the international board of the New Energy Coalition in Holland. His main research interest centers around decarbonization, hydrogen and geopolitics.

Andreas Østhagen is a Senior Research Fellow at the Fridtjof Nansen Institute. He is also an affiliated Senior Researcher at the High North Center for Business and Governance at Nord University, a Senior Fellow at the Arctic Institute and a Global Fellow at the Wilson Center in Washington DC. His work currently focuses on maritime disputes and resource management in the

North and beyond, under the larger framework of international relations and Arctic geopolitics.

Rolf Rødven, PhD, MBA, is the Executive Secretary of the Arctic Monitoring and Assessment Program (AMAP). Rødven holds a PhD in Northern Populations and Ecosystems and an MBA in Strategic Leadership and Finance from UiT – the Arctic University of Norway. He has been authoring several scientific papers on Arctic socio-ecological systems. His previous positions include Research Director and Director at the Norwegian Institute of Agricultural and Environmental research – Northern department, and Research Director at the Norwegian Institute of Bioeconomic research, as well as leading positions in environmental management.

Elena Zhurova Sæther is an adviser at the High North Center for Business and Governance at Nord University Business School. She was raised in Saint Petersburg and graduated from the Baltic State Technical University. She holds a PhD in Business Economics from Nord University. Her research interests include sustainability and cooperation models in the oil and gas business.

Yngvar B. Steinholt is Associate Professor of Russian Culture at the Institute of Language and Culture, UiT – The Arctic University of Norway. Dr. Steinholt is cofounder of the research group Russian Space: Concepts, Practices, Representations (RSCPR) at UiT. He teaches Russian literature, history, media and contemporary culture. Author and coauthor of numerous articles and the books *Rock in the Reservation* (MMMSP 2005), *Punk in Russia* (Routledge 2014) and *A War of Songs* (Ibid. 2019), Steinholt's many research interests include Soviet and Russian (as well as Ukrainian and Belarusian) popular music, literature, contemporary art, art activism and Norwegian-Russian cultural relations.

Marit Sundet is full-time Professor of Sociology at High North Center for Business and Governance, Nord University. For many years, she was a member of the Academic Leadership Team of University of the Arctic. Sundet's main scholarly interest is international cooperation between higher education institutions (HEIs), which has captivated her for nearly 20 years. Her empirical studies in this field have been published in various journals, revealing that such cooperation is rooted in very different political, social, economic and cultural conditions and that these conditions are prerequisites for the individual participating in international networks.

Roman Vakulchuk is a Senior Research Fellow at the Norwegian Institute of International Affairs (NUPI) in Oslo. He also works as a researcher at Nord University (Norway) focusing on energy transition and foresight methodology. He holds a PhD degree in Economics from Jacobs University Bremen in Germany. His major research areas are economic transition, trade, energy, climate change and investment policy in emerging markets. Vakulchuk worked at the Shell oil company in Germany and has more than ten years of research experience with scenario-building methodology. In 2013, Vakulchuk was awarded the Gabriel Al-Salem International Award for Excellence in Consulting.

Jan-Gunnar Winther is the Director of the Centre for the Ocean and the Arctic at UiT – The Arctic University of Norway. Dr. Winter has more than 30 years of experience in polar exploration, science and management. He has served on a large number of national and international committees, boards and delegations, including deputy chair of the Norwegian Government's Expert Committee on Northern Regions Policies, lead author on IPCC's fifth assessment report, the World Economic Forum, the UN, China Council for International Cooperation on Environment and Development, the High Level Panel for a Sustainable Ocean Economy and the Explorers' Club.

Note from the Editors

This book was written during the year 2021 by a cross-disciplinary group of 34 Arctic researchers based in Norway, Russia, Finland, Canada and US Alaska. At present, the war in Ukraine causes huge uncertainty to the whole world and indeed the Arctic. None of the contributors to this book could imagine this could happen. We do not know what is awaiting us, but for sure the world would never be the same. This cannot be ignored in our book as it deals with international cooperation and future scenarios for the Arctic. In this regard, we invited all the contributors to write ex-post reflections to their original chapters. These reflections are placed at the end of the chapters.

We kindly ask our readers to appreciate the fact that this book was written under geo-political conditions which are no longer as they were. Nevertheless, we firmly believe that discussions on the Arctic past, present and future are still crucial. Maybe they will become even more crucial in new yet unclear realities. Climate change, the fragility of the environment and the natural resources of the Arctic remain urgent issues. International cooperation is becoming even more difficult than before. We hope that the analytical approaches and findings presented in this book will inspire interesting thoughts and discussions.

In hope for peace in Ukraine,
Editors
Andrey Mineev, Anatoli Bourmistrov, Frode Mellemvik
May 30, 2022

Acknowledgments

We wish to express our gratitude to all authors of these chapters for sharing their knowledge, patience and hard work. In addition to writing excellent chapters, the authors participated in a scenario workshop in Bodø in September 2021 (physically and digitally). This workshop provided a valuable foundation for Chapter 1 in this volume. We are profoundly grateful to the two dedicated anonymous reviewers for their excellent work with constructive and insightful comments to the authors and editors. Furthermore, we highly appreciate the work done by Alena Nelaeva – who served as a technical editor of the manuscript, Virginia Mattila – for language proofing and Ilpo Koskela for the illustrations of the scenarios. We would also like to thank the Open Access Fund at Nord University for partial coverage of the publication costs. Last but not least, we wish to express our gratitude to the excellent experts at Routledge for the professional help, support and for guiding us toward publication.

Introduction

Andrey Mineev, Anatoli Bourmistrov and Frode Mellemvik

This book gives a view of the Arctic as a key region for global development in the 21st century. It starts with a chapter presenting major scenarios for international cooperation in the Arctic up to 2035. Further, this book unfolds to 14 thematic chapters presenting major trends and future prospects for international cooperation in the Arctic in various sectors: energy, business and economy, transportation and logistics, climate change, diplomacy and security, culture, innovations, higher education and research. The concluding chapter of the book provides an analytical summary of the preceding chapters, offers a cross-disciplinary conceptual framework of international cooperation in the Arctic and discusses the implications of this framework for global development. The book is written by a cross-disciplinary group of 34 Arctic researchers based in Norway, Russia, Finland, Canada and the US Alaska.

This book challenges conventional and rather utilitarian approach to the Arctic as a prospective area for either development or protection of the natural resources. Discussing main trends and using scenario methodology, this book demonstrates that the Arctic can serve as a learning example for the development of cooperation to solve global sustainable development challenges. The four scenarios proposed in this book highlight and problematize underlying present and future trends shaping the development of the Arctic.

This book moreover bridges the gap in analytical understanding of international cooperation. Various academic disciplines contributed to development of somewhat fragmentary approaches to international cooperation. In practice, cooperation between nation states, political bodies and industry, industry and the general public, and even business cooperation is often viewed and dealt with as a zero-sum game (one party's gain is another party's loss). Even in cases of "successful" international cooperation, parties manage to cooperate to maintain stability or the status quo (e.g., in the area of international relations). Cooperative development for solving global challenges, also represented in the Arctic (e.g., climate change, sustainable development, societal and humanitarian development), is extremely difficult. This book offers a cross-disciplinary conceptual framework for understanding international cooperation and exemplifies a case of a self-reinforcing cooperation model found in the Arctic as a viable alternative to cooperation based on zero-sum premises.

The first chapter, "International Cooperation in the Arctic 2035 – The Four Scenarios," authored by Andrey Mineev, Jan Dietz, Petter Nore, Roman Vakulchuk and Anatoli Bourmistrov, presents four scenarios which explore the context for international cooperation in the Arctic up to 2035: *Klondike Arctic, Tech Arctic, Chinese Arctic* and *EU Arctic*. These four scenarios are based on the assumption that major uncertainties for future cooperation in the Arctic are associated with international responses to climate change and attitudes to natural resources in the Arctic. *Klondike Arctic* is a high-powered global race for resources in the Arctic where nation states and global corporations both compete and cooperate as they extract hydrocarbons, biomass and other resources on a massive scale. In the *Tech Arctic* (*Technological Arctic*), the Arctic has turned its back on traditional resource exploitation and become a test bed for new green solutions and the scene of new technology-driven rivalry. In the case of *Chinese Arctic*, China becomes the dominant force in the Arctic, using cooperation with Russia to expand its influence. Then, China has wide access to Arctic infrastructure and resources, and is also eager to project itself as the chief global architect of "Net Zero". The *EU Arctic* assumes the EU to have taken the lead in turning the Arctic into a "Northern Sanctuary" where the extraction of oil, gas, minerals and other important resources is expressly forbidden. Then, the Arctic becomes a global beacon of hope and at the same time a source of great frustration for many of the actors located there.

Chapters 2 and 3 discuss international cooperation in the fields of international governance and security in the Arctic. In their chapter "International Governance Facilitating Sustainable Development in the Arctic – The Arctic Council as a Multi-Role Actor and Forum" (Chapter 2), Stefan Kirchner and Timo Koivurova analyze the work of the Arctic Council and the member states' historic and current use of this cooperation institution. The chapter looks at the evolution of the Arctic Council since its inception and at its recent role between notable successes and ongoing challenges, with a particular emphasis on the issue of sustainable development, taking into account pan-Arctic challenges and obstacles to the full realization of the Sustainable Development Goals (SDGs) as well as current trends in international Arctic governance.

The third chapter, "Unboxing Arctic Security Relations and Dynamics" by Andreas Østhagen, questions several assumptions underpinning recent work on military security in the Arctic. The chapter moves away from the dichotomous view of the Arctic as a region where great power rivalries or resource wars are likely or as a part of the world defined by cooperative traits and shared security interests. By making distinctions between the international, national and regional interactions that take place, the chapter advances the way to describe and understand security dynamics in the Arctic.

Chapters 4–8 deal with issues of Arctic cooperation in the fields of economy, business and innovations. The fourth chapter, "Global Arctic Economic Development Scenarios" by Andrey Krivorotov, addresses some long-term economic trends in Arctic development, with a special focus on the contemporary and potential future drivers. The main idea is that although the Arctic regions were until recently profoundly separated with regard to political and value chain respects, there have always been a number of common features in their destinies and ways of life originating from the need to meet similar challenges of surviving under the extreme natural conditions of the Circumpolar North. However, with the recent shifts in the market and policy, which may well persist until 2035, the Arctic will face new challenges to its economic resilience, posing rigorous requirements for managers' and politicians' creative thinking. In this respect, the chapter outlines some key challenges and policy choices and also presents the results of previous exercises in charting partial scenarios.

Petter Nore in his chapter "Gas Bridges and Geo-Economics of the Arctic" (Chapter 5) reviews changing energy relations between international partners in the Arctic. The "Russian-Norwegian Gas Bridge" lasted from 1990 to 2014. A combination of external and internal forces brought the cooperation to a halt. Russia's attention then turned to the East in the form of the "Sino-Russian gas bridge". The chapter, using the perspective of geo-economics, offers examples of these developments toward the future in the form of two narratives. The first narrative is of how the "Sino-Russian gas bridge" by 2060 is declining in importance and becoming largely irrelevant. The second narrative describes how gas is consolidating its position in China's energy mix and gas imports from Russia continue to increase.

Chapter 6, "Business Cooperation in the Arctic: Learning Points from the Russia-Based Oil and Gas Projects" by Andrey Mineev and Elena Zhurova, builds upon three case studies of Russia-based Arctic oil and gas projects: Kharyaga, Shtokman and Yamal. Considering these projects and their development over time, the authors identify various forms of cooperation between the Russian state and foreign investors from the West and the East. In each of these still workable forms of cooperation, there are potential contradictions embedded. Discussing this experience, the chapter gives future outlook for international cooperation in the oil and gas projects based in the Russian Arctic.

The next chapter, "Arctic Innovation and the Potential for the Creation of a Circumpolar Innovation Ecosystem" (Chapter 7) by Ken Coates and Carin Holroyd, highlights the needs and activities of the governments and business leaders to create an innovation sector in the Far North. There are obvious needs for innovation in areas as diverse as food supply and energy, health care delivery and the operation of northern mining sites. Yet few Circumpolar communities have what are believed to be the core requirements for the successful and sustainable commercialization of technology: advanced research facilities, a highly skilled workforce, access to investment capital, supportive consumer markets, and robust government support and infrastructure systems. This chapter suggests that Circumpolar-wide innovation and commercial collaboration is required if the region hopes to overcome the formidable barriers that stand in the way of global competitiveness.

Chapter 8, "Smart City Dialog in the Arctic: Opportunities and Challenges" by Evgenii Aleksandrov, Elena Dybtsyna, Nadezda Nazarova and Igor Khodachek, investigates so-called smart city initiatives in the Russian and Norwegian Arctic, revealing and comparing how the smart cities dialog unfolds across governance actors. The findings demonstrate that smart city dialog formation has so far faced many challenges. Particular attention in both countries should be paid to existing governance traditions and national/international players' roles. Without intervention, there is a risk of smart city dialog ending up a monolog of the dominant voices from outside the Arctic, challenging Arctic sustainability. Yet, in future projections, the chapter demonstrates that current developments present opportunities across the two countries for joint competence building, networking and research.

The next three chapters are devoted to international cooperation in the Arctic in the sectors of science, education and culture. Chapter 9, "Arctic Climate Change – Perspectives on International Scientific Cooperation" by Jan-Gunnar Winther, Larry Hinzman and Kim Holmén, describes the development of key processes of international Arctic scientific cooperation over the past 20 years. It also addresses changes in thematic focus and how the nature of cooperation has evolved. The chapter attempts to identify areas where cooperation in the Arctic has had global relevance and global impact. The authors argue that Arctic climate change is a forceful driver for developing sustainable, climate-friendly business solutions. Yet the right balance between production and protection must be found. In this regard, more work is needed to advance the scientific understanding of the Arctic.

The chapter "Internationalization of Higher Education in the High North: Purposes and Strategies", by Marit Sundet, Leif-Kristian Monsen and Elena Dybtsyna (Chapter 10), is based on two studies: an empirical study of the internationalization strategies at four universities (the Northern Arctic Federal University, the University of Oulu, Luleå University of Technology and Nord University) and a study of an educational program developed in collaboration between three universities (the Moscow State Institute of International Relations (University), East China Normal University and Nord University). The chapter discusses internationalization purposes and strategies at higher education institutions in the High North. It highlights the criteria that seem appropriate to include as preconditions for future projections on the internationalization of higher education in the Arctic and the High North.

Chapter 11, "Challenging Boundaries from Below: Cross-Border Culture in the Arctic" by Bjarge Schwenke Fors and Yngvar B. Steinholt, focuses on Norwegian-Russian cultural cooperation. In so doing, the chapter contributes to the discussion about the impact of culture on international cooperation and sustainable development. The chapter is based on a study of two cases of cultural cooperation between the two countries: a cultural institution in the border area and a large-scale cultural event. By distinguishing between border-affirmative cultural cooperation implemented by the authorities (from above) and border-transcending cultural cooperation emerging from the population in the borderland (from below), the authors demonstrate how the latter interacts with and influences the former. The findings presented suggest that cultural initiatives from below have played a more prominent role than commonly perceived.

This helps explain how and why cultural cooperation has been so important and why it will remain a decisive factor in the future development of the countries' relationship.

Chapters 12–15 are devoted to international cooperation in the spheres of environment, shipping and emergency response. Chapter 12, "Issues of Environmental Monitoring and Management in the Arctic" by Alexei Bambulyak, Lars-Henrik Larsen, Rolf Rødven, Denis Moiseev and Salve Dahle, looks at the environmental monitoring and assessments carried out in the Arctic and presents the roles of the six working groups organized under the Arctic Council. Enhanced industrial activity in the Arctic is likely to increase the environmental risk associated with industrial developments. The authors argue that experiences form the environmental cooperation in the Barents Sea (The Barents Sea Large Marine Ecosystem) can be a driving force for the introduction of ecosystem-based management in a pan-Arctic perspective. In the Barents Sea, Norway and Russia share and manage biological fisheries and other resources and carry out industrial activities, including oil and gas exploration and production, which entails environmental impacts and risks in the cross-border context.

The next chapter, "International Shipping and the Northern Sea Route" by Björn Gunnarsson and Arild Moe (Chapter 13), analyses development trends for the Northern Sea Route. According to the authors, traffic on the Northern Sea Route along Russia's Siberian coast has increased since 2010. Yet international transits between the Pacific and the Atlantic have not flourished as anticipated. Instead, the transportation of energy and mineral resources – mainly liquefied natural gas – from fields in Russia's Arctic has come to dominate the sea route. The chapter argues that Russian protectionist measures have limited the role of international shipping companies, but alliances between Russian and foreign companies are expected to become important. Russia aims to build infrastructure facilitating year-round usage of the route, but international freight and commodities markets are likely to be decisive for the future volume of shipping.

The chapter "Polar Ship Design and Operations: Past, Present, and Future" by Ove Tobias Gudmestad (Chapter 14) reviews historical highlights related to sailings into polar seas, particularly the Barents Sea region. This chapter identifies how ship design and operations for the polar seas have evolved over the centuries and describes the role of international cooperation in this process. The discussion leads to a review of the present requirements for the fleet of polar vessels carrying cargo, containers and products from the oil and gas industry as well as for cruise ships. Thereafter, the chapter presents possible future requirements for ships operating on polar seas. The role of the International Maritime Organization is highlighted with a call for launching additional relevant international norms as the maritime activities in the Arctic are on the increase and the technology is progressing toward autonomous ships.

Andreassen and Borch, in their chapter "International Cooperation in Emergency Response in the Arctic Sea Areas" (Chapter 15), discuss cross-border cooperation in maritime search and rescue (SAR) operations. The discussion builds upon experiences from two incidents in the Arctic Ocean and the Barents

Sea, namely the search for the Russian MI-8 helicopter that disappeared in the Isfjorden near Barentsburg at Spitsbergen and the rescue of the crew from the grounded dry cargo vessel *Victor Koryakin* on the Rybachiy Peninsula in the Barents Sea. This chapter focuses on the key actors, their roles and the coordination of a cross-border maritime emergency response. It sheds light on the nature of the processes of international cooperation in maritime SAR and discusses the factors that facilitated the successful cooperation practice. Finally, historical and more recent experiences in cross-border emergency response and their potential impacts in the Arctic seas are discussed.

Chapter 16, "International Cooperation for Global Development: What Can We Learn from rhe Arctic?" by Anatoli Bourmistrov, Frode Mellemvik and Andrey Mineev, integrates and concludes this book. The chapter presents an analytical summary of all the preceding thematic chapters. This chapter offers a so-called Arctic self-reinforcing cooperation model based on common patterns of cooperation observed throughout the book chapters. Self-reinforcing cooperation can be found in various sectors: climate cooperation, maritime search and rescue, culture, higher education, environmental protection, security and governance. The self-reinforcing model is in many ways based on the colocation and mutual dependence of the Arctic actors, but also on bottom-up initiatives. Such cooperation is motivated by the real-life paradoxes to which the actors are exposed. Once big economic interests (e.g., extractive industries) come into the picture, cooperation may develop in a different, non-self-reinforcing way.

1 International Cooperation in the Arctic 2035 – The Four Scenarios

Andrey Mineev, Jan Dietz, Petter Nore, Roman Vakulchuk and Anatoli Bourmistrov

Introduction to the scenarios

The Arctic has always fascinated people; its history, its present, and its future. The future of the Arctic has increasingly become a subject of academic research and the application of scenario methodology. Scenarios can be defined as prospective storytelling (Schoemaker, 1993), presenting a set of plausible, contrasting images of the future (Schatzmann et al., 2013), and indicating what alternative futures might look like (Amer et al., 2013). Studies offering scenarios of future development of the Arctic include Brigham (2007), Myllylä et al. (2016), Lazariva et al. (2021), Petrov et al. (2021), Haavisto et al. (2016), and Bourmistrov et al. (2015); see also the chapter by Krivorotov in this volume. The farther we look ahead, the more uncertain the future appears. There can never be full consensus on what major trends and driving forces will have the greatest impact on the future. But precisely for this reason, any kind of structured thought experiment, such as scenario development, is valuable and can add new knowledge and shared understanding.

Works dedicated to Arctic scenarios so far have largely focused on resource extraction, climate change, geopolitics, and economic and social development as key factors shaping the Arctic's future. Our chapter adds to this body of knowledge by giving more weight to the dynamics of international politics and cooperation, including the pressures for a green transition. In our analysis, we treat the Arctic as an object of interest to global society and a topic of growing importance in international affairs. Namely, we present four scenarios that describe how the context for international cooperation in the Arctic might change in the years leading to 2035.

The time frame chosen for our scenarios is the 15 years between 2021 and the end of 2035. This is a time horizon that gives us enough space to elaborate on plausible developments and capture the big picture for international cooperation in the Arctic. At the same time, we can be concrete enough as 15 years is a future which is not too far away, at least in our perception. We believe that most of the trends that will shape the Arctic in the coming 15 years are already in place.

We have identified a set of certain, already evolving trends which will significantly influence international cooperation in the Arctic the next 15 years (2035)

DOI: 10.4324/9781003246015-1

and beyond, although many of the implications are difficult to untangle. They appear, indirectly or directly, in the four scenarios we have elaborated. These trends form the backdrop of our analysis and are valid across all scenarios:

- Non-Arctic actors will have a greater say in making rules for the Arctic
- Climate change will be a key driver
- Pressure for a green transition will mount
- Technological development will continue to accelerate
- Arctic demography is a permanent concern
- Russia will remain heavily dependent on fossil fuel resources
- China will strive to increase its global power
- Geopolitical tensions will remain high

Having established the predictable and even predetermined factors and trends, it is important to outline the *key uncertainties*. These uncertainties define the contrasts between the scenarios, and they therefore play out differently in each story. We have identified two major uncertainties:

- ***Fragmented versus coordinated response to climate change***
- ***Arctic lockdown versus Arctic resource extraction***

The first relates to international cooperation to combat climate change. Given the myriad of national, economic, and institutional actors with different interests in the Arctic, the future response to climate change is profoundly uncertain. Will there be a coordinated response and established efficient international institutions to handle climate issues for the best of the planet? Or, in contrast, will we move into a future characterized by fragmented, anarchic responses undertaken by a variety of actors? The second major uncertainty concerns generally accepted public and political attitudes to Arctic resources. Will resources be extracted to meet growing world demands for energy and food? Or will they be subject to a formal or de facto lockdown, implying that they will be highly regulated and/or preserved? There are too many factors which can influence both sides, yet the outcome remains highly uncertain.

When combined, the two key uncertainties outlined above lead to four contrasting outcomes – the scenarios (Figure 1.1).

Klondike Arctic – this is a high-powered global race for resources in the Arctic. In 2035, nation-states and global corporations both compete and cooperate as they extract hydrocarbons, biomass, and other resources on a massive scale.

Tech Arctic – the Arctic has turned its back on traditional resource exploitation and become a test bed for new green solutions and the scene of new technology-driven rivalry. In 2035, the main actors in the Arctic are tech companies, newly set-up national Arctic ministries, the EU, and indigenous groups.

Chinese Arctic – China has become the dominant force in the Arctic, using cooperation with Russia to expand its influence. In 2035, China has wide access

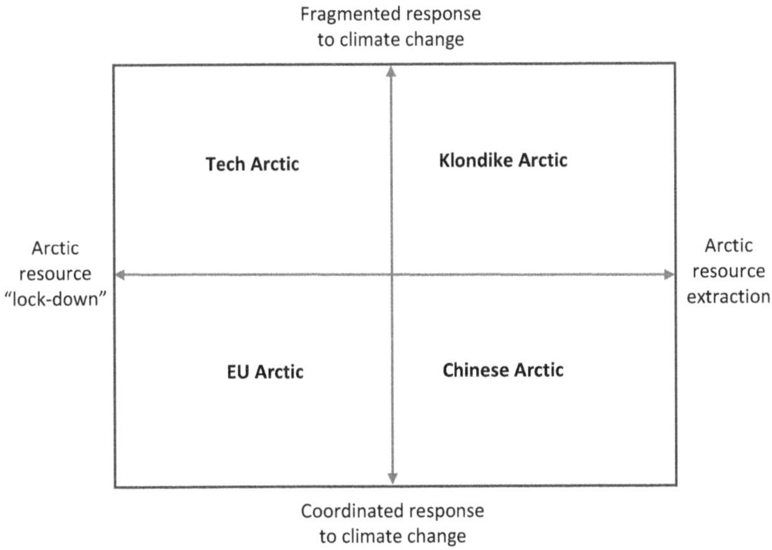

Figure 1.1 The four scenarios for international cooperation in the Arctic in 2035.
Source: Authors.

to Arctic infrastructure and resources, and is also eager to project itself as the chief global architect of "Net Zero".

EU Arctic – the EU has taken the lead in turning the Arctic into a "Northern Sanctuary" where the extraction of oil, gas, minerals, and other important resources is expressly forbidden. In 2035, the Arctic has become a global beacon of hope and at the same time a source of great frustration for many of the actors located there.

Approach

Scenarios are not attempts at forecasting or simple projections; rather, the purpose is to identify alternative possible pathways in acknowledgment of the fact that the future interaction between multiple factors is impossible to extrapolate. The interaction can and must be *imagined* by constructing coherent stories, that is, scenarios. Their true value lies in that they enable the reader to grasp how and why the world might change beyond recognition. Thus, the reader can learn more about how the present can evolve into radically different futures and better understand the possible threats and opportunities we are going to face. The aim, of course, is to challenge prevailing mindsets of the present and to become better prepared for the future.

There is a bewildering number of approaches, methods, and techniques that can be used to construct scenarios, for example, intuitive logics methodology, the French school ("La prospective"), real options, integrated risk management, probabilistic trends methodology, and prospective methodology (Bradfield et al., 2005; Burger-Helmchen, 2008; Miller and Waller, 2003). We have been inspired by the intuitive logics school of thought pioneered by Royal Dutch Shell (Schoemaker, 1993, 1995) and have sought to construct scenarios based on causal analytical reasoning and the determination of plausible cause-and-effect relationships between hypothetical events and possible future outcomes. Specifically, we have used the classical 2 × 2 matrix technique.

Following this methodology, opinions about the future must be grounded in basic assumptions about the world. Some trends and events form our assumptions because we think that they are significantly more probable than many other developments or even predetermined. Further, we were especially conscious of key uncertainties, that is, driving forces that are both unpredictable and decisive for future outcomes. Key uncertainties are factors, trends, or driving forces that could easily tip developments one way or the other. Building on this analysis, we have pieced together four scenarios based on thought experiments where we have set up different, contrasting combinations of these assumptions and uncertainties.

Before creating the scenarios and writing the storylines, we conducted quite extensive preparatory work. First, all the authors of the thematic chapters in this book were asked to provide specific inputs for the scenarios in terms of specifying assumptions, uncertainties, and wild cards. The factors were sorted and analyzed by the authors of this chapter, who made a comprehensive list summing up insights. Second, we organized a two-day workshop where authors of the thematic chapters as well as representatives of different stakeholders not involved in the book project were invited to pool ideas with us. The discussions on assumptions and uncertainties were organized into several groups moderated by the coauthors of the chapter. Each group produced and presented a set of scenarios. Third, those different sets of scenarios were further discussed among the authors of this chapter in a series of intensive internal scenario-building meetings in which the scenarios were refined. Finally, also drawing on inputs from authors of the thematic chapters, wild cards were discussed and presented. Wild cards are low-probability events which may have a very high impact and may dramatically change the course of events and even invalidate all scenarios.

Eight basic assumptions

Non-Arctic actors will have a greater say in making rules for the Arctic

The world is increasingly becoming a "global village", to borrow Marshall McLuhan's famous phrase (McLuhan, 1962), meaning that there is a growing interdependence between the world's economies, cultures, and populations. The globalization of the Arctic can be seen as an example of the interconnectedness that is emblematic of our era. International institutions like the UN and the EU and even some non-Arctic states (e.g., Asian countries) are becoming involved in making the rules for the Arctic. Arctic-related issues (e.g., climate) and opportunities (e.g., resources) are gradually recognized as having a global nature (Heininen and Finger, 2018; Kristoffersen and Langfelle, 2017). A growing number of actors express an interest in the Arctic[1]. We assume therefore that the Arctic increasingly will be seen as a global concern in the future, implying a greater involvement of non-Arctic actors in Arctic affairs and perhaps new constraints on the sovereignty of the Arctic nations, including militarization and security concerns (see, e.g., chapter by Østhagen in this volume). The interplay of Arctic and non-Arctic actors is discussed by Kirchner and Koivurova, also in this volume.

Climate change will be a key driver

Discernible changes in the climate in the Arctic are already underway and will most likely become more severe. Changes in the patterns of the weather are accelerating, affecting the oceans, land surfaces, glaciers, and ice sheets in the Arctic, often in an irreversible manner. Everywhere in the Arctic, ice is melting, and one important consequence is that shipping activity in the Arctic will increase. Even though we may expect more liquefied natural gas (LNG) transport, the Northern Sea Route (NSR) will not be a key transport route between Europe and Asia as early as 2035 (for more details about shipping in the Arctic, please refer to chapters of Gunnarsson and Moe, and Gudmestad in this volume). Environment and energy issues, which are directly affected by climate change, will become ever more important to both national and international policymakers (please see Bambulyak et al. in this volume). Energy scarcity is bound to become a major worry in many parts of the world. It is safe to assume that climate change in the Arctic will be high on the agenda for political and scientific cooperation. For instance, increased shipping activity will require heavy investments in search and rescue infrastructure to prevent loss of life and to handle environmental impacts (Andreassen and Borch, in this volume). Climate change will also prompt innovation efforts and business development. The chapter by Winther et al. in this volume is devoted to scientific cooperation on climate change.

Pressure for a green transition will mount

We expect continued pressure for a more rapid green transition, both from national actors around the Arctic and from international institutions such as the UN, the EU, and the Organisation for Economic Co-operation and Development (OECD). Europe is set to become climate neutral by 2050, according to the European Green Deal (EU Commission, 2019). The EU Arctic policy launched in October 2021 literally means that unexploited Arctic oil, gas, and coal resources must be left permanently in the ground. Green parties – once seen as radical outsiders – have increasingly claimed a place in mainstream politics, especially in Western Europe (McBride, 2021). Further, youth climate movements are gaining influence. For example, on September 25, 2020, some 3,500 climate strikes took place in 154 countries, from the Arctic to South Africa, from the Pacific to Latin America, "to demand climate justice now".[2] Perhaps we are on the brink of a generational upheaval involving a wholly new attitude toward environmental protection. Proposals for a truly low-carbon economy may jeopardize many existing or planned economic activities in the Arctic, as these are heavily based on raw material extraction. Investors may want to avoid projects that could appear controversial. The growing interest in protecting biodiversity could further slow down current plans in the Arctic.

Technological development will continue to accelerate

Rapid technological development is a global megatrend. Digitalization and robotization could be extensively used in the Arctic to make it easier to work in harsh weather conditions and to extract resources that are difficult to access. Increased use of advanced technology could reduce the need for manpower and minimize the risk involved. Artificial intelligence and machine learning may lead to breakthroughs in several fields that are relevant to the Arctic and to sub-Arctic areas (e.g., fisheries and environmental monitoring). However, the Arctic would probably lag southern and urban centers in technological innovation. For example, Smart Cities in the Arctic will likely be developed using ideas and innovations originating outside the Arctic (Alexandrov et al. in this volume). Historically, innovations have developed in larger metropolitan areas far away from the Arctic, but it is also conceivable that the Arctic could inspire more Arctic-specific technologies (Coates and Halroyd in this volume).

Arctic demography is a permanent concern

The world population is growing overall but not in the Arctic, where the trend is negative or flat in most areas. Demography and the robustness of local communities will always be an issue in the Arctic. Developing and maintaining infrastructure in remote, far-flung settlements is costly. According to studies made by Business Index North (Nord University in Bodø, Norway),[3] urban areas in the Arctic tend to be stable, while rural areas experience depopulation and an outflux of youth. Cross-border cultural and education cooperation has always been important to make the Arctic an attractive place to live in (Dybtsyna et al.; Fors and Steinholt, both in this volume). Nevertheless, population in the Arctic can be strengthened through big projects, yet these can also present a challenge as they tend to rely on commuting specialists and workers who move into the region for some time but who do not settle there for good. Extensive commuting adds little or no value to local communities in the Arctic – or anywhere else for that matter. Arctic communities are, however, especially vulnerable in demographic, economic, social, and cultural terms.

Russia will remain heavily dependent on fossil fuel resources

Russia will continue to have access to the world's largest fossil fuel resources, and we assume that there will be a continued Russian willingness to exploit these resources. Natural resource extraction (gas, oil, coal, metals) makes up about 13.5% of Russian gross domestic product (GDP). The three other largest industries – manufacturing, wholesale trade, and transportation and storage – are inextricably linked to the same natural resources: processing of the resources, sales of products made from the resources (e.g., fuel), and delivery to customers. In times of crisis such as that brought about by the COVID-19 pandemic in 2020, the government of Russia uses earnings from natural resources to support citizens and subsidize other sectors. Huge investments are made in the development of the NSR and the resource deposits along this route (natural gas, oil, minerals). The abundant energy resources in the Russian Arctic and the growing energy demand in China is an important explanation for the strategic alliance between the two countries. For more detail on Chinese-Russian cooperation in the Arctic, please refer to the chapters by Nore and by Mineev and Zhurova (both in this volume).

China will strive to increase its global power

China has emerged as the world's largest energy market, and we expect that China will forge ahead to extend its global influence and economic presence in the Arctic in the years to come. In 2017, the Arctic area and the NSR were added to the geographical scope of the Chinese Belt and Road Initiative (BRI). The BRI is a Chinese attempt to introduce a global governance concept and to give China a more prominent role on the world stage. Increasing its role in the Arctic, China has invested in or indicated its interest in joining large oil and gas projects like Yamal LNG and Alaska LNG, and to becoming involved in major infrastructure developments such as the Kirkenes-Rovaniemi rail route and the NSR (Krivorotov, 2018).

Geopolitical tensions will remain high

Clearly, globalization has not reduced the potential for geopolitical unrest. As pointed out by Deutsche Welle analysts (Schacht and Koschyk, 2019), wars have become more complex: until the beginning of the 2000s, only two or three *external parties*, on average, participated in any given conflict. In the following years, this average rose to between four and five. The war in Syria, for example, has involved at least ten major external parties since it started in 2011, according to various estimates. Involvement of external parties can, inter alia, be in the form of sending troops or supplying weapons, expertise and training, and in staging campaigns in both mass and social media. This means that modern military conflicts tend to have a ripple effect, affecting the security of other countries and regions as well as having an economic, political, and social impact.

Professor Samuel P. Huntington argued that conflicts between civilizations (cultures) rather than interstate conflicts would dominate world politics after the end of the Cold War (Huntington, 1996). He pointed out that we now live in a multipolar, multicivilizational world. According to Huntington, the power of the West will be contested. Efforts of the West to promote its values of democracy and liberalism, to maintain its military predominance and to advance its economic interests, will engender countering responses from other civilizations (Huntington, 1993, 29).

If Huntington is right, it is highly unlikely that the underlying tensions between the West and non-Western cultures and societies (in particular, Russia, China, and Islam) will disappear in the next 15 years. When much is at stake, cultural affinity and loyalty, based on historical and religious identity, tend to override other factors in international politics. Geopolitical tensions, therefore, will influence the Arctic heavily in the years to come, perhaps even more so than in the past, as the Arctic is set to play a more prominent role in global politics.

Two key uncertainties

Fragmented versus coordinated response to climate change

It seems obvious that the most difficult and pressing issues related to climate change cannot be resolved at local or even at national level. The Arctic is an excellent example of both shared interests and divergent interests, of both common ground and a lack of common denominators. The Arctic is also a frontier territory, not quite a no man's land, but not a highly regulated area either.

This means that understanding and combating climate change in the Arctic is a truly complex and challenging undertaking. Given the many and varied national, economic, and institutional actors involved, the future response to climate change in the region and on a global level is profoundly uncertain. We do not know how policies will be formulated, to what extent they will be coordinated, and how effective they will be. To complicate matters, the four major powers – Russia, China, the EU, and the US – can be seen as rivals in a game to shape the future of the Arctic region. In some quarters doubts persist with respect to whether the world is facing a climate emergency.

Given that the Arctic region is extremely exposed and vulnerable to climate change and at the center of global concerns over melting ice, we arrive at a fundamental uncertainty: what kind of actions to mitigate climate change will be taken by Arctic nations, non-Arctic actors, and international institutions over the next 15 years (by 2035)? On the one hand, the growing threat of climate change may increasingly serve to unite Arctic and non-Arctic states in joint and well-coordinated efforts to stem and adapt to climate change. On the other hand, although agreeing in principle on the need to fight climate change in a coordinated manner, Arctic states may choose to follow their own agendas in the Arctic. Other considerations, which have little or nothing to do with the Arctic, may make binding international commitments impossible to achieve.

At one end of the spectrum, it is possible to envisage a coordinated supranational response to escalating climate problems in the Arctic and other areas. The issue of climate change could, in theory, be dealt with by a transnational organization with undisputable legitimacy and authority. Such an organization would employ universal principles of environmental protection, taking the perceived common good of the planet as its point of departure. Regarding climate change, it could override the economic and political interests of any given country or business corporation. At present, there are no supranational organizations in this pure form. Still, one can argue that clear signs of supranationalism can be found in the EU.[4] Seeds of supranationalism can also be found in the UN Sustainable Development Goals (SDGs), which are defined in universal (global) terms.[5] Another example of emerging supranationalism could be the growing willingness to pool resources using the World Health Organization (WHO) and other instruments in the wake of the COVID-19 pandemic.

There seems to be a growing awareness that global problems require global solutions and some degree of global authority. If the effects of climate change and associated problems become truly global and severe, one can hypothesize that a more binding approach to international, multilateral cooperation may emerge in the next 15 years. We may even see the beginnings of a supranational model of governance in the Arctic. International Maritime Organization's Polar Code and the Arctic Investment Protocol by WEF (World Economic Forum) can be viewed as such attempts.

At the other end of the spectrum, one can imagine a situation where countries with a strong stake in the Arctic pursue their own policies and economic interests and simply ignore calls for more international action. Their primary concern could be vital national interests. Countries like Russia, China, the US, Canada, and Norway have a strong interest in the development of Arctic resources and may actively resist attempts by politicians in the EU to forbid the extraction of hydrocarbons, coal, and metals from the Arctic. By the same token, the prospect of extracting tangible benefits from the Arctic resources may contribute to fragmented national responses to climate change in general.

States pursuing national interests may not consider climate change in the Arctic a separate issue, but rather as a part of a much broader problem. They may take the view that they are working to solve climate problems in a more realistic way through unilateral action. One can also reasonably imagine the evolution of a new set of bilateral and multilateral agreements that seek to coordinate action to combat climate change. Groups of likeminded countries may come together to improve existing bilateral and multilateral cooperation. In this case, the countries will maintain full formal sovereignty and can choose the terms of their participation. In case of dissent or conflict, they may either withdraw from the agreement[6] (which may be a costly alternative) or exercise their right of veto. Multilateral agreements can have a global, continental, or regional scope, leaving formal sovereignty untouched. How effective these arrangements are, and how much real sovereignty individual states are left with, is a different matter. It is decidedly unclear how the interplay of strong national and economic interests will unfold in the years to come.

All in all, the response to climate change in the Arctic presents a fundamental uncertainty: will Arctic actors and the world at large respond to climate change in the Arctic by moving toward coordinated supranational solutions or will nations mostly "go it alone"?

Arctic lockdown versus Arctic resource extraction

The second major uncertainty zooms in on Arctic resources and public and political attitudes to their future use. One the one hand, the world is moving toward a low or non-carbon economy. Finite resources need to be used more carefully, and few will dispute that nature and wildlife must be afforded greater protection. On the other hand, the global population continues to grow and may increase by more than one billion before 2030. The world needs more resources in almost every form – water, food, and energy. Today, the world gets 80% of its energy from fossil fuels. These realities combine to create a global dilemma or conundrum: will hitherto largely unexploited Arctic resources be extracted, using either traditional or new technologies, or will pressure to preserve these resources completely prevail? This is a polar yes/no question and both outcomes seem plausible in a long-term perspective. One may, however, also foresee a situation in which resources are somehow both extracted and preserved within a set of strict environmental rules. The very nature of these rules remains uncertain. There are two legal maxims:

A Everything which is not forbidden is allowed
B Everything is forbidden unless it is permitted

The idea underlying maxim A, which is fundamental to liberal democracies, is that we are inherently and naturally free to do anything, so long as it is not expressly prohibited by law.[7] Maxim B may work in authoritarian settings but also in critical situations such as the COVID-19 pandemic, where there is a recognized common, overall danger. The strong call to "stay at home" during the lockdowns has been used in both democratic and authoritarian countries. One may not leave home unless an exception can be justified. The same approach could be taken to Arctic resources: in general, they are preserved and extraction is banned, but it could be possible to utilize some of them if, for example, the proper technology is applied or if a distinction is made regarding the resources taken (e.g., food resources are allowed while hydrocarbons are not). Traditionally, the Arctic has been developed based on maxim A: extraction of natural resources is allowed, but some restrictions apply (e.g., national and international environmental protection regulations).

If maxim A still applies in 2035, the Arctic will be *a place characterized by extensive resource extraction*. If we have a shift to maxim B, then the Arctic will be subject to a *resource lockdown*, although possibly with some exceptions. These are two fundamentally different situations. For example, in the case of a resource lockdown, one can envisage the EU playing a leading role, setting the rules for conservation and the introduction of a more circular economy in Europe. Such a role would fit the environmental ideals and the high level of multilateral coordination in the EU. A resource lockdown in the Arctic would probably be part of a much wider change in thinking. Only "smart" cities and communities would then be encouraged in the Arctic and, as a rule, only smart (green) technologies would be allowed in the industry. Large tracts of the Arctic would, however, become a nature reserve or a museum.

A resource lockdown might be met with both enthusiasm and fierce resistance. There could be a backlash where national and commercial actors strive to have the Arctic recognized as the opposite: a resource base of global importance, essential to meeting the growing demand for energy and food in a world where resources are becoming increasingly scarce.

The four scenarios

Klondike Arctic 2035

This is the story of a tense global race for resources in the Arctic. In 2035, nation-states and global corporations both compete and cooperate as they extract hydrocarbons, biomass, and other resources on a massive scale.

Big picture in 2035

"Klondike Arctic" can in many ways be seen as a logical continuation of current trends and conventional, convenient perceptions of how the world functions: "Business as usual" is preferable, the detrimental effects of climate change can be controlled through gradual improvements in technologies, the Arctic is becoming increasingly accessible, and the world urgently needs a more abundant supply of resources.

In 2035, climate change issues are primarily handled through national policies and to some extent, through loose international agreements. Countries with strategic interests in the Arctic retain a high degree of sovereignty and are free to pursue their own interests. As the Arctic evolves into a global focal point for the extraction of natural resources, it also becomes an arena for tough international competition marked by political contradictions. Besides economic and technological capacities, a strong military capacity is an important prerequisite for success in the competition for resources.

Arctic resources are desperately needed. By 2035, 80% of world consumption stems from the 7.2 billion people living in Southeast Asia and Northwest Africa (out of a total world population of nine billion). Asian and African regions are the main destinations for shipments of goods, while the main destination of LNG shipments is China. Global energy use is now 25% higher than in 2020. Because the energy mix includes a greater share of renewables, CO_2 emissions have only grown by 6% in the same period. In 2035, hydrocarbon-based fuels remain the principal source of energy. Nevertheless, the damage to the Arctic environment caused by climate change is not seen as a significant problem – the common perception is that the advantages outweigh the disadvantages. Both the North-West Passage and the North-East Passage have become ice-free and are now viable alternatives to shipping via Suez.

What has happened?

Since 2025, the Canadian government has been committed to opening the North-West Passage to global trade, making essential investments in the development of its infrastructure, comparable to Russian investments in the NSR – the largest part of the North-East Passage in 2015–2025. Previously controversial areas in the Arctic have been opened for exploration and development of resources: Baffin Bay between Canada and Greenland, the Barents Sea between Norway and Russia, the Arctic National Wildlife Refuge (ANWR) in the US, and Chukchi Sea between Russia and Alaska (US). Modern coastal and offshore sea farming has been developed extensively along the circumpolar Arctic, alongside petroleum and renewable energy activities. Non-Arctic countries such as China, Japan, South Korea, and India are active through their governmentally backed companies, in addition to companies based in Arctic countries and transnational corporations.

In 2025, the opening of the ANWR area in Alaska for drilling and other activities and the continued production of tar sand in sub-Arctic Alberta (Canada) resulted in massive protests led by young environment activists and representatives of indigenous communities. The protests, which took place in the US, Canada, and Brussels, were quickly quelled by the police. In the wake of the protests, however, governments introduced stricter environmental regulations.

Since 2030, Arctic countries have discussed the establishment of a new intergovernmental institution – the Arctic Economic Union (AEU). The AEU discussions have achieved no results due to disagreements between the EU, Russia, and the US. In 2027, Greenland initiated the development of oil deposits in the Baffin Bay area in cooperation with Canadian and American companies. In 2030, Greenlanders tried to arrange a referendum on political independence from Denmark but were stopped in their tracks by the Danish government with the strong support of the EU, which was also skeptical of Canadian and American moves.

China has to a great extent managed to implement its Polar Silk Road initiative in the Russian Arctic and elsewhere in the period of 2020–2030. Western corporations have found ways to participate in the development of the Russian Arctic. The Russian government and national corporations must deal with Chinese and Western investors who, in turn, have somewhat opportunistic motives. Relations between Russia and the West have changed: both sides understand that there is a growing global need for resources from the Russian Arctic. A new progressive government in Russia is intent on expanding trade and modernizing Russian infrastructure.

The Arctic has witnessed substantial investments, for example, in a grid of 5G transmitters, fiber-optic infrastructure, data centers, and development of *Internet of Things* (IoT) technology. Shipping operations, land transportation, mining, sea farming, and extraction of hydrocarbons are mainly unmanned and robotized. This has led to a loss of jobs and to further depopulation. In 2035, only the Scandinavian Arctic and the European part of Northern Russia have sizeable populations. People cluster in a small number of urban centers while the rural areas have declined noticeably. Arctic settlements along the circumpolar coastline are relatively small and far apart. Only highly skilled and educated employees involved in knowledge-intensive services live there with their families, enjoying high levels of economic well-being.

Svalbard has attracted increased attention. In 2025, large businesses in Russia, the US, and China demanded that Norway change its position and open the Svalbard shelf for exploration and development of marine biomass resources, hydrocarbons, and subsea minerals. EU Member States have pressed for increased fishing and snow crab quotas. Meanwhile, China has managed to establish a town in Svalbard devoted to climate research and the development of commercial Arctic tourism. China has also pulled off a coup by buying Hurtigruten, the renowned Norwegian Coastal Steamer company.

In 2032, an American-Russian joint venture company was established to explore minerals under the seabed in the Arctic, and a huge discovery of uranium ore was made in the so-called Svalbard box. In 2035, the Norwegian parliament was presented with an ultimatum by China, Russia, the US, and India. Norway was forced to interpret the Svalbard treaty in a nondiscriminatory way, meaning that the whole Svalbard box area should be a special economic zone open for international exploration and development.

In brief

The global race for natural resources has accentuated the geopolitical and economic importance of the Arctic, leading, for instance, to the gradual deterioration of the ANWR in Alaska and to challenges to Norwegian authority on Svalbard. In 2035, the Arctic is no longer a neglected, secluded place but an arena of fierce competition. Convenient perceptions of "business as usual" (from the early 2020s) do not match with the reality of "business not as usual" in 2035. Illustration of the "Klondike Arctic" scenario is given on the Figure 1.2.

Tech Arctic 2035

This is the story of how the Arctic has turned its back on traditional resource exploitation and become a test bed for new green solutions and the scene of new technology-driven rivalry. In 2035, the main actors in the Arctic are hi-tech companies, newly set-up national Arctic ministries, the EU, and indigenous groups.

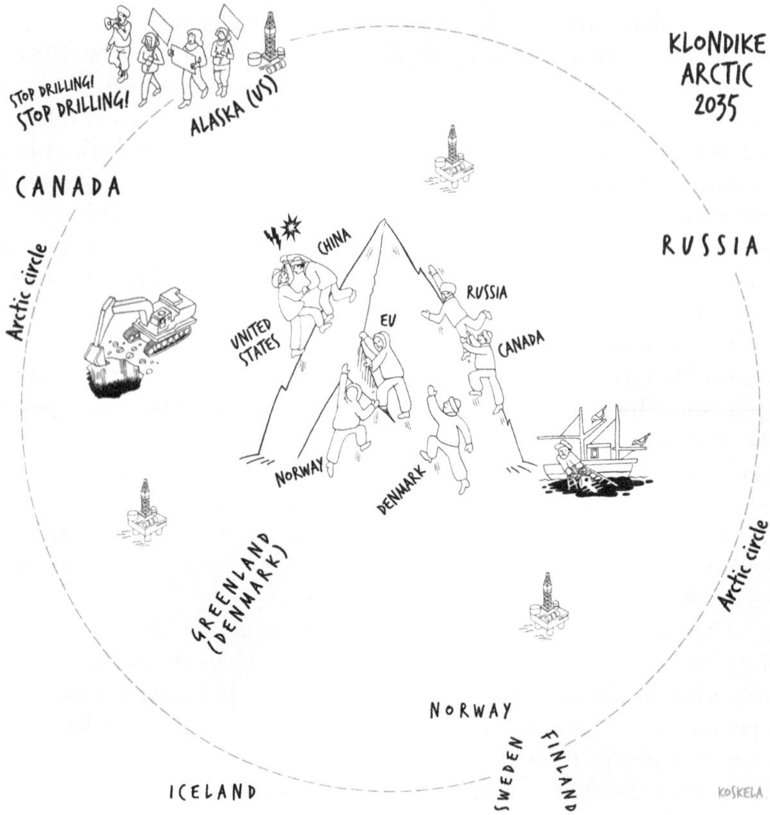

Figure 1.2 Illustration of the Klondike Arctic 2035.
Source: Authors.

Big picture in 2035

The Arctic has almost been turned upside down. New technologies and industries have emerged, making traditional extraction and production outdated, and old business models ineffective and obsolete. The oil and gas industry has had its Kodak moment[8]: the assets are stranded, and investments cannot be recouped. Advances in wind and solar energy and storage technologies have reduced the need for new fossil fuel and mineral resource extraction in the Arctic and elsewhere. Mineral resources produced before the green wave were reused multiple times through circular economy infrastructure. The global response to climate change has been indecisive and fragmented, something which paradoxically has increased the significance of the Arctic. The Arctic with its vast unpopulated areas has become extremely attractive to green energy entrepreneurs and investors. In the Arctic, they have more freedom than in the more densely populated and regulated south; wind and solar companies have access to vast stretches of cheap land and they rarely need to worry about the effects on local communities. Similarly, conditions are ideal for data centers and data mining.

Contrary to expectations, the overall security situation in the Arctic has improved. Competing agendas and interests have made the Arctic politically more vulnerable but have also increased interest in keeping security problems at bay. The Arctic exemplifies complex interdependence, a concept formulated by Robert Keohane and Joseph Nye (1973). Relations between the Arctic and non-Arctic states have become close, multifaceted, and complex, leading to beneficial interdependence and improved security. No country is willing to risk escalation by advancing their interests and prestige too hard. The green agenda also imposes limits as to how far countries can promote selfish or belligerent foreign policies.

In 2035, following "peak oil" and the collapse of "business as usual", Arctic and non-Arctic states have reached a gentleman's agreement that hydrocarbon resources, as a rule, should be left in the ground and that only renewable resources should be developed. In many parts of the Arctic, indigenous communities have become direct owners and beneficiaries of electricity produced using wind and solar power.

What has happened?

The Arctic has become a place for political and commercial competition, characterized by a multitude of policies and views on the implementation of green change as well as changes in the roles Arctic countries play in the development of their northern territories. In this Arctic future, two surprising events have occurred. First, Greenland became independent in 2033 and expressed a strong interest in participating in Arctic international affairs. Second, the rise in the global sea level has forced inhabitants in some areas in Asia to relocate to the Arctic and other regions of the world. Some highly skilled eco-migrants from Bangkok, Jakarta, Manila, Tokyo, and other Asian cities – representing the first wave of Asian eco-migration – have moved to the Arctic and expanded the Arctic talent pool, especially as regards the Information Technology (IT) sector.

The EU has taken a strong pro-climate position and introduced a unilateral moratorium on the development of Arctic resources. The moratorium, which is rooted in UN Goals for a Green Arctic, and rigorous policies of major Western banks and rating agencies has influenced both Arctic and non-Arctic states, including Russia, the US, and China. However, a formal, binding agreement on a resource lockdown in the Arctic has been impossible to achieve. Interests differ too much, and the EU finds it difficult to exercise global leadership. The fragmented responses to climate change internationally have complicated matters: there is little agreement about the pace and scale of achieving carbon neutrality, meeting the needs of the developing world, and saving the ecosystems. In this situation, the Arctic has become a prime example of the need to "think globally and act locally".

Canada was the first country to establish a Ministry of the Arctic in 2027, signaling the growing importance of the Arctic as a valuable natural habitat and an area in need of environmental protection. Denmark, Iceland, Norway, Russia, and Sweden soon followed suit by establishing similar ministries for the Arctic. The interaction and communication between Arctic states largely takes place through informal and formal channels in the Arctic Council, intergovernmental agencies, ministries, networks of towns and cities, and councils consisting of representatives of local indigenous communities.

China has become a major investor in Arctic infrastructure and received informal approval from the Arctic Council. Since 2025, Norway has invested a fortune in the transformation of its oil industry, with offshore wind development north of Finnmark, and new battery and hydrogen plants becoming a cornerstone of green policies. Russia has started to look into large-scale offshore wind power development, slowly reacting to pressures for a green shift. Norwegian companies have found ways to participate in the development of the Russian Arctic.

The Arctic has become the major location of data centers and colocation centers in Denmark, Finland, Iceland, Norway, and Sweden. Back in 2021, the Arctic was already home to 40 distinct data centers in Greenland, Iceland, and Norway (Sovacool et al., 2022). By 2035, the number of data centers and colocation centers has quadrupled. Also, by 2035, the number of high schools and universities in the Arctic has tripled. Growing migration to the region, the emergence of data towns and environmental eco-centers spur the demand for high-quality education. Most of the educational institutions have merged with technological hubs and professional centers. The Arctic has seen the emergence of a new digital class of entrepreneurs and service providers. However, the region has also provided ample opportunity for professionals in fields such as green infrastructure and construction, renewable energy maintenance, and environmental protection.

Ultimately, technological breakthroughs and new business models rather than political agreements have paved the way for the resource lockdown. Oil, gas, and coal are hardly exploited and are no longer essential to economic growth. Many of the traditional producers were unable to foresee and adapt to the dramatic change in the business environment caused by the green transformation. The untapped resources, which seemed so attractive, became a trap. Extracting mineral resources in remote Arctic and sub-Arctic areas is too costly and risky. The looming climate crisis has accelerated the transition to a complete lockdown of mineral resources: everyone understands that the Arctic is especially vulnerable.

In brief

Life in the Arctic is more technologically driven, more hectic, and even more sophisticated in 2035 than before. In contrast to the situation in the 2020s, when the Arctic faced a brain drain problem with the best minds leaving for opportunities elsewhere, in 2035 the Arctic states are experiencing a sizeable inflow of highly skilled young labor to the region. Illustration of the "Tech Arctic" scenario is made on the Figure 1.3.

Chinese Arctic 2035

This is the story of how China has become the dominant force in the Arctic, using cooperation with Russia to extend its influence. In 2035, China has wide access to Arctic infrastructure and resources, and is also eager to project itself as the chief global architect of "Net Zero".

Big picture in 2035

Through patience and perseverance, China has managed to change the power dynamics, cooperation patterns, and commerce in the Arctic. In 2035,

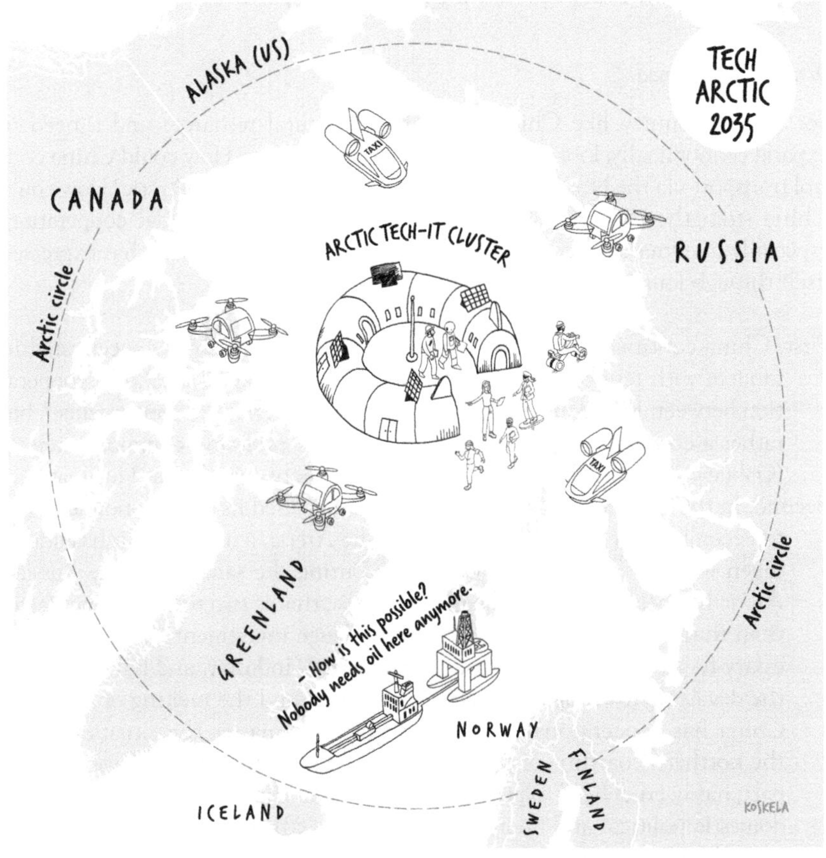

Figure 1.3 Illustration of the Tech Arctic 2035.
Source: Authors.

international arrangements in the Arctic increasingly favor the Chinese-Russian axis. At the same time, China has become the world's largest economy. The key to the country's success has been a shift in emphasis from traditional industry and capital-intensive export to technology, services, and a commitment to "Net Zero". China has become the world's leading superpower, and the Arctic has given the country new networks, revenues, and options to shape events.

By 2035, China has maneuvered itself into a position as the de facto hegemon in the Arctic. From both an economic and an institutional perspective, the writing is on the wall. And the Chinese have managed to achieve this without firing a shot or threatening military action, even if the country by that date maintains a sizeable military presence in the region. The Chinese have managed to fulfill the dictum of their historic strategist Sun Tzu, who in the 5th century BC stated: "To subdue the enemy without even fighting is the supreme act of war".

In 2035, however, Russia has second thoughts. Russia is on the verge of severing its strong ties with China and turning toward other partners.

What has happened?

For a large country like China with limited natural resources and a need to expand economically, looking to the north was necessary. How could China control transport via the NSR and exploit natural resources in the Arctic? How could China strengthen its strategic position and role in international cooperation, especially in climate matters? China's increasing role in the Arctic has expressed itself through four factors:

First, China continued close cooperation with Russia, which has accelerated in tandem with tensions between the West and Russia. By 2022, the cooperation between Russia and China is no longer a "marriage of convenience", but rather a coming together of the two countries' profound common interests (Gabuev, 2021). In 2035, cooperation is even more complex and intimate.

Second, in the years leading up to 2035, China has used its vast economic power to become the economic superpower of the Arctic, a move that it has undertaken with the concurrence of Russia. During the same time, the Russian Arctic has been subject to a temperature rise that is two to three times faster than that found in the rest of the world. Huge investments have been necessary to protect Russian Arctic infrastructure, industry, and buildings from the devastation wrought by high temperatures and the melting of methane. China has concentrated on helping Russia to master the situation along the northern coastline of Siberia, where the NSR would pass. Russia, for its part, has been obliged to prioritize protection of the interior of the Arctic for domestic political and financial reasons.[9] In 2031, huge offshore deposits of critical minerals needed for the development of renewable energy were found in the Laptev Sea at the mouth of the Lena River. China offered Russia to take charge of developing these resources. Russia accepted the offer from its closest strategic partner, while China ended up with a majority share in the project. China initially also tried to invest in projects in Western countries but was rebuffed for "security reasons", largely on the initiative of the US. Around 2030, however, as the US retreated into increasing isolation, there was a new rapprochement between China and the EU, and such investments were again accepted.

Third, China supported the idea of "Net Zero" and broadly adhered to its principles in contrast to Russia. The reason was simple: many of the negative climate consequences affecting China had their origins in the Arctic. Concretely, this meant that while China was working diligently to open the Arctic for both mineral and some fossil exploitation, these projects would only be carried out under the strictest supervision, ensuring truly sustainable, cutting-edge outcomes. The first blue hydrogen and ammonia projects using CCUS (Carbon Capture Utilization and Storage) in the Arctic were organized by Chinese companies, which also invested heavily in wind power. The Russians viewed these projects with interest but chose not to make major investments.

Fourth, in parallel with China's investments, China slowly, methodically, and with great skill strengthened its cooperation with other countries through international institutions. At the same time, China officially always paid lip service to the principles of the Law of the Sea.

In 2029, China managed to obtain full membership of the Arctic Council, arguing that its government and businesses have control over more than 50% of all investments in the Arctic. From then on, it managed to initiate a series of initiatives that favored the Sino-Russian axis, such as strengthening the Arctic dimension of the Shanghai Group. China also slowly tightened its grip on the NSR and invested in the newly independent Greenland. Last, but not least, China took full advantage of its formal position as one of the signatories of the Svalbard treaty. By 2030, several countries started a strong "pushback" against Norway and the way it was exercising its sovereignty over Svalbard. This gave China a new firm foothold in how the Arctic was governed.

However, by the mid-2030s, rifts between China and Russia started to appear. "Net Zero", which until then had relatively minor consequences for the actual export of oil and gas from Russia, suddenly "kicked in" (see, e.g., Gustafson, 2021, p. 210, and the chapter by P. Nore in this volume). And what many traditional Russian analysts interpreted as yet another "temporary setback", proved to be a turning point. Dramatically decreasing sales of oil, gas, and coal ensued. And Russia's key partner China was leading the change that would have profound consequences for state income, economic growth, and social stability. The realization that almost all other countries had accepted ten years before finally hit Russia. And what was possibly even worse: with respect to diversification of its economic structure, Russia found itself in almost the same position in 2035 as it did in 2022. No strong alternatives to the dominance of the fossil fuel industry had been developed, especially not in the Arctic, while the brightest and best educated young people in Russian society for many years had led the exodus out of the country.

The first post-Putin government understood much more clearly than the previous regime that Russia faced a truly existential choice. And it did not hesitate. The new government which took office in the autumn of 2035[10] recognized that the very close relationship with China, while important in many respects, stood in the way of necessary change. Russia therefore broke with China and started to look for other global partners.

In brief

The objective of "Net Zero" has given China legitimacy and influence in the Arctic. China uses its sway over the NSR to foster trade and to impose its will on Arctic countries. However, by 2035, the Sino-Russian alliance is effectively

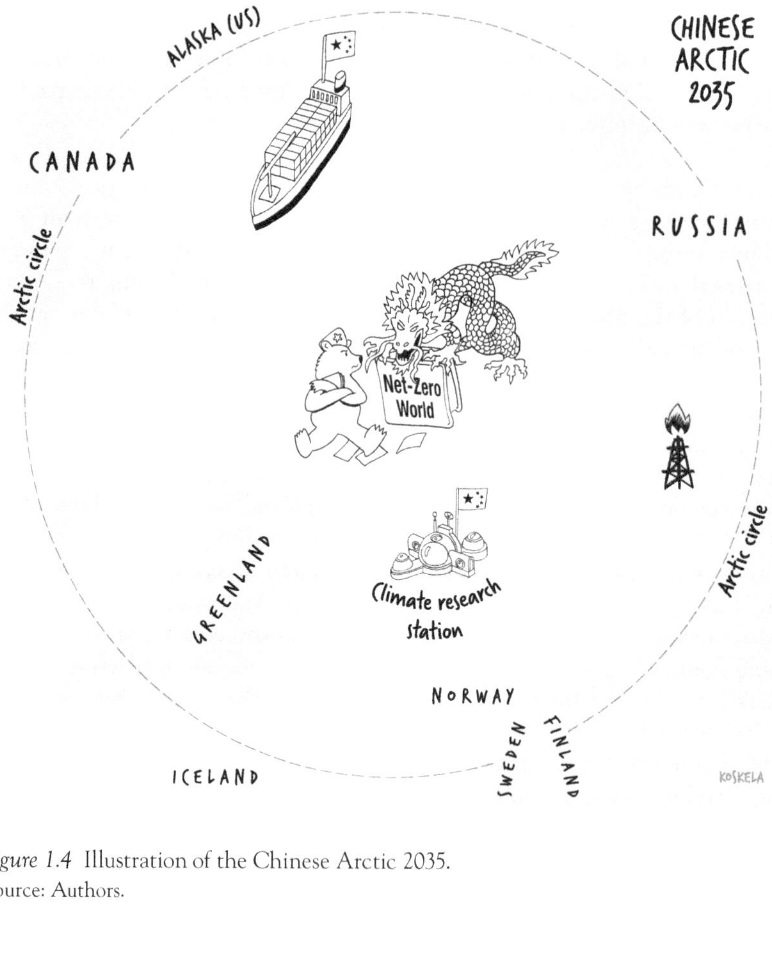

Figure 1.4 Illustration of the Chinese Arctic 2035.
Source: Authors.

over, and no one can rule out that Russia may attempt to take back control of the NSR and actively oppose Chinese policies. Illustration of the "Chinese Arctic" scenario is presented on the Figure 1.4.

EU Arctic 2035

This is the story of how the EU has taken the lead in turning the Arctic into a "Northern Sanctuary" where extraction of oil, gas, minerals, and other important resources is expressly forbidden. In 2035, the Arctic has become a global beacon of hope and at the same time a source of great frustration for many of the actors located in the Arctic.

Big picture in 2035

The resource lockdown is appealing because it marks a clean break with the sins of the past: No further efforts to explore and extract the chief source of CO_2 emissions – fossil minerals – will be allowed. The ban is unambiguous and "smart", at least on the surface. In 2035, the EU provides funding for small, digitally advanced "smart" communities in the circumpolar coastal areas and for measures to protect or recreate the unique biodiversity found in Arctic and sub-Arctic areas. The EU also promotes eco-friendly maritime transport in the Arctic Ocean. The new "smart" communities are cool in both senses of the word, and they attract a small band of artists, recluses, and researchers from all over the world. The original inhabitants in Arctic and sub-Arctic areas, however, tend to be skeptical of these developments.

While the resource lockdown is a source of pride for EU politicians, by 2035 it is increasingly contested politically. Sometimes complaints and civil cases are brought before national and international courts. The lockdown is also quietly ignored by various corporations, oligarchs, and indigenous groups. One strategy is to complicate the monitoring of the oceans undertaken by EU research agencies by sabotaging subsea surveillance systems. Some obstruct the EU's efforts by jamming signals and creating streams of false data. Seen from a distance, the sanctuary appears to be functioning well and the bureaucratic procedures of the EU seem to be comprehensive and foolproof. In 2035, other geographical areas and political problems are in the spotlight in Brussels. Researchers and environmentalists despair because the lockdown, ironically, diverts attention from the Arctic.

By 2035, the Arctic Council, which was sidelined by the EU in discussions on the future status of the Arctic, has become a forum for dissent and at times angry resistance to the resource lockdown. New technology has emerged, and experts maintain that energy and minerals can be extracted in a much gentler way than before. Mining in the Arctic could help secure precious, rare minerals. Russian and Nordic actors argue that the rules and regulations should be more flexible and lenient, while the EU Commission responds that exceptions are dangerous because they would undermine respect for the "Northern Sanctuary" and generally weaken resolve in the global struggle for carbon neutrality and a just green transition.

What has happened?

In the mid-2020s, the demands for visible, radical action to combat climate change reached fever pitch. No politician could ignore the demands for a dramatic reduction of CO2 emissions. However, few organizations were able to offer concerted action on a grand scale, and in most instances, strong interest groups could block progress and play for time by insisting on special transitional arrangements and by asking for further scientific studies. Storing CO2 underground using Carbon Capture and Storage (CCS) technology was increasingly seen as a ruse and too risky. In this situation, the EU found that an unequivocal resource lockdown in the Arctic would cut the Gordian knot. The "Northern Sanctuary" could be used as a rallying point for environmentalists, politicians, scientists, and young people wishing to make a difference. It should also be noted that continental European countries, the EU Commission, and the European Parliament had little to lose in economic, social, and political terms, in contrast to the Arctic states, communities, and companies.

So it was that a string of qualified majority decisions was pushed through in the EU in 2025. The Arctic was considered "easy prey" and a chance for the EU to assert itself as a global champion of sustainability, biodiversity, and justice for marginalized, indigenous peoples.

The EU Commission persuaded and coerced the Nordic countries outside the EU – Norway, Greenland, and Iceland – into accepting the lockdown: the "Northern Sanctuary" was presented as an example of Nordic values and environmental ideals put into practice. Russia, which was keen to escape from the political and economic isolation that it had suffered for several years, paid lip service to the idea of the new "Northern Sanctuary". So long as Russia could retain control of the Northwest Passage in the face of growing international interest, this would be an acceptable concession. Canada and the US, which could not be seen to obstruct efforts to fight global warming in any shape or form and which also wanted to be recognized as allies of indigenous peoples in the north, quickly gave their assent. Even China found it difficult to object to the resource lockdown as it was firmly committed to a policy of "Net Zero".

The EU immediately initiated programs and projects to strengthen development in the Arctic regions. Far north in Greenland, a special reservation for polar bears was set up. Here, vulnerable yet obviously quite dangerous polar bears could be viewed by well-protected tourists with expensive cameras. Data center facilities on Svalbard and Novaya Zemlya were supported. However, transport and popular tourism were strictly regulated, all activities having to comply with an increasing number of EU directives. Only a few smart communities in the circumpolar coastal areas seemed to flourish after the initial EU honeymoon in the Arctic.

The lockdown was heavily criticized, especially by organizations, industries, and communities in Norway, Russia, and Greenland, for stifling natural growth and encouraging reliance on subsidies and funding from outside sources. On top of this, European and national programs were often seen as incoherent, contradictory, and even too theoretical and ambitious. With permafrost thawing and forest fires adding to the worries of Arctic communities, infrastructure and economic growth should be given priority, not "eco villages". Critics argued that the Arctic cannot be protected in the same sweeping way as Antarctica, not only because the resources in the Arctic are plentiful, valuable, and highly needed, but also because the Arctic is home to indigenous populations and old settlements.

The strongest criticism, however, was voiced over the rigid fisheries policies and food, health, and safety procedures that came on the heels of the new ban on the extraction of subsea minerals imposed by the EU in 2032. Fishing interests in Norway asserted that the EU had overreached itself and that the new policy played directly into the hands of Russian and Chinese pirate trawlers.

In brief

The "Northern Sanctuary" seemingly heralded a new age of sustainability and a commitment to a greener and more prosperous Arctic. But, in truth, not much is happening in the region. In 2035, emissions have gone down but temperatures are still rising and the infrastructure is still deteriorating. Quite possibly, EU policies will be met with open defiance and the sanctuary be rolled back in the years to come. Illustration of the "EU Arctic" scenario is presented on the Figure 1.5.

Eight wild cards

Nobody knows what the world and the Arctic will look in 15 years' time. That is why we need both scenarios and "wild cards". "Wild cards" are commonly defined as low-probability, high-impact events. Wild card events occur suddenly, seemingly out of the blue, and they tend to have irreversible effects. Frequently, they are exogenous to the system and the trends that underpin the main scenarios, yet sometimes they are tipping points of an underlying trend (Overland et al., 2015). Examples of wild cards are the fall of the Berlin Wall and the terrorist attacks against the US in September 2001. Wild card events divide history into a "before" and an "after". Although difficult to imagine, wild cards are highly useful in scenario projects such as this one because they help to stretch thinking about what could plausibly happen.

We have conceived of eight wild cards that would significantly change developments in the Arctic at some point in the next 15 years. Please note that the wild cards are game changers that break with the logic in the scenarios and that they therefore stand alone. We invite the reader to reflect on the possible implications of the wild cards.

1 A sudden and dramatic acceleration in climate change

A sudden and massive release of methane from the permafrost in Siberia occurs, leading to environmental, economic, and infrastructural devastation

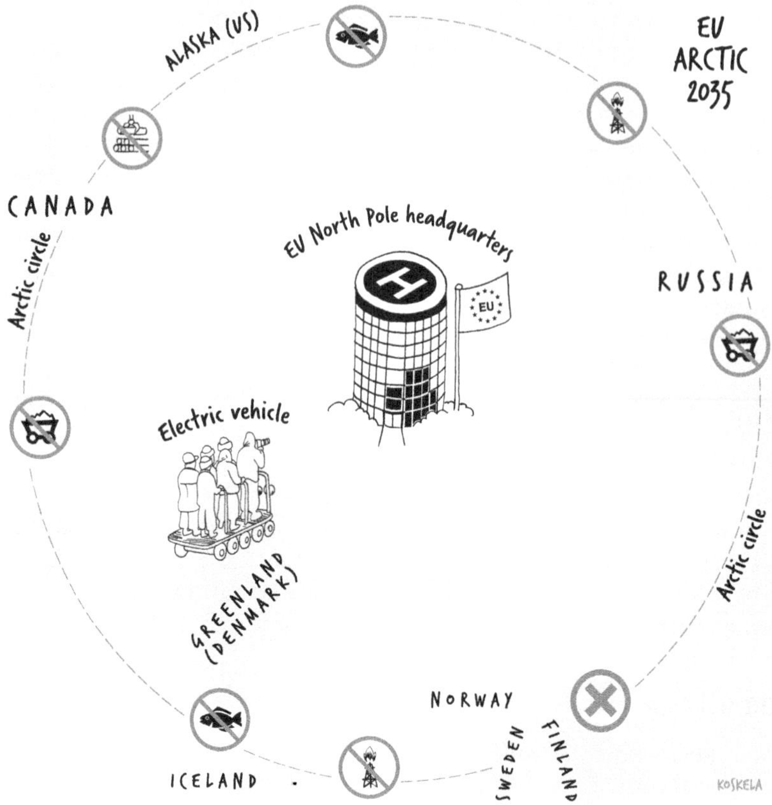

Figure 1.5 Illustration of the EU Arctic 2035.
Source: Authors.

in Artic and sub-Arctic areas and to a radical change in both regional and global weather patterns. Key ecological processes are altered and the degradation becomes irreversible.

2 A pandemic that originates in the Arctic

A dangerous and highly contagious virus is released as old carcasses thaw in the Siberian permafrost. The virus, which is of an unknown kind, takes both the scientific community and authorities by surprise. Life in the Arctic becomes even more precarious.

3 A new global financial crisis

Following a collapse in the Chinese property market, the global financial system implodes. This crisis is much more severe than that of 2008, as it pulls out the rug from underneath global supply lines. Basically, only local and regional barter trade is possible in the Arctic.

4 The Suez Canal is closed
A radical Islamist group seizes control of Egypt. International trade is actively discouraged, and the Suez Canal falls into disrepair. This dramatic turn of events necessitates new alternatives and makes the NSR critically important. Accelerating infrastructure projects in the Arctic becomes urgent.

5 Russia turns to the West and goes green
Russia becomes a state that respects European liberal values. New policies are implemented to strengthen the rule of law, compliance with international law, and human rights after the first post-Putin new Russian government is elected. The new government surprises electors by giving priority to a close alliance with the EU and a massive political swing in a green direction. Much of Russia's oil, gas, and coal production is shut down.

6 Deep-sea metal mining alters the economy
Vast deposits of copper, zinc, cobalt, and other valuable metals are discovered in Norwegian and Russian waters. These metals are crucial for the green shift. Deep-sea metal mining alters the economy in Norway and Russia and hastens the end of oil and gas.

7 China's economic and political model weakens, and the country turns inward
The Chinese Communist Party (CCP) tries to regain full control of the economy but fails. Propping up the property market, monitoring the internet, and micromanaging entrepreneurs proves impossible. The CCP loses its grip on power and the Chinese state descends into chaos. Arctic countries push back against the Chinese Arctic initiatives.

8 The Arctic becomes highly militarized
Geopolitical tensions mount, and new technologies dramatically increase the importance of surveillance. A breakdown in trust between the West, Russia, and China leads to an intense arms race and a struggle for dominance in the Arctic. Arctic and sub-Arctic areas witness a massive military build-up.

December 2021

(Subsection *Geopolitical tensions will remain high* and wild card 8 *The Arctic becomes highly militarized* were included in May 2022)

Ex-post reflections

The original chapter on Arctic cooperation scenarios was finished in December 2021. The Russian invasion of Ukraine on February 24, 2022, was a game changer, seemingly a "wild card" event that has divided history into a before and after. As of today, there is no consensus on how to stop the ongoing Ukraine war by diplomatic means. The world has changed and may be on the verge of even more dramatic geopolitical changes.

In our chapter, we have presented two fundamental uncertainties about Arctic cooperation: attitudes to the Arctic resources (fragmented vs. coordinated response to climate change) and response to climate change (Arctic lockdown vs. Arctic resource extraction). These uncertainties remain valid and important and should be included in further scenario work. However, the war in Ukraine has brought to the forefront one more fundamental uncertainty – the future security architecture in Europe. This war is unfolding in the heart of Europe, and most European countries are indirectly yet heavily involved (by introducing serious, often double-edged economic sanctions and by providing military equipment and weapons to Ukraine). The outcome is uncertain, and there is always the risk of escalation.

Are we moving toward a new security architecture in Europe? Will we see a shift from a fairly stable security situation to one that is (perhaps for a long time to come) more precarious? The future architecture will be heavily influenced by whether a relatively clear resolution of the Ukraine war can be achieved, and further, by how this resolution comes about. A big question is if a peace deal that is satisfactory to all the major parties can and will be achieved. In this case, the security architecture will be stable even though tensions probably will persist. The resolution can also come as a result of the decisive victory of one of the parties and the strategic defeat of the other. What happens then is extremely unclear. If, on the other hand, no clear resolution is achieved, we may be entering an era of a continuous hostility, marked by more and less active phases of fighting and subversion on shifting fronts. We may experience a confusing blend of political, psychological, economic, and military warfare.

Most of the parties directly and indirectly involved in the Ukraine conflict have interests and ambitions in the Arctic. As has been pointed out throughout this volume, the Arctic region is unique in various ways, also in the sense that it is vulnerable. That is why this third major uncertainty – will we have a stable or unstable security architecture in Europe? – must be taken into consideration in further work on Arctic scenarios. The uncertainty we have briefly touched upon here will affect Arctic geopolitics, including prospects for keeping the Arctic as a region of low tensions. With mounting uncertainty, the urge to start a "Scramble for the Arctic", even backed with military tensions, may grow. We recommend that readers keep this additional uncertainty dimension in mind while considering our four basic scenarios.

Finally, we would express the hope that building on successful cooperation of major powers in the Arctic after 2000, a new interstate cooperation in the Arctic during these turbulent and uncertain times will facilitate peace and security, building measures around which the European and global security will start consolidating and improving.

May 20, 2022

Notes

1 For example, the Arctic Circle Assembly, which holds annual conferences in Iceland, is attended by heads of state and government, ministers, members of parliaments, indigenous leadership and representatives, officials, experts, scientists, entrepreneurs, business leaders, environmentalists, students, and activists.

2 Source: fridaysforfuture.org – a youth-led and organized movement that began in August 2018, after 15-year-old Greta Thunberg and other young activists sat in front of the Swedish parliament every school day for three weeks, to protest against the lack of action on the climate crisis. She posted what she was doing on Instagram and Twitter, and it soon went viral. In 2021, FridaysForFuture had 14,000,000 supporters from all continents.

3 BIN (Business Index North) project aims at raising awareness of business opportunities and development challenges in the Arctic. Please refer to the website of the project for more information and reports (https://businessindexnorth.com/).

4 The European Parliament and the Council of the European Union are, to varying degrees, empowered to make and execute laws at a European level in the areas of trade, business, foreign policy, and security. Supranational EU influence is also felt through the EEA (European Economic Area), to which Norway and Iceland are tied.

5 The SDGs are conceptualized as the world's shared plan to end extreme poverty, reduce inequality, protect the planet, and so on by 2030. The plan has gained considerable support worldwide, not least because it offers an almost universally accepted view of a world where "nobody is to be left behind". Many countries and institutions are committed to the SDGs and have made efforts to integrate them into their policies.

6 For example, consider the US decision made under the Trump administration to withdraw from the Paris agreement, a decision which was subsequently overturned by the Biden administration in 2021.

7 Please refer to "Everything is forbidden unless it is permitted" by Christine Van Geyn – an interesting discussion of these two maxims in case of Canada (https://theccf.ca/everything-is-forbidden-unless-it-is-permitted/).

8 The "Kodak moment" refers to a situation that occurs when a business fails to foresee disruption in the environment and sticks to a "business-as-usual" strategy despite warning signals. The well-known company Kodak, which produced analog cameras and was considered highly successful, was overtaken by digital picture producers. The Kodak moment signifies a sudden and total collapse of the very foundations of business.

9 This story of "two Arctics" is inspired by Gustafson (2021) "Klimat", Harvard University Press.

10 Vladimir Putin, according to the Russian Constitution, can be President until 2036, but this scenario assumes that he resigns one year before that date "for health reasons".

References

Amer, M., Daim, T.U. and Jetter, A. (2013). A review of scenario planning, *Futures*, 46, pp. 23–40.

Bourmistrov, A. et al. (eds.) (2015). *International Arctic Petroleum Cooperation: Barents Sea Scenarios*. London and New York: Routledge.

Bradfield, R., Wright, G., Burt, G., Cairns, G. and van der Heijden, K. (2005). The origins and evolution of scenario techniques in long range business planning, *Futures*, 37(8), pp. 795–812.

Brigham, L.W. (2007). Thinking about the Arctic's future, *The Futurist*, Sept–Oct, pp. 27–34.

Burger-Helmchen, T. (2008). From investment rules of thumb to routines: A real option approach, *Problems and Perspectives in Management*, 6(1), pp. 64–73.

EU Commission. (2019). The European Green Deal – Communication from the Commission to the European Parliament, the European Council, the Council, the European Economic and Social Committee and the Committee of the Regions (December 11, 2019).

Gabuev, A. (2021). The emergence of a pax sinica including Russia could draw new dividing lines over Eurasia. *Carnegie Moscow Center*. Retrieved 17 December 2021 from: https://carnegiemoscow.org/commentary/84135.

Gustavson, T. (2021). *KLIMAT: Russia in the Age of Climate Change*. Cambridge, MA: Harvard University Press.

Haavisto, R., Pilli-Sihvola, K., Harjanne, A. and Perrels, A. (2016). Socio-economic scenarios for the Eurasian arctic by 2040, *Finish Meteorological Institute*, 1, pp. 1–65.

Heininen, L. and Finger, M. (2018). The "global arctic" as a new geopolitical context and method, *Journal of Borderlands Studies*, 33(2), pp. 199–202. https://doi.org/10.1080/08865655.2017.1315605

Hjort, J., Karjalainen, O., Aalto, J., Westermann, S., Romanovsky, V.E., Nelson, F.E., Etzelmüller, B. and Luoto, M. (2018). Degrading permafrost puts arctic infrastructure at risk by mid-century, *Nature Communications*, 9(1), pp. 1–9. ISBN 978-0-8020-6041-9

Huntington, S.P. (1993). The clash of civilizations? *Foreign Affairs*, 72(3), pp. 22–49.

Huntington, S.P. (1996). The *Clash of Civilizations and the Remaking of World Order*. New York: Simon & Schuster.

Keohane, R.O. and Nye, J.S. (1973). Power and interdependence, *Survival*, 15(4), pp. 158–165.

Kristoffersen, B. and Langhelle, O. (2017). "Sustainable development as a global-arctic matter: Imaginaries and controversies." In: Keil, K. and Knecht, S. (eds.). *Governing Arctic Change*. London: Palgrave Macmillan. https://doi.org/10.1057/978-1-137-50884-3_2

Krivorotov, A. (2018). China's polar policy: Gaining momentum, obtaining scope, multiplying vectors, *China in the Global and Regional Politics: History and Nowadays*, 23, pp. 183–199.

Lazariva, A., Kalinin, A., Middleton, A., Nilssen, F. and Belostotskaya, A. (2021). *Arctic 2050: Mapping the Future of the Arctic*. Moscow: SKOLKOVO Institute for Emerging Market Studies.

McBride, J. (2021). How green-party success is reshaping global politics. *Council on Foreign Relations* (www.cfr.org). Retrieved 26 October 2021 from: https://www.cfr.org/backgrounder/how-green-party-success-reshaping-global-politics.

McLuhan, M. (1962). *The Gutenberg Galaxy*. Toronto, Canada: University of Toronto Press.

Miller, K.D. and Waller, H.G. (2003). Scenarios, real options and integrated risk management, *Long Range Planning*, 36(1), pp. 93–107.

Myllylä, Y., Kaivo-oja, J. and Juga, J. (2016). Strong prospective trends in the arctic and future opportunities in logistics, *Polar Geography*, 39(3), pp. 145–164.

Overland, I., Bambulyak, A., Bourmistrov, A., Gudmestad, O., Mellemvik, F. and Zolotukhin, A. (2015). "Barents sea oil and gas 2025: Three scenarios". In: Bourmistrov, A. et al. (eds.). *International Arctic Petroleum Cooperation: Barents Sea Scenarios*. London and New York: Routledge.

Petrov, A.N., Smith, M.S.R., Krivorotov, A.K., Klyuchnikova, E.M., Mikheev, V.L., Pelyasov, A.N. and Zamyatina, N.Y. (2021). The Russian arctic by 2050: Developing integrated scenarios, *ARCTIC*, 74(3), pp. 306–322.

Schacht, K. and Koschyk, M. (2019). How war has changed since 1989. *Deutche Welle*. Retrieved 23 May 2022 from: https://p.dw.com/p/3SZ2K

Schatzmann, J., Schäfer, R. and Eichelbaum, F. (2013). Foresight 2.0 – Definition, overview and evaluation, *European Journal of Futures Research*, 1(15).

Schoemaker, P.J.H. (1993). Multiple scenario development: Its conceptual and behavioral foundation, *Strategic Management Journal*, 14(3), pp. 193–213.

Schoemaker, P.J.H. (1995). Scenario planning: A tool for strategic thinking, *Sloan Management Review*, 36(2), pp. 25–50.

Sovacool, B.K., Monyei, C.G. and Upham, P. (2022). Making the internet globally sustainable: Technical and policy options for improved energy management, governance and community acceptance of Nordic datacenters, *Renewable and Sustainable Energy Reviews*, 154, pp. 111793.

Part I

International Governance and Security

2 International Governance Facilitating Sustainable Development in the Arctic – The Arctic Council as a Multi-Role Actor and Forum

Stefan Kirchner and Timo Koivurova

Introduction

The challenge of Arctic sustainability

Given the small number of people living in the Arctic, with approximately 4 million living along and above the polar circle, the Arctic takes up a disproportionately large amount of international attention, in particular on the part of policymakers. More and more states, including many non-Arctic nations, are creating Arctic-specific policy documents (Heininen, 2020). The interest in the Arctic can be explained in part by the global impact of the effects of climate change on the Arctic (from a governance perspective cf. Moe, 2020), such as the rise of sea levels or climate feedback loops (cf. Goosse et al., 2018), but also by a sense of "opportunities for commerce and business in a warmer Arctic" (Dodds and Nuttall, 2016, p.119). The latter approach perceives the Arctic as a source region for natural resources, such as hydrocarbons (Dodds and Nuttall, 2016, p.123 et seq.), which themselves are a key driver of climate change. In the Arctic itself, the tension between the need for economic development and the protection of the natural environment, on which many in the Arctic depend for their livelihoods, can be felt most acutely. Here, sustainable development is not a lofty goal but essential for the continuation of many communities that are facing unprecedented challenges. Alongside several other efforts aimed at enhancing the sustainability of the economic development of the Arctic, which is sparsely populated and often characterized by decoupled economies (Durfee and Johnstone, 2019, p.114) and limited alternative income streams, the Arctic Council (AC) plays a key role in the efforts to achieve and promote sustainability in the Arctic. Sustainability is a challenge, but it is a challenge that Arctic communities or states do not need to face alone. This chapter aims to provide a contribution to the ongoing discourse on sustainable development in the Arctic within the AC.

Purpose and outline of this chapter

The purpose of this chapter is to attempt to assess the future potential of the AC in advancing sustainable development in the Arctic. We will look back at

DOI: 10.4324/9781003246015-3

earlier experiences of the AC, its successes, and some of the challenges it has been facing, but also at potential future developments. To this end, we will first look at the long-term challenges the Arctic is facing today and which are likely to remain relevant until 2035. The two major challenges which fit this description are globalization and climate change. In the next step, we will look at sustainability and in particular sustainable development in the Arctic as it is today. While a complete overview of the issue of sustainability in the Arctic would exceed go beyond the scope of this chapter, we will focus on different sustainable development goals (SDGs) to provide the reader with snapshots of the status quo in the Arctic. The focus here will be on the situation of the people who live in the Arctic. From this close-up and human-centered approach, we will then take a step back and look at governance in the Arctic and in particular at the institutions involved in international Arctic governance. Here, we will pay particular attention to the AC and its work on sustainable development. This will provide the basis from which we will approach the core task of this chapter, when we will attempt to look into the future and venture an estimate of the future role of sustainability in the AC before concluding with a brief summary and outlook.

Expecting the unexpected

As the COVID-19 pandemic has shown, any attempt at forecasting will be limited by the possibility that hard-to-predict events will occur. Events that appear unlikely just a short time before they actually happen, such as the pandemic or the end of the Cold War, can have high impacts, which can fundamentally alter the (geopolitical) landscape on which the AC operates. Some of these "wildcards" can be predictable to a certain degree, although their likelihood at any given time and their specific actual impact can be difficult to predict. Examples can include extreme natural events, like the 2004 and 2011 earthquakes and tsunamis, or events which are likely but the details of which can matter a great deal. The likelihood of such disturbances may be lower in the Arctic than in other parts of the world, and the history of the AC shows that cooperation despite severe political differences is one of its key strengths. One reason for this is that the attention of the AC has so far been focused on issues that are not very sensitive from a political perspective (e.g. scientific research cooperation) and which have a clear practical value for everybody living in the Arctic (e.g. search and rescue operations).

The binding international treaties concluded by Arctic states under the auspices of the AC, the 2011 Agreement on Cooperation on Aeronautical and Maritime Search and Rescue in the Arctic, the 2013 Agreement on Cooperation on Marine Oil Pollution Preparedness and Response in the Arctic, and the 2017 Agreement on Enhancing International Arctic Scientific Cooperation (see Arctic Council, 2021) have also been accepted easily because they were not created in a vacuum. Instead, they have a solid basis in the work done by many experts from the different member states in working groups and task forces. These treaties, but also the Central Arctic Ocean Fisheries Agreement (CAOFA; formally the Agreement to Prevent Unregulated High Seas Fisheries in the Central Arctic Ocean; see Vylegzhanin et al., 2020), have their roots in nonbinding, soft law norms, such as guidelines. The international legal community of the Arctic, states and non-state actors, in particular indigenous representative organizations, has shaped these norms. This increases the acceptance of the AC as a forum not only for cooperation but also for the future development of international norms of Arctic Law, notably norms which are as politically unproblematic as those which have already been moved from the soft law to the hard law level.

Key challenges for the Arctic

Climate change and its impact on the Arctic

Climate change is the key factor influencing the Arctic today. While it is possible that action will be taken against some short-lived climate forcers, such as black carbon (cf. Arctic Council, 2019, p.9), and that it is not too late to act against climate change (cf. Herring and Lindsey, 2020), it can be assumed that climate change will continue to be the core challenge in the Arctic in 2035. The impact of climate change is not restricted to the loss of sea ice cover and volume in the Arctic Ocean. In the year 2035, for which this chapter is intended to provide an educated attempt at forecasting, the Arctic Ocean may already be largely ice-free during the summer months. The ice sheets are losing both size and volume and permafrost and glaciers in the northern hemisphere are disappearing rapidly. Already today, the melting of permafrost is causing infrastructure problems, affecting roads and settlements in Canada and Russia, often cutting off local communities and increasing their vulnerabilities. This, in turn, can lead to more migration from rural to urban areas and to outmigration from the Arctic – and to a further loss of services for those left behind. Communities across the Arctic are also impacted by climate change due to their dependence on the natural environment for food security and livelihoods: fish stocks migrate poleward as the oceans' temperatures rise, providing some economic opportunities, as in the case of the mackerel fisheries in Iceland (Astthorsson et al., 2012). Often, however, these opportunities will be only short-lived and shifting between different marine spaces, opening up the possibility of conflicts between coastal and flag states, for example, in the fisheries industries.

Globalization and its impact on the Arctic

Ensuring the sustainability of economic development in the Arctic is already difficult because local economies in the Arctic often face structural disadvantages. Traditional livelihoods remain relevant not merely out of concern for the preservation of local, including in particular indigenous, cultures, but also out of necessity, especially in fields that are directly related to the production of food, such as hunting, fishing, and herding. Food available in stores in remote parts of the Arctic is often less healthy and very expensive when put in relation to the locally available purchasing power. Many Arctic communities have a strong focus on extractive industries and tourism, a trend which in some parts of the Arctic is likely to continue as the region is perceived from the outside as both a source of natural wealth and an "exotic" travel destination. But while the nature on which local communities (and in the context of tourism also local cultures and the communities themselves) depend is part of the value chain, only a part of the profits generated by tourism or the extractive industries actually remains in the Arctic, for example, in the form of local tax revenue or profits by locals who provide goods or services. There is no uniform picture across the Arctic and situations may vary significantly between different Arctic nations due to differences in tax legislation and economic development approaches. The negative consequences in the form of environmental impact, however, will continue to be borne and felt by local communities for many years.

This is felt most acutely in the high Arctic, where nature needs a long time to regenerate. In many communities, the lack of alternative economic opportunities contributes to a loss of population due to outward migration, away from the rural Arctic. More recently, the widespread availability of high-speed internet in some parts of the Arctic and sub-Arctic and the perception of these regions as the last holdouts of cold temperatures (and hence lower cooling costs when compared to more southerly regions) have made some Arctic locations attractive destinations for the information technology sector, for example, data centers. Similarly, the emphasis on telework due to the COVID-19 pandemic and the relatively good internet connectivity in the Arctic allows highly trained workers to remain in or return to their respective Arctic communities. This, however, only applies to a very small part of the Arctic: the more densely populated and economically more developed areas.

The second contemporary impact felt in the Arctic is globalization. Like climate change, it is perceived in multiple ways, as an opportunity and as a problem. Arctic economies depend on and benefit from globalization, with extractive industries, tourism, and remote work only among the most obvious examples. Because the Arctic does not form a coherent economic unit, the situation in Arctic regions is similar to that of export-dependent nations with limited domestic markets. Arctic economies are affected by disruptions of globalization, especially if they cannot tap into the economic potential of domestic markets. This problem can be mitigated somewhat by participation in free trade arrangements. International trade can, however, be disrupted due to factors that have their origins outside the Arctic, such as political disputes leading to the imposition of trade barriers or the ongoing pandemic. Globalization is a double-edged sword for Arctic economies. It leads to more opportunities, but not all of these opportunities provide for sustainability in development. Some actually impair the ability to engage in other, more sustainable, ways of generating income, for example, mining, which limits the ability to engage in other forms of land use, such as traditional reindeer herding. Outside economic interests may lead to economic development in the Arctic but not necessarily in a sustainable manner. An increasing outside interest in the Arctic, therefore, does not remove the need for sustainability.

Reinforcing effects of climate change and globalization

Climate change and globalization are mutually reinforcing each other. Climate change renders the Arctic Ocean more readily accessible, enabling an increase in trade. Today, many of the ships operating in the Arctic either transport raw materials such as hydrocarbons from the Arctic, use Arctic locations as tourist sites, or bypass Arctic coastal communities completely. More often than not, no income is generated in the local communities along the coast from ships that just pass by, but they bear the negative impacts such as the risk of oil spills and the adverse health effects of air pollution. These negative environmental and health impacts, in turn, put other sustainable economic endeavors at risk. The same is true of the extractive industries and tourism themselves: climate change allows for more extractive industries in parts of the Arctic, but the economic benefits for local communities are limited both in terms of the time during which income is generated and in terms of the amount received by local communities, but there are significant long-term costs, for example, if land or wetlands can no longer be used anymore after mining or due to pollution (mine runoff to rivers, etc.). Tourism is often seen as a cleaner alternative to extractive industries or as complementary to them. In many locations in the Arctic, tourism remains a seasonal business, the continuation of which results in long-term impacts on the environment, in particular in mountain and fell areas. In addition, tourism is a rather volatile business because it depends on external factors that are outside of local control, as has become apparent again during the pandemic.

Challenges for the sustainability of economic development in the Arctic

Extreme climates, a limited number of income streams, especially the reliance on extractive industries, great distances, and low population densities are just some of the challenges to sustainable economic development in the region (see also Durfee and Johnstone, 2019, p.114 et seq.). They are amplified by climate change and globalization. While it will not be possible to entirely overcome these challenges, a number of tools can be identified which can support Arctic communities on the local, national, and international levels to enhance the sustainability of economic development. In the following, we will briefly look at sustainability indicators in the Arctic before moving on to the role international governance institutions can play in supporting Arctic stakeholders and decision-makers.

Sustainable development goals and the Arctic

Sustainability is notoriously difficult to define or to measure, but the SDGs provide indicators about different potential facets of sustainability. We therefore use the SDGs as indicators for sustainability. In the Arctic, the SDGs are imperfectly realized, with some SDGs being more difficult to achieve than others. One important fact about the Arctic that has to be kept in mind is that there is not one Arctic. While the Arctic is rich in resources but "much of the wealth leaks out of the region altogether" (Durfee and Johnstone, 2019, p.100), some communities in the Arctic experience extreme poverty. There are many similarities, but also significant differences between different parts of the Arctic, for example, between urban locations such as Murmansk or Rovaniemi, rural regions with limited infrastructure, and the extremely remote regions of the High Arctic. There is no one-size-fits-all solution when it comes to efforts to achieve the SDGs in the Arctic, but it is possible to identify a few areas of particular concern across all SDGs: Climate change (SDG 13) and pollution have significant impacts on land and on water (SDGs 14 and 15). The aforementioned impact of climate change on food security in remote communities, for example, in Qaanaaq, Greenland, remains an obstacle to achieving SDG 2, "zero hunger", a problem to which private enterprises also contribute. Closely related to this are relative poverty (SDG 1) and the high cost of living in many parts of the Arctic. The January 2021 Tuluksak incident (Zidak, 2021) also showed the wider public that even access to clean water (SDG 6) cannot necessarily be taken for granted in a region characterized by a large number of wetlands. Access to health care services (SDG 3) is not only limited due to great distances and lack of infrastructure (SDG 9), but also because of limited human resources, such as physicians speaking indigenous languages. Here, digital services can provide some support for rural communities in the form of telemedicine. Education (SDG 4) is a field in which significant progress has been made in recent decades, but the options for higher education in indigenous languages remain limited. All too often, accessing higher education is associated with high cost, a need to move, and so on. These inherent limitations are due to great distances and lack of infrastructure. Education, however, provides low-hanging fruit when it comes to realizing the benefits of digitalization. Gender equality (SDG 5), on the other hand, is an SDG characterized by a significant discrepancy between the situation on paper and in reality. Many political leaders in Arctic countries are women and in theory, gender equality is a highly ranked value in many Arctic communities. These ideals contrast sharply with the actual reality of domestic violence, which remains a serious problem in many, including economically highly developed parts of the Arctic (cf. Burman and Svensson, 2018), which relates the issue to SDG 16, justice.

While there is no uniform picture across the Arctic (cf. SDG 10), some aspects are encountered across the Arctic, such as a relative lack of services (cf. SDGs 7 and 8), disconnected economies (cf. SDG 12), and outmigration (cf. SDGs 8 and 11). One of the most notable challenges regarding the realization of SDGs in the Arctic is that traditional livelihoods, which are of enormous practical importance for many Arctic communities, in particular indigenous peoples, are often not measured in economic terms. This means that they hardly, if at all, appear in statistical data on the economies of Arctic regions. In addition to the aforementioned emphasis on states, noncommercial traditional livelihoods mark another blind spot that limits the utility of the SDGs.

Solutions are already being pursued, in particular through cooperation (cf. SDG 17). In the next steps, we will ask what such (international) cooperation to achieve the SDGs in the Arctic looks like and what it might look like in the future.

Such cooperation is facilitated within regional, Arctic-specific frameworks as well as through global fora. In addition, there are a number of regional institutions and agreements which are not Arctic specific, such as the EU, the US-Mexico-Canada Agreement, or the Comprehensive Economic and Trade Agreement between the EU and Canada, which also facilitate cooperation with regard to the Arctic. Arctic-specific fora include, among others, the Northern Sustainable Development Forum (under the Northern Forum), the Barents Euro Arctic Council, the Barents Regional Council, the Nordic Council, and the Arctic Council. The AC has emerged as the key institution which brings together all Arctic states, while the five coastal nations of the Arctic Ocean (commonly referred to as A5) have occasionally cooperated with each other. In the following, the focus will be exclusively on the latter and its role in the promotion of sustainable development in the Arctic.

Sustainable development at the Arctic Council

The AC was established in 1996 and can trace its origins back to 1989, when talks started between the Arctic nations which would eventually lead to the adoption of the Arctic Environmental Protection Strategy in 1991. The eight Arctic states, the US, Canada, Denmark (of which Greenland and Føroyar are autonomous territories), Iceland, Norway, Sweden, Finland, and Russia, are members of the AC. In addition, six representative organizations of indigenous peoples, the Aleut International Association, the Arctic Athabaskan Council, the Gwich'in Council International, the Inuit Circumpolar Council, the Russian Association of Indigenous Peoples of the North, and the Saami Council are Permanent Participants within the AC. A number of non-Arctic states as well as intergovernmental organizations and nongovernmental organizations have sought and obtained observer status. This signals a growing international interest in the Arctic. This includes nations with long histories of Arctic research, such as Italy and Germany, mountain nations that share some of the challenges experienced in the Arctic, such as Switzerland and India, but also nations with an economic interest in the Arctic, such as the PRC.

The bulk of the work of the AC is organized beyond the governance framework. Monitoring, assessments, and recommendations are among the science-based tools available to the AC (see in more detail Durfee and Johnstone, 2019, p.74 et seq.). Generating science-based knowledge and disseminating it are among the most effective approaches used by the AC. In 1998, it established the Sustainable Development Working Group (SDWG). The SDWG consists of a Social, Economic, and Cultural Expert Group as well as an Arctic Human Health Expert Group, thereby covering key sustainability issues. The AC's 2017 SDWG Strategic Framework (Arctic Council Sustainable Development Working Group, 2017) covers a range of areas with which the SDWG in particular is concerned, including the reduction of inequalities, human health, education, culture and heritage, economic assessments, infrastructure, transportation, energy, water, sanitation and also the role of businesses, science, and research. This broad outlook already highlights the parallels between the 2017 SDWG Strategic Framework and the SDGs.

But is it possible to extrapolate from these currently used working methods how the SDWG will work in the future? Given the achievements across the spectrum of fields in which the SDWG and the AC's other working groups are active, the expert groups method is very likely to continue. This science-based approach has been proven successful in tackling issues of importance for the Arctic states and is accepted by the member states, permanent participants, and observers in the AC. The issues the SDWG is currently focusing on under the 2017 Strategic Framework are likely to remain relevant in the future (as well). In particular, energy and infrastructure will attract significant attention, especially with the increasing global interest in moving away from hydrocarbons as energy sources. This will have positive environmental impacts for the Arctic overall, but can lead to short-term challenges in those parts of the Arctic which are heavily dependent on the extraction of hydrocarbons. Tourism is very likely to make a comeback as soon as the worst challenges of the pandemic have been overcome. This, however, depends on conditions that can hardly be influenced by the communities which depend on Arctic tourism.

One issue which is already gaining more attention, and which will remain topical for the foreseeable future, is human health. In several Arctic countries, there is already an increasing focus on the concept of "One Health" (cf. Ruscio et al., 2015), a growing interest that has been accelerated by the COVID-19 pandemic (cf. de Garine-Wichatitsky et al., 2020), understanding environmental protection, animal health, and human health as part of the same set of challenges. The AC has long been pursuing human health from multiple perspectives and the SDWG has already been active in the field of mental health and suicide prevention, which makes it well placed to also approach human health issues in a holistic manner, and at the time of writing, there is already a One Health project at SDWG.

Another area of concern of growing importance for sustainable development in the Arctic is digitalization. Providing fast and fair access to the internet everywhere is essential for the participation of Arctic residents in the global digital economy. The work of the Arctic Council Task Force on Improved Connectivity in the Arctic is a good example of how the AC can react to new challenges. This will enable the AC to advise on related issues such as cybersecurity. Through this work, providing information to its member states, the AC can contribute to the equitable provision of access to high tech (artificial intelligence, space applications, etc.) also for small and medium-sized enterprises, which continue to play an important role for local economies in many parts of the Arctic. This is a good example of how the AC creates real value for Arctic communities. Like the inclusion of indigenous peoples' voices in decision-making processes, the emphasis on knowledge sharing is part of the core identity of the AC, which already provides access to high-value scientific information. It has neither the infrastructure nor budget to actually create high tech. In so far, it differs from technical institutions such as the European Space Agency (ESA), which cooperates with the EU and provides open access Copernicus satellite data for commercial use (European Space Agency, 2021), which, in turn, is also being used in Arctic contexts, for example, when determining sea ice coverage, in the monthly reports published by the European Centre for Medium-Range Weather Forecasts, an independent organization which utilizes the Copernicus data (see, e.g., European Centre for Medium-Range Weather Forecasts, 2021).

The AC's working groups not only provide technical information or raw data but reports which, according to Durfee and Johnstone, "are highly regarded as authoritative and impartial assessments of the state of the Arctic" (Durfee and Johnstone, 2019, p.75) and support the Senior Arctic Officials who meet in the AC with recommendations for policies (ibid., p.74).

The AC is not an intergovernmental organization (ibid., p.78), nor does it have international legal subjectivity (ibid., p.78), but it has great practical relevance for the governance of the region (ibid.). Within the limitations of its mandate, in particular concerning the sharing of knowledge, the AC can make contributions to the sustainable development efforts by different Arctic actors. The emphasis on knowledge is also important for building a basis fundament for future generations in the Arctic by facilitating high-level education and access to information. While the creation and operation of educational institutions are seen as a national and local task, educational institutions in the Arctic already have a tradition of cooperation, for example, through the University of the Arctic. Digitalization and cooperation across borders can contribute to closing the gap which still persists in the non-urban parts of the Arctic. Education, one of the SDGs (SDG 4), is essential for ensuring the sustainability of development across different fields. Today, utilizing educational opportunities usually requires physical relocation to cities. Digitalization makes education accessible of location, thereby reducing the risk of outward migration away from rural and remote areas (which, in turn, impacts sustainability efforts). Through its efforts in the field of digitalization, the AC therefore also has a positive influence on access to education. Also, in addition to the explicitly sustainability-related work of the SDWG, the work accomplished in other working groups, for example, concerning the protection of the natural environment, is closely connected to sustainable development. Sustainability requires a view of the big picture and a capacity to pay attention to detail when and wherever needed. Not all activities of the AC related to sustainable development are labeled as such.

Potential future roles for the Arctic Council

This holistic way of addressing sustainability issues is also promising when it comes to potential future developments. By providing space for different aspects, topics, and perspectives, the AC maintains the inclusiveness it already has in terms of its organization, for example, by including indigenous representative organizations. By opening the door to non-Arctic observer states, the AC recognizes the interconnectedness of the Arctic and the interests of non-Arctic states without relinquishing control over its internal working processes.

In recent years, there has been a trend toward a "legalization" of Arctic governance and soft law guidelines have been turned into legally binding hard law norms. This can be seen not only in the context of the Polar Code or the CAOFA, both of which had nonbinding soft law predecessor texts. Binding norms were created outside the "Arctic Council system", a term coined by Molenaar (2012), both of which had nonbinding, soft law, predecessor documents, but also within the work of the AC. The three treaties associated with the AC have been created on the scientific basis within the aforementioned system provided by the AC. These norms were not created by the AC but were negotiated by the member states under the umbrella of the AC, benefiting from the input by permanent participants and observers. This strategy is not new but has been built over the course of several chairmanships of the AC in the first two decades of the 21st century. The international treaties created by the state based on the work undertaken within the AC system have not been made by the AC but by the member states. This step-by-step approach, building consensus to create legal norms on the fundament provided by scientific research, which includes not only states but also permanent participants and allows for input from observer organizations and observer states, appears to be sustainable insofar as it is independent of day-to-day politics and, in fact, also largely independent of the organizational structure of the AC.

However, it seems unlikely that binding treaties will be used any time soon to advance sustainability in the Arctic. Barring major political and economic disruptions on the same scale as the integration of Western European economies in the decades after World War II, which currently appear highly unlikely, it seems reasonable to assume that there will not be a binding international treaty explicitly forcing change toward sustainability as such. In order to be meaningful in practice, sustainability will have to be realized on the local level. Top-down efforts can support the move toward more sustainability but actual change has to be implemented locally. Most importantly, economic development is a sensitive issue for nation-states. The states of the Arctic are no exception in this regard and this view is also supported by international law; the right to self-determination of peoples also contains a right to economic self-determination. It is highly unlikely that Arctic states will accept binding international legal rules having an explicit direct impact on their domestic economic development. What does appear feasible in the next 15 years is the development of more specific sets of soft law norms which contribute to sustainability in economic development in the Arctic without imposing strict obligations. Such a development, which would likely build on existing soft law tools such as the SDGs, would be in line with prior developments in the Arctic: similar to the earlier processes in the Arctic which led to the treaties created under the auspices of the AC, the Polar Code or the CAOFA, such soft law norms may eventually lead to the adoption of specific treaties. As far as economic development is concerned, the adoption of hard norms seems highly unlikely, but there is a potential to advance from soft law to hard law if the member states feel that the political conditions for this are right.

Hard law, however, may not even be needed to advance sustainability in the Arctic. This is evidenced by the fact that the AC's own working methods regarding sustainable development are in themselves sustainable. In principle, there is no reason to assume that the current present methods would not also work in the future, especially because the AC has long enabled cooperation despite significant political differences and under difficult circumstances. Barring a truly massive change of circumstances in the form of highly unlikely events with a very serious impact, the AC's success story is likely to continue. Such a "black swan", to use a term made prominent by Taleb (2007), events could include an armed conflict in the Arctic or a failed state situation in an Arctic state. Such events are theoretically possible but highly unlikely. What cannot be excluded is that conflicts elsewhere might have spillover effects on the Arctic. While such cooperation, despite differences frequently practiced in the Arctic, is politically easier in areas in the common interest and not very politicized, for example, scientific cooperation, economic and political dimensions of sustainability may to some extent restrict the willingness to cooperate. As sustainability leads to better well-being, security, safety, and prosperity of the people living in the Arctic, this is an area in which agreement is possible. As the differences concerning the origins and consequences of climate change at the 2019 Arctic Council Ministerial in Rovaniemi (cf. Koivurova, 2020) have shown, governments have different visions for the economic future of the Arctic and sustainability is not necessarily always the first priority. Sustainability requires a long-term outlook. Even a major disruption of the current global economic system on a scale far exceeding the impact of the 2008/2009 economic crisis or the COVID-19 pandemic, maybe one triggered by faster-than-imagined extreme climate disruptions, would not make the issue of sustainability irrelevant – quite the contrary. Sustainability remains a challenge for future generations.

For the mid-term, until around 2035, such major disruptions seem unlikely. The AC's previous work has continued despite major challenges such as Russia's wars in Georgia and Ukraine, the challenges of the former administration in Washington, and significant disagreements over climate change. While the AC continues to function, there are lessons to be learned for the future (cf. Koivurova, 2020). While the upcoming Russian chairmanship of the AC may lead to new questions, these are unlikely to have an impact across the board but might be limited to issues, for example, security. Despite major political differences and "challenges" (Kalinichenko et al., 2019, p.108, cf. also ibid., p.111) concerning the implementation of international law by the Russian Federation, it is unlikely that the AC will be irreparably damaged during the Russian chairmanship of the AC. By emphasizing soft law (Schramm, 2020, p.50) and focusing on what Wilson Rowe calls "low-political" (Wilson Rowe, 2018, p.93) issues, the Arctic states have created what might be considered a failsafe mechanism that has the potential to protect the international cooperation in the Arctic and the work happening under the umbrella of the AC.

As long as this minimum commitment by all Arctic states is a given, there is reason for optimism about the future of the AC. From almost informal (Schramm, 2020, p.49), environmental, and scientific (Hønneland, 2020, p.154) beginnings, the AC has evolved into a forum that begins to resemble a more classical intergovernmental organization, albeit without actually becoming one while maintaining a strong role for nongovernmental participants (Koivurova and Heinämäki, 2006). The limited role of the organization itself, the space it gives to members and permanent participants, but also the inclusion of non-Arctic actors in debates, but their exclusion from decision-making can be seen as an organizational advantage: while the AC as a forum is fairly open, for example, to observer states, binding rules are not made by the AC but in the traditional form of multilateral treaties between the Arctic states which are the member states of the AC.

Conclusions and outlook

As an institution, the AC itself gives every indication of functioning as effectively as before long beyond 2035. It not only has a future as a forum, but it also has the potential to grow as a decision-making entity, provided it remains true to its established modus operandi and that it is neither exhausted by demands of Arctic or non-Arctic actors nor hampered by lack of commitment on the part of Arctic states and permanent participants. The same positive outlook applies to the way the AC addresses sustainability issues. Its approach, science-informed governance through soft and hard law with a focus on issues that are practically relevant for the people who live in the Arctic and which transcend political considerations, can allow the AC to be a role model or at least an example for regional cooperation elsewhere, in particular in parts of the world where cooperation has not met expectations despite shared concerns.

Sustainability will remain a key concern for the AC also by 2035. Efforts to achieve and promote sustainable development in the Arctic will continue at that time. While many challenges will be similar to the way they present themselves now, some will hopefully have been overcome while new challenges are likely to emerge, in particular as climate change continues to impact the Arctic. The AC has the tools needed to address these challenges. Neither today nor in 2035, however, should it be expected to be able to solve all problems. The AC will continue to contribute to Arctic governance, not just in the form of governance through regulation but also by providing governance through the sharing of scientific knowledge. It will, however, only be as strong as the commitment of its member states. In the future, as today, the AC's effectiveness will depend on them.

December 2021

Ex-post reflections

When we wrote the chapter "International Governance Facilitating Sustainable Development in the Arctic – The Arctic Council as a Multi-Role Actor and Forum" in 2021, it was with a conception of the AC as a success story for international cooperation. Since then, the escalation of Russia's war against Ukraine has dramatically impacted international cooperation, also in the Arctic. As a result of Russia's invasion, the AC has put on hold all meetings since early March 2022. Since then, relations between Russia and the West have deteriorated markedly, already leading to Finland and Sweden seeking to join *North Atlantic Treaty Organization* (NATO). This does not mean that international cooperation for Arctic governance has come to an end; we consider it highly likely that cooperation between the seven Western Arctic states is going to continue, but the future *form* of international Arctic governance is, for the moment, uncertain. While cooperation with Russia does not appear feasible at the moment, it is possible that this future cooperation will happen within the framework of a revised version of the AC. What is certain is that sustainable development will continue to remain a challenge for the Arctic. As one of the concerns that are shared throughout the circumpolar north, sustainable development suggests itself as an issue for cooperation across borders and the AC, in particular its SDWG has made significant contributions to the advancement of sustainable development in the region. This knowledge, these experiences, and the connections that have been forged over years are too valuable for Arctic states to give them up at this time.

May 24, 2022

References

Arctic Council (2019). *Expert Group on Black Carbon and Methane Summary of Progress and Recommendations 2019*. Accessed from: http://hdl.handle.net/11374/2411 (all URLs are up to date as of 7 March 2021).

Arctic Council (2021). *International Cooperation in the Arctic*. Accessed from: https://arctic-council.org/en/explore/work/cooperation/.

Arctic Council Sustainable Development Working Group (2017). *The Human Face of the Arctic - Strategic Framework 2017*. Accessed from: http://hdl.handle.net/11374/1940.

Astthorsson, O.S., Valdimarsson, H., Gudmundsdottir, A., and Óskarsson, G.J. (2012). Climate-Related Variations in the Occurrence and Distribution of Mackerel (Scomber scombrus) in Icelandic Waters, *ICES Journal of Marine Science*, 69, pp. 1289–1297. Accessed from: https://doi.org/10.1093/icesjms/fss084.

Burman, M., and Svensson, E.-M. (2018). Women's Human Rights in the Governance of the Arctic – Gender Equality and Violence Against Indigenous Women, *Yearbook of Polar Law Online*, 9, pp. 53–78. Accessed from: https://doi.org/10.1163/22116427_009010004.

de Garine-Wichatitsky, M., Binot, A., Morand, S., Kock, R., Roger, F., Wilcox, B.A., and Caron, A. (2020). Will the COVID-19 Crisis Trigger a One Health Coming-of-Age? *The Lancet Planetary Health*, 4, E377–E378. Accessed from: https://doi.org/10.1016/S2542-5196(20)30179-0.

Dodds, K., and Nuttall, M. (2016). *The Scramble for the Poles*, Cambridge: Polity.

Durfee, M., and Johnstone, R.L. (2019). *Arctic Governance in a Changing World*, Lanham: Rowman & Littlefield.

European Centre for Medium-Range Weather Forecasts (2021). *Sea Ice Cover for January 2021*. Accessed from: https://climate.copernicus.eu/sea-ice-cover-january-2021.

European Space Agency (2021). *Sentinel Online*. Accessed from: https://sentinel.esa.int/web/sentinel/sentinel-data-access.

Goosse, H., Kay, J.E., Armour, K.C., Bodas-Salcedo, A., Chepfer, H., Docquier, D., Jonko, A., Kushner, P.J., Lecomte, O., Massonnet, F., Park, H-S., Pithan, F., Svensson, G., and Vancoppenolle, M. (2018). Quantifying Climate Feedbacks in Polar Regions, *Nature Communications* 9, pp. 1919. doi: 10.1038/s41467-018-04173-0. Accessed from: https://www.nature.com/articles/s41467-018-04173-0.pdf.

Heininen, L. (2020). An Arctic Boom of Policies & Strategies: 32 and Counting, *Arctic Yearbook*. Accessed from: https://arcticyearbook.com/images/yearbook/2020/Briefing-Notes/6_Heininen.pdf.

Herring, D., and Lindsey, R. (2020). Can We Slow or Even Reverse Global Warming? *NOAA Climate.gov*. Accessed from: https://www.climate.gov/news-features/climate-qa/can-we-slow-or-even-reverse-global-warming.

Hønneland, G. (2020). *Russia and the Arctic - Environment, Identity and Foreign Policy*, London: I. B. Tauris.

Kalinichenko, P., Petrov, R., and Karliuk, M. (2019). International Law in Russia, Ukraine, and Belarus: Modern Integration Projects, *Russian Law Journal*, 7, pp. 107–133. Accessed from: https://doi.org/10.17589/2309-8678-2019-7-3-107-133.

Koivurova, T. (2020). Lessons from the Finland's Chairmanship of the Arctic Council: What Will Happen with the Arctic Council and in General Arctic Governance, *Yearbook of Polar Law Online*, 12, pp. 195–209. Accessed from: https://doi.org/10.1163/22116427_012010013.

Koivurova, T., and Heinämäki, L. (2006). The Participation of Indigenous Peoples in International Norm-Making in the Arctic, *Polar Record*, 42, pp. 101–109.

Moe, A. (2020). "The Dynamics of Arctic Development", in: Svein Vigeland Rottem and Ide Folkestad-Soltvedt (eds.), *Arctic Governance, Volume I, Law and Politics*, London: I. B. Taurus, pp. 9–21.

Molenaar, E.J. (2012). Current and Prospective Role of the Arctic Council System within the Context of the Law of the Sea, *International Journal of Marine and Coastal Law*, 27, pp. 553–595. Accessed from: https://doi.org/10.1163/15718085-12341234.

Ruscio, B.A., Brubaker, M., Glasser, J., Hueston, W., and Hennessy, T.W. (2015). One Health – a Strategy for Resilience in a Changing Arctic, *International Journal of Circumpolar Health*, 74, pp. 27913. Accessed from: https://doi.org/10.3402/ijch.v74.27913.

Schramm, S.O. (2020). "Pros and Cons of Institutional Complexity: The Case of Arctic Governance", in: Svein Vigeland Rottem and Ide Folkestad-Soltvedt (eds.), *Arctic Governance, Volume I, Law and Politics*, London: I. B. Taurus, pp. 45–71.

Taleb, N.N. (2007). *The Black Swan: The Impact of the Highly Improbable*, New York City: Random House.

Vylegzhanin, A.N., Young, O.R., and Berkman, P.A. (2020). The Central Arctic Ocean Fisheries Agreement as an Element in the Evolving Arctic Ocean Governance Complex, *Marine Policy*, 118, pp. 104001. Accessed from: https://doi.org/10.1016/j.marpol.2020.104001.

Wilson Rowe, E. (2018). *Arctic Governance - Power in Cross-Border Cooperation*, Manchester: Manchester University Press.

Zidak, J. (2021). Multiagency Response and Recovery Efforts Continue After Tuluksak Water Plant Fire, *Alaska Native News*, 26 February 2021. Accessed from: https://alaska-native-news.com/multiagency-response-and-recovery-efforts-continue-after-tuluksak-water-plant-fire/54385/.

3 Unboxing Arctic Security Relations and Dynamics

Andreas Østhagen

Introduction

Few places have given rise to so much speculation, hype, and sweeping generalizations as the Arctic region at the start of the 21st century. Propelled onto the agenda by flag plantings and resource appraisals a decade ago, the Arctic continues to lure researchers and journalists to venture northwards to "the next great game" (Dadwal, 2014).

However, ideas of the Arctic as an arena for political competition and rivalry are often juxtaposed with the view of the Arctic as a region of harmony and shared interests. Underpinning cooperation in the Arctic is a desire to ensure stable operating environments for extracting costly resources far away from their prospective markets, and the foreign ministries of the Arctic states repeatedly highlight cooperation (Heininen et al., 2020; Lavrov and Støre, 2010; Rahbek-Clemmensen, 2017). Scholars point to the different layers of Arctic cooperation and emphasize that the Artic has generally remained a zone of cooperation, even after the deterioration in relations between Russia and the West after 2014 (Byers, 2017; Elgsaas, 2019; Østhagen, 2016; Stephen and Knecht, 2017).

The common point in these two diverging views on Arctic political relations is the tendency to describe dynamics in the entire circumpolar region with one stroke of the brush. With rhetoric about Arctic security threats intensifying over the past decade, security challenges are seen as coherent across the circumpolar North (Jegorova, 2013; Lanteigne, 2016; Padrtová, 2017), and scholars and media alike increasingly refer to the Arctic as one region, where various types of state security interests are inherently intertwined (Borgerson, 2008; Huebert, 2013; Weber, 2015).

Security studies offer multiple approaches to the study of specific regions. An underlying assumption has been that the security concerns and priorities of states located within a region are interlinked and overlapping. Regional relations between actors may compound over time, giving rise to patterns that may not make sense from a purely systemic point of view (Frazier and Stewart-Ingersoll, 2010; Kelly, 2007). The case of the Arctic is well suited to examining the idea of a "security region." What are the characteristics of the Arctic in terms of military and state security (for more on definitions of security, see, e.g., Hoogensen Gjørv

DOI: 10.4324/9781003246015-4

et al., 2014)? Does the Arctic, as a region, share security interests and concerns—and why should that matter?

This chapter unpacks the nuances of traditional security concerns and dynamics in the Arctic in order to better understand recent developments and questions some of the assumptions underlying the concept of (security) regions more broadly. Moreover, by introducing a "level of analysis"—or, in other words, by making distinctions between state interactions that take place at different levels in the international arena (e.g., Singer, 1961; Soltani, 2014; Waltz, 1959)—we can move away from broad, sweeping generalizations on regional relations and advance the way we understand and describe security dynamics in the Arctic at different levels (for more, see Østhagen, 2021).

Unpacking the different levels of Arctic (Geo)politics

To understand how the various security region concepts fit with the Arctic in the 21st century, it is purposeful to separate them into three different levels of analysis. Naturally, these are not watertight divisions, with each level influencing the others. Yet they help tease out some of the nuances of Arctic geopolitics and unearth the security dynamics that are prevalent at different levels of international politics.

The regional (Arctic) level

As the Cold War's systemic overlay faded away, regional interaction and cooperation in the North flourished. Furthermore, as the melting ice at the turn of the millennium opened up opportunities for greater maritime activity (shipping, fisheries, oil and gas exploration and exploitation), the Arctic states began to look northwards in terms of investments as well as presence. Climate change was leading to accelerated ice melting in the north, which, coupled with high oil prices and positive estimates of the region's hydrocarbon resources (Hobér, 2011; United States Geological Survey, 2008), resulted in portrayals of the Arctic as the world's new energy frontier and northern "shortcut" to Asia (Ho, 2011; Humpert, 2013).

In particular, Russia's ambitions concerning the Northern Sea Route require presence as regards both military and civilian infrastructure and capacity (Konyshev and Sergunin, 2014; Sergunin and Konyshev, 2014; Wilson Rowe and Blakkisrud, 2014). The other Arctic states have been following suit; with more and more of their northern waters remaining ice-free for longer periods, establishing a forward presence through coast guards, patrol aircraft, and exercises has become a challenge and priority for all Arctic littoral states (Østhagen, 2020).

As the Arctic states—Canada, Denmark (via Greenland), Finland, Iceland, Norway, Russia, Sweden, and the US—placed the North on their domestic and foreign policy agendas, and non-Arctic states like China, France, Japan, South Korea, and the UK expressed interest in the north, predictions foresaw the region as the next arena for "geopolitical" conflict (Borgerson, 2008; Dadwal, 2014; Grindheim, 2009; Sale and Potapov, 2010). However, a range of studies have pointed out that many Arctic predictions have proved inaccurate, whether made before or after the deterioration in relations with Russia and the drop in oil prices in 2014. Over the past decade, scholars have produced more balanced accounts of the dynamics within the region as a whole and among the actors with stakes in the Arctic (e.g., Dodds and Nuttall, 2016; Greaves and Lackenbauer, 2016; Łuszczuk, 2016; Rahbek-Clemmensen, 2017; Tamnes and Offerdal, 2014).

In particular, the Arctic states are recognized as mutually dependent in creating a political environment favorable to investments and economic development (Østhagen, 2018). In response to the outcry and concerns about the "lack of governance" in the Arctic spurred by the growing international awareness of the region, political representatives of the Arctic states have continued to declare the Arctic to be a region of cooperation through venues such as the Arctic Council (Jacobsen, 2018). Foreign ministries in the Arctic states actively emphasize the "peaceful" and "cooperative" features of the region (Heininen et al., 2020; Wilson Rowe, 2020). The deterioration in relations between Russia and the other Arctic states that started in 2014 has not changed this (Byers, 2017; Østhagen, 2016).

The emergence of the Arctic Council as the primary forum for regional affairs in the Arctic plays into this setting (Graczyk and Rottem, 2020). The Arctic states have shown a preference for a stable political environment in which they maintain dominance in the region. This is supported by the importance attributed to the Law of the Sea and issue-specific agreements signed under the auspices of the Arctic Council. These developments benefit the Northern countries in particular, while also ensuring that Arctic issues are generally dealt with by the Arctic states themselves.

The international (systemic) level

What happens *in* the Arctic is one thing, but politics *over* the Arctic are another. During the Cold War, the Arctic held a prominent place in the political and military standoffs between the two superpowers. It was important not only because of interactions in the Arctic itself, but also because of its strategic role in the systemic competition between the US and the Soviet Union. Norway was one of only two *North Atlantic Treaty Organization* (NATO) countries (the other being Turkey) that shared a land border with the Soviet Union. Alaska was also in proximity to the far-eastern region of Russia, albeit separated by the Bering Strait. Greenland and Iceland held strategic positions in the North Atlantic, and the Kola Peninsula was—and still is—central in Russian military planning, given its unrestricted access to the Atlantic.

With the end of the Cold War, the Arctic was transformed from a region of geopolitical rivalry to one where Russia would be included in various cooperative arrangements with its former adversaries (see discussion above). Subsequently, although interaction among Arctic states and Arctic peoples increased in this period, the region disappeared from the geopolitical radar and lost its systemic importance beyond its significance to these Northern countries themselves.

Over the last 15 years, the strategic importance of the North has risen again. Recalling some of the dynamics of the Cold War, the strategic importance of the Arctic has evolved primarily because Russia is intent on reestablishing its military power at large, and the Arctic is one domain where it can do so basically unobstructed (Hilde, 2014, pp. 153–155). This comes not necessarily because of the Arctic itself, but because of Russia's dominant position in the North, with its Northern Fleet based on the Kola Peninsula, base for the strategic submarines essential to the county's status as a nuclear power on the world stage (Sergunin and Konyshev, 2014, p. 75).

Furthermore, unlike during the Cold War, China has now emerged as an Arctic actor. With Beijing continuing to assert its influence on the world stage, the Arctic is one of many regions where China's presence and interaction are components of an expansion of power in both soft and hard terms (e.g., Bennett, 2017; Guo and Wilson, 2020; Ye, 2014). China has described itself as a "near-Arctic state" as a way of legitimizing involvement from Beijing (Koivurova et al., 2020, p. 26). This is linked partly to Chinese interests, such as research and investments, but also to its position as an emerging superpower (see Koivurova and Kopra, 2020; Sun, 2014). Safeguarding Chinese interests, which range from businesses to opinions on developments related to the Law of the Sea, is part of this expansion of power (Willis and Depledge, 2014; Ye, 2014).

Although China is not an Arctic state, its growing global stature has triggered challenges, particularly from the US. Marking a shift in the cooperative Arctic rhetoric, in May 2019, US Secretary of State Pompeo lambasted both Russia and China in a speech held before the ministerial meeting of the Arctic Council (United States Department of State, 2019), and one month later the US Department of Defense (DoD) heavily criticized the same states in its updated Arctic Strategy (Office of the Under Secretary of Defense for Policy, 2019, p. 5). Pompeo's warning—that Beijing's Arctic activity risks creating a "new South China Sea" ("US warns Beijing's Arctic activity risks creating 'new South China Sea'", 2019)—shows how the US sees the Arctic as yet another arena where the emerging systemic competition between the two countries is intensifying (e.g., Tunsjø, 2018).

In other words, much of the increase in tension that we have witnessed in the Arctic—be it between NATO and Russia since 2014 or between the US and China since 2018/2019—has little do with events *in* the Arctic and everything to do with relations between these actors globally. The Arctic plays a role in these increasingly competitive relationships due to its military importance for Russia and to Chinese global economic interests in the North.

The national level

One can describe the overarching Arctic security environment in sweeping, general statements, depicting it as either driven by strategic interests and competition or dominated by regional cooperation and shared interests. However, it is important to probe deeper into the metrics of the Arctic security concerns of each actor. These are, naturally, informed by the two levels already outlined. Still, to disentangle the security dynamics of the Arctic region, we must consider how the Arctic states actually interact on a regular basis.

Central here is the role the Arctic plays in considerations of national defense. This varies greatly across the Arctic, with vast divergence in what each country chooses to prioritize and target in its northern areas in terms of national security and defense (Hilde, 2014). For Russia, the Arctic is integral to national defense considerations (Sergunin, 2014). Although these are—as described—chiefly linked to developments elsewhere, investments in military infrastructure in the Arctic have an Arctic impact, particularly for countries located close to Russia (in essence, Finland, Norway, and Sweden). For the Nordic countries, the Arctic is fundamental to national defense policy, precisely because this is where Russia—as a great power—invests considerable military capacity (Jensen, 2017; Saxi, 2011).

In North America, the Arctic arguably does not play the same seminal role in national security considerations. The Arctic has primarily been the location for missile defense capabilities, surveillance infrastructure, and a limited number of strategic forces (Østhagen et al., 2018). Many commentators argue that the most immediate concerns facing the Canadian Arctic are not defense capabilities, but the social and health conditions in Northern communities and the poor rates of economic development (Greaves and Lackenbauer, 2016). Alaska has a somewhat more prominent role in US defense policy, given its proximity to the Russian region of Chukotka across the Bering Strait; however, this cannot be compared to the role of the Russian land border in Norwegian (and NATO) security concerns (e.g., Østhagen et al., 2018).

A geographical dividing line falls between the European Arctic and the Arctic, in tandem with variations in climatic conditions. The Northern Norwegian and the Northwestern Russian coastlines are ice-free during winter, but ice—even though it is receding—remains a constant factor in the Alaskan, Canadian, and Greenlandic Arctic. Due to the sheer size and inaccessibility of the region, the impact of security issues on either side of the dividing line is, in turn, relatively low. Despite rhetoric to the contrary, Russian investments in Arctic troops and infrastructure have had little impact on the North American security outlook at large. Flyovers by Russian bombers and fighter planes may cause alarm, but the real threat to the North American states in the Arctic is limited (Lasserre and Têtu, 2016).

It is therefore difficult to generalize about how Arctic countries themselves perceive and respond to their security interests and challenges across the circumpolar North. Security—and essentially defense—dynamics in the Arctic remain anchored to the subregional and bilateral level. Of these, the Barents Sea and European Arctic stand out. Here, bilateral relations between Russia and Norway are especially challenging in terms of security interactions and concerns. Norway, a small state and NATO member, borders on a Russia intent on investing in the Arctic for regional and strategic purposes.

Since 2014, defense aspects have made relations increasingly tense, with bellicose rhetoric and a surge in military exercises (Friis, 2019; Norwegian Intelligence

Service, 2020). With Russia intent on reestablishing the prominence of its Northern Fleet, primarily for strategic purposes (albeit also with an eye toward regional development), Norway—whose defense posture is defined by the situation in its northern areas—faces a more challenging security environment (Sfraga et al., 2020).

However, bilateral dynamics—as in the case of Norway-Russia—are always multifaceted. The two states also engage in various types of cooperation, ranging from the management of fish stocks to search-and-rescue operations and border crossings. In 2010, Norway and Russia resolved a four-decades-long maritime boundary dispute in the Barents Sea, partly in order to be able to initiate joint petroleum ventures in the disputed area (Moe et al., 2011). From 2012, Norwegians and Russians living less than 30 kilometers from the border have been able to travel across the border without a visa. These cooperative arrangements and agreements have not been revoked after the events of 2014 (Østhagen, 2016; Rowe, 2018), a clear indication of the complexity of one of the most challenging bilateral relations in the Arctic.

The Arctic: an emerging security region?

In light of the above review of the three levels as well as the concept of security regions and regionalism more broadly, how can we better understand security dynamics in the Arctic? Some scholars have argued that we are witnessing the emergence of an Arctic security region, even a regional security complex, where military security interests are increasingly overlapping and intertwined (Lanteigne, 2016, p. 4; Padrtová, 2017, p. 1). The idea is that the security interests among Arctic states have become interlinked—that is, the actions of one actor impact the others—on a regional (Arctic) level.

Others argue that the foundation of the Arctic cooperative environment is not military security interests and overlap as it would be in a traditional security region; it is rather the *absence* of these concerns from general Arctic affairs—as with the specific exclusion of military security issues from the Arctic Council—that ensures peace and stability (Grønning, 2016; Rottem, 2017). As stated by Exner-Pirot (2013, p. 120), "the Arctic, fundamentally, is a regional security complex built around interdependence on environmental and ocean issues." According to Steinberg and Dodds (2015, p. 108), the Arctic is "increasingly a region that…has an institutional structure that encourages cooperation and consultation among states so as to facilitate commerce." Byers (2017, p. 394) notes that the Arctic "is of interest because Russian–Western relations in that region have been insulated, to some degree, from developments elsewhere." Keil (2013, p. 252), albeit writing before 2014, even moots the idea of a nascent Arctic security community.

During the Cold War, the entire region was subjected to superpower rivalry. The overlay of the systemic level overrode the concerns of regional players as the Arctic was turned into a frontline, complete with nuclear submarines and bombers. With the end of the Cold War, this systemic overlay receded and regional politics emerged as a driving force in the region. Canada and Finland took the lead in founding the Arctic Council to promote their focus on environmental issues (Rottem, 2017), while the global hegemon—the US—became relatively disentangled from the region.

Today, Russia has reemerged as the most active Arctic state, investing in its Arctic capabilities for both military and civilian purposes. The US was initially a reluctant Arctic actor, but it has sharpened its focus on the region—at least rhetorically—since 2019 (Conley et al., 2020). If things were to change further, the US would be able to project its power into the Arctic. Furthermore, China is engaging in Arctic issues. China's focus comes not from a position of geographic proximity, but as a consequence of its general global outreach and engagement. In other words, in the case of being influenced by *systemic* developments and rivalry, the Arctic is not only similar to most parts of the world, but is also increasingly characterized by a so-called geostrategic competition that has very little to do with the Arctic in and of itself.

Where the idea of the Arctic as a traditional security region encounters problems is with *proximity*. The importance of the Arctic to national security and defense policies differs considerably from region to region *within* the Arctic. For example, looking at Canada and Norway, the contrasts stand out. Located on different continents, these two states are arguably only loosely connected (if at all) when it comes to national security interests. The border with Russia dominates Norwegian security concerns, but Norway's security concerns and neighbor relations do not stretch across the Atlantic or the Arctic to Canada (Østhagen et al., 2018). At best, the wider security context can be said to include the North East Atlantic, specifically Iceland and Greenland, which, along with the UK, were known during the Cold War as the GIUK gap (Smith et al., 2017). The basic principle that geographical proximity spurs mutual threat conceptions—what Buzan and Wæver (2003) call "interlinkages"—does not seem to hold up *across* the Arctic. This is a result of one simple but relevant fact: the distance between Norway and Canada is far too great, and Russia is also too far removed from Canada.

Furthermore, is it possible that the Arctic is bound into a single region by a security *externality*? Barring the existential threat posed by climate change, which falls outside the scope of this article's emphasis on state and military security, the most likely candidate would be a militarily resurgent Russia. With its annexation of Crimea, its investments in military installations across the Arctic, and its increasing number of military exercises in the North (Expert Commission, 2015, p. 20; Norum, 2018), might Russia be the shared security externality that forms an Arctic security region?

However, here again we see the dividing line defined by geography and proximity. As outlined by Østhagen et al. (2018), the countries' respective positions on NATO are indicative of differing threat perceptions. If concern about Russian behavior and investments is the key factor, then this security region would also include countries outside the Arctic, including most NATO members. Moreover, it does not make sense to have a "security region" where half of the geographic domain in question—Russia—is not part of the shared security externality, but rather the source of it.

Turning to the different, yet linked, ideas of developing the Arctic into a "region," this approach seems the most fruitful for explaining *why* the Arctic is sometimes depicted as a security region despite the logical pitfalls outlined above. Foreign ministries in the Arctic countries (Wilson Rowe, 2020) as well as officials working with issues pertaining to the Arctic Council or other Arctic-specific entities seem to have had an interest in portraying the Arctic as a zone of cooperation (Heininen, 2012; Heininen et al., 2020).

The 2008 Ilulissat Declaration from the five Arctic littoral states, which was repeated in 2018, signaled to the world the explicit intention to solve potential disputes between states through diplomacy within the framework of the UN Convention on the Law of the Sea (Jacobsen, 2018). The rebranding of the Arctic Council with the establishment of a permanent secretariat in Tromsø in Northern Norway, and the attendance at meetings by all ministers of foreign affairs from all Arctic countries in 2008–2009 (Rottem, 2014; Steinberg and Dodds, 2015), indicated such a pathway toward an Arctic "community" region.

Scholars have further gone on to highlight the cooperative features and the "uniqueness" of the Arctic region's amicable cooperation, while relations between the same actors deteriorated elsewhere (Berkman and Young, 2009; Byers, 2017; Rahbek-Clemmensen, 2017). These views are still held, despite the post-2014 souring of relations with Russia (Østhagen, 2016; Raspotnik, 2018; Stephen and Knecht, 2017).

The idea of a security region that appears most relevant in the Arctic context is consequently that of a normative region or a "constructed" region (after Neuman, 1994)—constructed, or built, by those actors engaged in Arctic studies, Arctic policy-making, and Arctic governance (see Keskitalo, 2004, 2007). Crucial here, however, is the fact that military security discussions did not figure to a great extent in these region-building efforts. The Arctic might indeed be a "region" in terms of dealing with issues ranging from economic development to climate change research, but in terms of military security no such region-building efforts have occurred.

Exemplifying this, the most pressing challenge in the Arctic in the 2020s is indeed how to deal with and talk about Arctic-specific (military) security concerns, which are excluded, for example, from the Arctic Council. The debate over what mechanisms are best suited to further expand security cooperation has been ongoing for a decade (Conley et al., 2012), with discussions about whether the Arctic Council should acquire a security component (Graczyk and Rottem, 2020; Grønning, 2016); others look to the Arctic Coast Guard Forum or more ad hoc venues (Østhagen, 2020; Sfraga et al., 2020). The Northern Chiefs of Defense Conference and the Arctic Security Forces Roundtable were initiatives established to this end in 2011/2012 (Depledge et al., 2019), but they fell apart after 2014.

In summary, descriptions of the Arctic that depict it as possessing its own regional security dynamics in the traditional sense clash with the realities of the region: The Arctic Ocean is simply too vast and remote. Security dynamics in the Arctic have remained anchored to other national and regional levels: the Barents area, the Northwest Atlantic, and the Bering Sea/Strait area. From a normative understanding of security regions, however, a different picture emerges. The concept of a nascent security community concerning the Arctic was mentioned in the period between 2008 and 2014. Efforts by the foreign ministers of Arctic countries as well as by Arctic governance scholars to depict the Arctic as a special or sheltered region have also fed the view of the Arctic as a security community. However, these conceptualizations never covered traditional military security concerns. Moreover, they have been fracturing since 2014, and suffered a severe blow in 2019, with the US noting the growing possibility of "great power politics" influencing relations in the North.

Concluding remarks

The Arctic is increasingly being referred to as a "region" in which the security concerns and interests of states are interlinked and overlapping. The "region" label is frequently used, but without a proper analysis of what this label means and how it is linked to the notion of the region in international studies. In terms of national security, the desire to see the Arctic as a coherent region does not correlate with empirical facts. As has been shown here with regard to the immediate security threats perceived by Arctic states and the defense posture that follows, the Northern European and North American security domains are only marginally aligned. This fact contradicts arguments that the Arctic is a typical security region—it is simply too vast and inaccessible to fit the various definitions of a security region.

This article has also unpacked the various, and at times contradictory, security dynamics in the Arctic. Some dynamics are best understood through the threefold distinction presented here: international competition (why the US is increasingly focusing on China in an Arctic context), regional interaction (why Arctic states still meet to sign new agreements hailing the cooperative spirit of the North), and national defense (why some Arctic states and not others invest heavily in their northern defense posture).

What does this all mean when looking at the Sustainable Development Goals (SDGs) toward 2035? We could envision that the Arctic states also engage in further region-building with a security focus. The SDGs, in particular number 13 on Climate Action and number 16 on Peace and Justice, might be used as relevant frameworks for joint efforts among the Arctic countries within the framework of the Arctic Council or otherwise. Moreover, SDG number 17 especially highlights international cooperation as a way of resolving many, if not most, of the issues that the goals target. This is also highly relevant for the Arctic states and how they deal with an array of security concerns, ranging from soft security to hard security. Indeed, more cooperation is needed, as Arctic political relations are fraying due to tense global security relations. Using the SDGs as umbrella mechanisms to spur on low-level cooperation—that, in turn, perhaps could have positive effects on larger pan-Arctic political relations—is a feasible option. In that sense, goal number 17 is perhaps as pertinent to the Arctic as anywhere else, if not even more so.

Moreover, looking to 2035, perhaps the increased focus on security in the north might actually spur the Arctic states to make efforts to tackle regional security matters. Yet, leaning on the different levels of analysis, questions would arise regarding the level upon which to focus. For example, should the focus be placed on national defense concerns or on international strategic competition? As shown throughout this chapter, it is difficult to pinpoint pan-Arctic *military* security concerns that include *all* Arctic states – apart from, perhaps, a shared code of conduct (e.g., Boulègue, 2019).

The difficulties encountered in trying to establish an arena for security discussions indicate that this issue is highly sensitive to, and influenced by, events elsewhere. Any Arctic security dialogue is fragile, and risks being interpreted through the lens of the increasingly tense NATO-Russia division in the Arctic. Paradoxically, progress in developing such an arena is tricky precisely because of what the arena is intended to achieve: hindering the spillover of tensions from other parts of the world to the Arctic. Nevertheless, looking to 2035 and beyond, matters of Arctic security and the need for dialogue, guidelines, and frameworks will not be less relevant or less in demand.

In turn, what these nuances imply is that simplistic, one-liner descriptions of "Arctic security" must be taken with a pinch of salt. This should inspire further studies on security politics in a region that is at least as complex as any other part of the world.

December 2021

Ex-post reflections

The Russian invasion of Ukraine starting on February 24, 2022, naturally also changes the way we view and analyse Arctic security relations. This chapter was written before those events. However, the main points made in it still stand. The chapter argues among other things that the primary security variable in the Arctic is Russia-NATO (or, if you will, the West) relations. With the invasion of Ukraine in 2022, this is further exemplified, as fears of a spillover to the Arctic materialized. At the time of writing (May 2022), this has not happened in terms of direct security operations or warfare. Still, cooperative mechanisms such as the Arctic Council and the Barents Euro Arctic Council have been suspended. Bilateral cooperation between Russia and the other Arctic countries over a range of issues (economic, political, research) have also been suspended. Another point made in this chapter is that it is not sufficient to generalize across the vast circumpolar region when discussing immediate security concerns in the North. The Ukraine invasion further exemplifies this, as the security concerns of Norway—bordering on Russia's Northern Fleet—are perceived as much more immediate than the security concerns of Canada or Greenland. Still, in all Arctic non-Russian spaces, discussion on defense and security emerged at the start of 2022. With Finland and Sweden deciding to join NATO, Arctic security relations at a systemic (global), regional (circumpolar), and national level will further change. No doubt, the Russian invasion of Ukraine was the final blow to the idea of Arctic "exceptionalism," that is, the Arctic is sheltered from security affairs elsewhere involving some of the same actors. However, this does not mean that the Arctic cannot be an area of cooperation and low tension, if Arctic states actively work toward that goal. Yet, looking toward 2035, the idea of security dialogue in the north involving Russia looks both more unlikely than before, and more needed.

May 21, 2022

References

Bennett, M. M. (2017). Arctic Law and Governance: The Role of China and Finland. *Jindal Global Law Review*, 8(1), pp. 111–116. https://doi.org/10.1007/s41020-017-0038-y

Berkman, P. A., and Young, O.R. (2009). Governance and Environmental Change in the Arctic Ocean. *Science*, *324* (17 April 2009). https://doi.org/10.1126/science.1173200

Borgerson, S. (2008). Arctic Meltdown: The Economic and Security Implications of Global Warming. *Foreign Affairs*, *87*(2), pp. 63–77.

Buzan, B., and Wæver, O. (2003). *Regions and Powers: The Structure of International Security*. Cambridge: Cambridge University Press.

Byers, M. (2017). Crises and International Cooperation: An Arctic Case Study. *International Relations*, *31*(4), pp. 375–402.

Conley, H. A., Melino, M., Tsafos, N., and Williams, I. (2020). *America's Arctic Moment: Great Power Competition in the Arctic to 2050*. https://csis-website-prod.s3.amazonaws.com/s3fs-public/publication/Conley_ArcticMoment_layout_WEB FINAL.pdf

Conley, H. A., Toland, T., Kraut, J., and Østhagen, A. (2012). A New Security Architecture for the Arctic: An American Perspective. *CSIS Report: Vol. January* 20. Center for Strategic & International Studies (CSIS).

Dadwal, S. R. (2014). Arctic: The Next Great Game in Energy Geopolitics? *Strategic Analysis*, 38(6), pp. 812–824. https://doi.org/10.1080/09700161.2014.952936

Depledge, D., Boulègue, M., Foxall, A., and Tulupov, D. (2019). Why We Need to Talk About Military Activity in the Arctic: Towards an Arctic Military Code of Conduct. *Arctic Yearbook*, *2019*. https://arcticyearbook.com/arctic-yearbook/2019.

Dodds, K., and Nuttall, M. (2016). *The Scramble for the Poles: The Geopolitics of the Arctic and Antarctic*. Polity Press.

Elgsaas, I. M. (2019). Arctic Counterterrorism: Can Arctic Cooperation Overcome its Most Divisive Challenge Yet? *The Polar Journal*, 9(1), pp. 27–44.

Exner-Pirot, H. (2013). What is the Arctic a Case of? The Arctic as a Regional Environmental Security Complex and the Implications for Policy. *The Polar Journal*, 3(1), pp. 120–135.

Expert Commission. (2015). *Unified Effort*. Norwegian Ministry of Defence. https://www.regjeringen.no/globalassets/departementene/fd/dokumenter/rapporter-og-regelverk/unified-effort.pdf

Frazier, D., and Stewart-Ingersoll, R. (2010). Regional Powers and Security: A Framework for Understanding Order Within Regional Security Complexes. *European Journal of International Relations*, 16(4), pp. 731–753. https://doi.org/10.1177/1354066109359847

Friis, K. (2019). Norway: NATO in the North? In N. Vanaga and T. Rostoks (Eds.), *Deterring Russia in Europe: Defence Strategies for Neighbouring States* (1st ed., pp. 128–145). Routledge.
Graczyk, P., and Rottem, S. V. (2020). The Arctic Council: Soft Actions, Hard Effects? In G. H. Gjørv, M. Lanteigne and H. Sam-Aggrey (Eds.), *Routledge Handbook of Arctic Security* (pp. 221–234). Routledge.
Greaves, W., and Lackenbauer, W. P. (2016, March 23). Re-Thinking Sovereignty and Security in the Arctic. *OpenCanada*. https://www.opencanada.org/features/re-thinking-sovereignty-and-security-arctic/
Grindheim, A. (2009). *The Scramble for the Arctic? A Discourse Analysis of Norway and the EU's Strategies Towards the European Arctic*. Fridtjof Nansen Institute.
Grønning, R. (2016, June 7). Why Military Security Should be Kept out of the Arctic Council. *High North News*. http://www.highnorthnews.com/op-ed-why-military-security-should-be-kept-out-of-the-arctic-council/
Guo, L., and Wilson, S. L. (2020, March 29). China, Russia, and Arctic Geopolitics. *The Diplomat*. https://thediplomat.com/2020/03/china-russia-and-arctic-geopolitics/
Heininen, L. (2012). Arctic Strategies and Policies - Inventory and Comparative Study. *Northern Research Forum*, April, pp. 1–97.
Heininen, L., Everett, K., Padrtová, B., and Reissell, A. (2020). *Arctic Policies and Strategies — Analysis, Synthesis, and Trends*. http://pure.iiasa.ac.at/id/eprint/16175/1/ArticReport_WEB_new.pdf
Hilde, P. S. (2014). Armed Forces and Security Challenges in the Arctic. In R. Tamnes and K. Offerdal (Eds.), *Geopolitics and Security in the Arctic: Regional Dynamics in a Global World* (pp. 147–165). Routledge.
Ho, J. H. (2011). The Arctic Meltdown and Its Implication for Ports and Shipping in Asia. In J. Kraska (Ed.), *Arctic Security in an Age of Climate Change* (pp. 33–43). Cambridge University Press.
Hobér, K. (2011). Territorial Disputes and Natural Resources: The Melting of the Ice and Arctic Dispute. *Oil and Gas Journal, 109*(6), pp. 54–60.
Hoogensen Gjørv, G., Bazely, D. R., Goloviznina, M., and Tanentzap, A. J. (Eds.) (2014). *Environmental and Human Security in the Arctic*. Routledge.
Huebert, R. (2013). Submarines, Oil Tankers and Icebreakers. *International Journal*, 66(4), pp. 809–824. http://www.jstor.org/stable/23104394
Humpert, M. (2013). *The Future of Arctic Shipping - A New Silk Road for China? Vol. November 2* (T. A. Institute (Ed.)). The Arctic Institute. http://issuu.com/thearcticinstitute/docs/the_future_of_arctic_shipping_-_a_n
Jacobsen, M. (2018, May 23). Ilulissat Declaration's 10-Year Anniversary. *High North News*. https://www.highnorthnews.com/en/ilulissat-declarations-10-year-anniversary
Jegorova, N. (2013). Regionalism and Globalisation: The Case of the Arctic. *Arctic Yearbook*.
Jensen, L. C. (2017). An Arctic 'Marriage of Inconvenience': Norway and the Othering of Russia. *Polar Geography*, 40(2), pp. 121–143. https://doi.org/10.1080/1088937X.2017.1308975
Keil, K. (2013). *Cooperation and Conflict in the Arctic: The Cases of Energy, Shipping and Fishing*. PhD Thesis: Berlin Graduate School for Transnational Studies. Freie University ät Berlin.
Kelly, R. E. (2007). Security Theory in the "New Regionalism." *International Studies Review*, 9(2), pp. 197–229.

Keskitalo, C. (2004). *Negotiating the Arctic: The Construction of Aan International Region*. Routledge.
Keskitalo, C. (2007). International Region-Building: Development of the Arctic as an International Region. *Cooperation and Conflict*, *42*(2), pp. 187–205. https://doi.org/10.1177/0010836707076689
Koivurova, T., and Kopra, S. (Eds.) (2020). *Chinese Policy and Presence in the Arctic*. Brill Nijhoff.
Koivurova, T., Kopra, S., Lanteigne, M., Nojonen, M., Smieszek, M., and Stepien, A. (2020). China's Arctic Policy. In T. Koivurova and S. Kopra (Eds.), *Chinese Policy and Presence in the Arctic* (pp. 25–41). Brill Nijhoff.
Konyshev, V., and Sergunin, A. (2014). Is Russia a Revisionist Military Power in the Arctic? *Defense and Security Analysis*, *30*(4), pp. 323–335. https://doi.org/10.1080/14751798.2014.948276
Lanteigne, M. (2016). Ties That Bind: The Emerging Regional Security Complex in the Arctic. *NUPI Policy Brief*, *4*, pp. 1–4.
Lasserre, F., and Têtu, P. (2016). Russian Air Patrols in the Arctic: Are Long-Range Bomber Patrols a Challenge to Canadian Security and Sovereignty ? *Arctic Yearbook 2016*, *2016*.
Lavrov, S., and Støre, J. G. (2010, September 21). Canada, Take Note: Here's How to Resolve Maritime Disputes. *The Globe and Mail*. http://www.theglobeandmail.com/opinion/canada-take-note-heres-how-to-resolve-maritime-disputes/article4326372/
Łuszczuk, M. (2016). Military Cooperation and Enhanced Arctic Security in the Context of Climate Change and Growing Global Interest in the Arctic. In L. Heininen (Ed.), *Future Security of the Global Arctic: State Policy, Economic Security and Climate* (pp. 35–54). Palgave Macmillan.
Moe, A., Fjærtoft, D., and Øverland, I. (2011). Space and Timing: Why Was the Barents Sea Delimitation Dispute Resolved in 2010? *Polar Geography*, *34*(3), pp. 145–162.
Norum, H. (2018, March 5). Russland Simulerte Angrep på Vardø-radar [Russia Simulated A ttack on Vardø-radar]. *NRK*. https://www.nrk.no/norge/_-russland-simulerte-angrep-pa-vardo-radar-1.13946450
Norwegian Intelligence Service. (2020). *FOCUS 2020: The Norwegian Intelligence Service's Assessment of Current Security Challenges*. https://forsvaret.no/presse_/ForsvaretDocuments/Focus2020-web.pdf
Office of the Under Secretary of Defense for Policy. (2019). *Report to Congress: Department of Defense Arctic Strategy*. https://media.defense.gov/2019/Jun/06/2002141657/-1/-1/1/2019-DOD-ARCTIC-STRATEGY.PDF
Østhagen, A. (2016). High North, Low Politics Maritime Cooperation with Russia in the Arctic. *Arctic Review on Law and Politics*, *7*(1), pp. 83–100.
Østhagen, A. (2018). Geopolitics and Security in the Arctic. In M. Nuttall, T. R. Christensen and M. Siegert (Eds.), *Routledge Handbook of the Polar Regions* (pp. 348–356). Routledge.
Østhagen, A. (2020). *Coast Guards and Ocean Politics in the Arctic*. Palgave Macmillan.
Østhagen, A. (2021). The Arctic Security Region: Misconceptions and Contradictions. *Polar Geography*, *44*(1), pp. 55–74.
Østhagen, A., Sharp, G. L., and Hilde, P. S. (2018). At Opposite Poles: Canada's and Norway's Approaches to Security in the Arctic. *Polar Journal*, 8(1), pp. 163–181.
Padrtová, B. (2017). Regional Security Patterns in the Arctic. In A. Bourmistrov, E. Dybtsyna and N. Nazarova (Eds.), *Management in the High North: Young Researchers'*

Contribution: Collection of Essays: Volume 3 (pp. 1–8). Nord University. https://brage.bibsys.no/xmlui/handle/11250/2432394

Rahbek-Clemmensen, J. (2017). The Ukraine Crisis Moves North. Is Arctic Conflict Spill-Over Driven by Material Interests? *Polar Record*, *53*(1), pp. 1–15. https://doi.org/10.1017/S0032247416000735

Raspotnik, A. (2018). *The European Union and the Geopolitics of the Arctic*. Edward Elgar.

Rottem, S. V. (2014). The Arctic Council and the Search and Rescue Agreement: The Case of Norway. *Polar Record*, *50*(3), pp. 284–292.

Rottem, S. V. (2017). The Arctic Council: Challenges and Recommendations. In S. V. Rottem and I. F. Soltvedt (Eds.), *Arctic Governance: Law and Politics. Volume 1* (pp. 231–251). I. B. Tauris.

Rowe, L. (2018). Fornuft Og Følelser: Norge Og Russland Etter Krim (Sense and Sensibility: Norway and Russia after Crimea). *Nordisk Østforum*, *32*, pp. 1–20.

Sale, R., and Potapov, E. (2010). *The Scramble for the Arctic: Ownership, Exploitation and Conflict in the Far North*. Frances Lincoln.

Saxi, H. L. (2011). Nordic Defence Cooperation after the Cold War. *Oslo Files: Vol. March 2011*. Norwegian Institute for Defence Studies.

Sergunin, A. (2014, November 25). Four Dangerous Myths about Russia's Plans for the Arctic. *Russia Direct*. http://www.russia-direct.org/analysis/four-dangerous-myths-about-russias-plans-arctic

Sergunin, A., and Konyshev, V. (2014). Russia in Search of its Arctic Strategy: Between Hard and Soft Power? *Polar Journal*, *4*(1), pp. 69–87. https://doi.org/10.1080/2154896X.2014.913930

Sfraga, M., Wilson Rowe, E., Sverdrup, U., Friis, K., and Hønneland, G. (2020). *A Governance and Risk Inventory for a Changing Arctic: Background Paper for the Arctic Security Roundtable at the Munich Security Conference 2020*. https://www.wilsoncenter.org/publication/governance-and-risk-inventory-changing-arctic

Singer, J. D. (1961). The Level-of-Analysis Problem in International Relations. *World Politics*, *14*(01), 77–92. https://doi.org/10.2307/2009557

Smith, J., Hendrix, J., and Kaplan, R. D. (2017). *Forgotten Waters - Minding the GIUK Gap*. https://www.cnas.org/publications/reports/forgotten-waters

Soltani, F. (2014). Levels of Analysis in International Relations and Regional Security Complex Theory. *Journal of Public Administration and Governance*, *4*(4), pp. 166–171.

Steinberg, P. E., and Dodds, K. (2015). The Arctic Council after Kiruna. *Polar Record*, *51*(1), pp. 108–110. https://doi.org/10.1017/S0032247413000557

Stephen, K., and Knecht, S. (Eds.) (2017). *Governing Arctic Change: Global Perspectives*. Palgave Macmillan.

Sun, K. (2014). Beyond the Dragon and the Panda: Understanding China's Engagement in the Arctic. *Asia Policy*, *18*(1), pp. 46–51. https://doi.org/10.1353/asp.2014.0023

Tamnes, R., and Offerdal, K. (2014). Conclusion. In R. Tamnes and K. Offerdal (Eds.), *Geopolitics and Security in the Arctic: Regional Dynamics in a Global World* (pp. 166–177). Routledge.

The Guardian. (2019, May 6). US Warns Beijing's Arctic Activity Risks Creating "New South China Sea." *The Guardian*. https://www.theguardian.com/world/2019/may/06/pompeo-arctic-activity-new-south-china-sea

Tunsjø, Ø. (2018). *The Return of Bipolarity in World Politics: China, the United States, and Geostructural Realism*. Columbia University Press.

US Department of State. (2019). *Looking North: Sharpening America's Arctic Focus*. Remarks. https://www.state.gov/looking-north-sharpening-americas-arctic-focus/

Waltz, K. N. (1959). *Man, the State, and War*. Columbia University Press.
Weber, B. (2015, April 22). Security Fears Growing in the Arctic Region: Survey. *CTV News*.
Willis, M., and Depledge, D. (2014, September 22). How We Learned to Stop Worrying About China's Arctic Ambitions: Understanding China's Admission to the Arctic Council, 2004–2013. *The Arctic Institute*. https://www.thearcticinstitute.org/china-arctic-ambitions-arctic-council/
Wilson Rowe, E. (2020). Analyzing Frenemies: An Arctic Repertoire of Cooperation and Rivalry. *Political Geography*, *76*(January). https://doi.org/10.1016/j.polgeo.2019.102072
Wilson Rowe, E., and Blakkisrud, H. (2014). A New Kind of Arctic Power? Russia's Policy Discourses and Diplomatic Practices in the Circumpolar North. *Geopolitics*, *19*(1), pp. 66–85. https://doi.org/10.1080/14650045.2013.789863
Ye, J. (2014). China's Role in Arctic Affairs in the Context of Global Governance. *Strategic Analysis*, *38*(6), pp. 913–916. https://doi.org/10.1080/09700161.2014.952938

Part II

International Cooperation in Economy, Business and Innovations

4 Global Arctic Economic Development Scenarios

Andrey Krivorotov

Economy of the north and its international dimension

The very special Arctic geography has always determined its economic development. On the one hand, the region is known for its extreme climatic conditions, vast distances, inaccessibility, and short weather windows. These result among others in scattered, isolated settlement patterns and in the characteristic, sustainable lifestyles of the Indigenous Peoples, finely balanced with the harsh nature and predominantly based on a precapitalist, solidarity economy.

On the other hand, the Arctic is rich in unique resources valuable enough to justify the enormous development and transportation costs. Various resources have played this driver role over the centuries: furs, gold, coal, military-strategic geographical position, oil and gas, rare earth elements, and so forth (Coates, 1985). Their importance, coupled with the exceedingly high upfront investments, has limited the number of economic actors and attracted the special interest of governments.

As a result, the economies of the Arctic regions feature three distinct and relatively separate sectors:

- Market-driven economic activities based on acquisition/extraction and export of valuable Arctic resources, plus local supporting industries.
- Traditional subsistence economy, predominantly in Indigenous communities.
- An extraordinary large transfer (public) sector, including critical infrastructure, government administration, and military installations (Kryukov and Kryukov, 2019, p 28; Larsen and Petrov, 2015, 2020).

While all the sectors are indispensable for the contemporary Arctic, the first one determines the overwhelming majority of its economic relations with the outside world, where the Arctic serves primarily as a supplier of unique, non-processed raw materials (AHDR, 2004; Glomsrød and Aslaksen, 2006, p. 18).

Arctic economies were for centuries dominated by powerful monopolists (like the old style chartered companies), which controlled huge territories, enjoyed earmarked privileges, and often performed some public functions on site. The Hudson's Bay Company (monopoly in the Canadian North 1670–1869), the

DOI: 10.4324/9781003246015-6

Russian-American Company in Alaska (1799–1867), the Royal Greenlandic Trade Company (1774–1979), or the Dalstroy Trust in the Soviet north-east (1931–1957) are typical examples of such entities, depicted as "super-organizations" by Dr. Alexander Pelyasov (Pelyasov and Zamyatina, 2019). While there was no super-organization in Northern Norway, its exports since the mid-14th century were similarly controlled by the oligopoly of Hanseatic merchants from Bergen (Christensen, 2009, pp. 8–10). This monopolism has left a lasting footprint on Arctic history and on its economic geography. While its resources, like *tørrfisk* (sun-dried cod from the Lofoten Islands) or North Russian furs, have circulated in foreign markets since the Middle Ages, the region as such was separated from them economically by large entities controlled from the south.

Similarly, in political terms, the Arctic used to be ruled from remote capital cities and was often accorded a lower formal status. Greenland only ceased to be a Danish colony in 1953, Alaska became a fully-fledged US state in 1959, and the Canadian northern territories are still not formally equated with the provinces (although this difference is becoming less obvious). The central government has always played a crucial role in the regional economies as the owner of resources, entrepreneur, regulator, and law enforcer.

The development of the Arctic has therefore been very unstable, depending upon market fluctuations and changing governmental priorities. For instance, Arkhangelsk, the first-ever Russian export harbor in the North, was founded by Ivan the Terrible in 1584 to establish a direct line of communication with the UK, but doomed in 1713–1718 by Peter the Great, who forced the conduct of Russian foreign trade via the newly established St. Petersburg. In the contemporary US, Republican Presidents lobby for petroleum activities in the Alaskan protected areas, while the Democrats block them for environmental and climate reasons, notably including Joe Biden on his inauguration day (Schreiber and Rosen, 2021).

Altogether, these features have created two traditional paradigms and popular perceptions of the Arctic (and, in Russia, also of Siberia, which has many similarities): "Resource Base" versus "Colony". Large outside investors, strongly supported by central authorities, tend to view the region as a complicated, yet immensely rich underdeveloped frontier area offering true adventure and opportunities for a civilizing mission vis-à-vis the Indigenous Peoples – an approach still visible in many governmental Arctic strategies (Heininen et al., 2020; Lasserre, 2021; Petrov, 2018). Many northerners, in turn, demonstrate a characteristic colonial identity, depicting their homeland as an underprivileged outpost, stripped of assets by outside businesses and overseen by central authorities (Table 4.1). Nikolay Yadrintsev, a Siberian scientist and public figure, raised this issue in relation to Asian Russia nearly 140 years ago (Yadrintsev, 1882). Over the past decades, researchers and politicians with northern background have applied (post)colonial narratives in relation to many Arctic regions worldwide (Brox, 1984; Coates, 1985; Körber and Volquardsen, 2020; Nergård, 2011; Silin, 1989; Standlea, 2006; Zubov and Inozemtsev, 2013).

Table 4.1 Arctic economy paradigms

Paradigm	*"Resource Base"*	*"Colony"*
Overall perception of the Arctic	Resource-rich frontier territory under development	Periphery area, abused by central government and outside investors
Impact on Arctic regional economy	Accelerated economic and social development	Resource exploitation, unequal exchange
Impact on local (including Indigenous) population	Modernization, improving living standards	Destruction of traditional lifestyles and habitats
Typical proponents	Central authorities Corporate headquarters Military Industrial economists Adventure seekers Clergy Teachers	Regional and local authorities Local SMEs Indigenous peoples Regional economists Environmental activists Ethnographers

Source: Author.

The two perceptions of the Arctic stand in sharp contrast to one another, highlighting various sides of the same, a sophisticated and incoherent process. But both recognize the Arctic as a distinct place compared to the core areas of the respective countries and focus strongly on the relations with capital cities rather than on an intra-North cooperation agenda.

Despite numerous common features, the Circumpolar North used not to be a holistic global macro region, due among other to its predominantly meridional, export-oriented system of transport routes, with little connectivity among the northern regions themselves. The Pomor Trade between the northern Norwegian and Russian merchants along the Barents and White Sea coasts was a very positive exception, although it also suffered from politically motivated bans, imposed at times in both countries.

The heavy militarization in the Cold War period, when the Arctic Ocean turned into an area of confrontation between the Soviet Union and five *North Atlantic Treaty Organization* (NATO) member states, was also an important economic driver. Johan Jørgen Holst, then Norwegian Minister of Defense, noted in 1991 that "the country's security policy [in the North] is also its regional policy in the highest grade" (Holst, 1991). In many respects, the military behaved similarly to the big corporations: both were large integrated organizations with immense human and financial resources, all controlled from headquarters. Their arrival dramatically changed lives in the arctic regions and settlements by improving the living standards, introducing new economic relations, goods, and services, but also caused them to be tied to a single investor and employer. Military secrecy significantly added to the isolation of arctic regions from one another, especially regarding cross-border contacts.

Liberal globalization and the emergence of pan-Arctic cooperation

This situation started to evolve rapidly in the 1990s. With the end of the Cold War and the unfolding globalization, the Arctic regions faced new challenges, but also discovered new, unseen cooperation opportunities.

On the one hand, the Arctic experienced a major setback with the spreading of economic liberalism, which gained new impetus after the collapse of the Soviet Union and the resulting demise of the ideas of governmental intervention and strong social policies. Monetarist concepts and Francis Fukuyama's ideas of the "end of history" with Western-type democracy as the "final form of human government" (Fukuyama, 1992) heavily dominated the political and economic thinking of the time.

The global Arctic was becoming exposed to increasingly liberalized international trade under predominantly low prices for raw materials (especially after the 1998 oil market slump) and accelerating international competition for investments. Arctic resources were often privatized at discount rates by major outside (therein foreign) investors. International oil major BP acquired ARCO, a key producer of Alaskan oil. In Norway, ship owners from the south secured large catch quotas for valuable fish species in northern waters free of charge (Grytås, 2014). Industrial giants of the Russian Arctic and Siberia were sold off to Moscow-based holdings (often registered in offshore jurisdictions) for a fraction of their real value. Besides, foreign investors established numerous joint ventures within extractive industries in the Russian North on very liberal terms (Gustafson, 2012). The Arctic became involved in new value chains and financial flows, very often with negative consequences.

In parallel, the military-strategic importance of the area diminished with the end of the bloc confrontation, as the Arctic states focused increasingly on NATO enlargement in Central and Eastern Europe, regional conflicts in the former Soviet Union and Yugoslavia, wars in Iraq and Afghanistan, and so forth. The closedowns of military installations in Northern Russia and Norway, Alaska or the Keflavik airbase in Iceland had dire negative side effects for local economies in terms of civilian jobs, supply contracts, and budgetary incomes.

Interregional disproportions were increasing all over the world, in keeping with the notions of "New Economic Geography". Mainstream economists and institutions like the World Bank welcomed this development, citing the success stories of large agglomerations, which created extra opportunities for business development and human well-being (World Bank, 2009). Remote territories with extreme natural conditions often evolved toward simpler economic structures and lifestyles, a growing reliance on subsistence economies, massive bankruptcies, and depopulation. The global Arctic was also becoming increasingly heterogeneous. The regions with large extractive industries (Northern Siberia, Alaska, and Northern Canada) were in the best position, while remote resource-poor regions abandoned by big companies and the military suffered most. In 2003, the variation in gross regional product per capita in the Arctic was 15-fold, between the

Northwest Territories in Canada and Evenkia in Russia (Glomsrød and Aslaksen, 2006, p. 22; Glomsrød, Duhaime and Aslaksen, 2021, p. 44). The list of Russian settlements with the biggest population decrease from 1990 to 2014 is headed by the northern towns of Ostrovnoy (−89.6%), Igarka (−71.6%), Susuman (−69.3%), and so on (Sokolov and Terentyev, 2015).

The liberalist thinking complemented the two traditional images of the Arctic economy with a third one: a redundant budgetary burden, a noncompetitive remote territory inhabited by welfare clients. In the post-Soviet Russia, it boiled down to the laconic 1992 formula by Yegor Gaidar, the key reforms architect, who dismissed the North as "obsolete and overly populated" (Logunov, 2001). This approach persisted until the mid-2000s.

On the other hand, the demilitarization led to a positive ideological shift toward a vision of the Arctic as an emerging new arena of global cooperation in light of new common problems. Multilateral initiatives on various levels soon emerged, notably including the Arctic Environmental Protection Strategy (1991), the Northern Forum (1992), the Barents Region (1993), the Arctic Council (AC, 1996), and the EU Northern Dimension (1997).

These initiatives actually pursued a dual aim:

- To fill the foreign policy vacuum and design a civil agenda for the Arctic
- To help northern regions adapt to tougher market conditions through cooperation

The efficiency of the above initiatives varied. Political dialog among the central and regional authorities was relatively intense and fruitful, especially within the AC and Barents Region. Indigenous Peoples' organizations established very good cooperation on a wide range of issues, notably including their empowerment in respect of extractive industries, co-decision and revenue sharing, promoting traditional arts and crafts, gender and labor relations, and so forth. Meanwhile, the business cooperation was generally lagging behind due to the structures of the Arctic economies (which are more competitive than complementary), the insufficient transportation routes, and not least, the hardships caused by the Russian radical reforms. Several initiatives for improving intra-Arctic connectivity were launched at that time, like simplified visa arrangements, especially for Indigenous Peoples, the Barents Euro-Arctic Transport Area (BEATA), the Norwegian idea of a special bilateral "Pomor Zone", or Arctic air and sea line projects between Canada and Russia (Maximova, 2015, pp. 209–213). During its first term as AC chair (2004–2006), Russia initiated broadening the AC agenda, which formerly concentrated on environmental and Indigenous issues, by including down-to-earth cooperation, like Arctic shipping and trade (Churkin, 2005, p. 7).

The early 21st century witnessed a dramatically increased interest in the Arctic among both politicians and businessmen (especially within oil and gas and shipping). The 2007 Russian private expedition, which planted the national titanium flag on the North Pole seabed, became the true trigger. Within a short time, the Arctic turned into a popular and "trendy" global issue. All the Arctic states and many non-regional actors, like the EU, the UK, or Asian countries, issued official Arctic policy papers.

The new contest for the Arctic was unfolding alongside dramatically intensified international cooperation (Wilson Rowe, 2018). The first three legally binding agreements were signed under the AC umbrella; numerous international organizations put the Arctic high on their agendas. The Canadian AC chairmanship (2013–2015) resulted in the establishment of the Arctic Economic Council. International investors and national champions jointly implemented breakthrough projects, like Snøhvit offshore Norway or Yamal LNG in Russia (Mitrova, 2019). Russia opened its Northern Sea Route (NSR) for international transit in 2010. The Arctic emerged as a new, dynamic region of international politics and global economy, with a distinct profile and specific rules formulated largely within the AC. This cooperation continued even under the overall setback in the Russia-West relations since 2014, giving rise to the concept of "Arctic exceptionalism" (Käpylä and Mikkola, 2019).

The growing interest and involvement of non-Arctic actors, especially China, became an unprecedented and important trend of the time. The opening up of the Arctic coincided with the rise of China as a global power, and the two processes proved mutually reinforcing (Brady, 2017). The Chinese Belt and Road Initiative (BRI) was expanded to cover the NSR and the Arctic at large within the "Ice Silk Road" framework. Chinese investors (mostly large state-controlled corporations, often working jointly with UK, Australia, or Hong Kong-based entities) acquired or expressed interest in industrial developments in nearly all Arctic countries regarding fossil fuels, minerals, rare earth elements, and so on. Chinese companies closely evaluated land-based transportation routes through the Russian Arctic and Northern Europe and were among the first foreign shipping lines to start transit navigation along the NSR.

Future macro trends and Arctic sustainability challenges

Over the past few years, however, the Global North has entered into a new transition phase characterized by multifaceted large-scale changes and complex policy choices. Characteristically, they are all relevant to the Sustainable Development Goals (SDGs), but often act in opposite directions. As Professor Jeremy Rayner (Canada) put it, the "comprehensiveness of the SDGs is going to be one of the most significant challenges, but also one of the most significant opportunities" (Menezes and Chater, 2018, p. 9). The following trends, among others, will frame the regional development in the medium- to long-term perspective.

First, the somewhat chaotic reconfiguration of the international economic space, occasioned originally by the global recession, the rise of China, and the drastically aggravated relations between Russia and the West after 2014. There have been numerous indications that the traditional models no longer function smoothly, including the failures of the US-sponsored transatlantic and transpacific trade partnerships, troubles with China's BRI, US-China tensions, including the trade war, a massive rise in protectionism, Brexit, and so forth. The process has gained serious additional momentum with the global COVID-19 pandemic, which within months transformed the entire world. It remains to be seen if humankind will be willing or able to "build back better", restoring globalization to what it was a decade ago or move toward a "Global Reset", as suggested by the World Economic Forum (WEF). Klaus Schwab, founder and Executive Chairman of the WEF, underscores that "the pandemic represents a rare but narrow window of opportunity to reflect, reimagine, and reset our world to create a healthier, more equitable, and more prosperous future" (Schwab, 2020). The Arctic has been less involved in these developments, but the reshaping of the global economic order whatever its outcome is sure to have strategic consequences for the northern rim.

Second, the long-term prospects will be closely tied to global warming, which is twice as rapid in the Arctic as elsewhere in the world on average. Its economic consequences include changes in the traditional ecosystems and habitats, a progressive opening up of high-latitude shipping routes, release of gas from the melting tundra permafrost, and degradation of Arctic waterfronts with a corresponding impact on coastal settlements (McPherson, 2015; Suter, Streletskiy and Shiklomanov, 2019; Zellen, 2009). Thus, SDG 13, on taking urgent action to combat climate change and its impacts, is going to be a major priority for all Arctic actors.

Global warming has also given rise to a growing awareness of climate among the general public and the real start of a transition to low carbon. As the renewable energy capacities expand dramatically and the leading CO2 emitters like the EU, the US, and China voice plans to become carbon neutral by the mid-century, analysts are abandoning the traditional visions based on M. King Hubbert's famous "peak oil" theory (King Hubbert, 1956). Instead, they have started to talk about an upcoming "peak oil demand", arguing more about its specific timing; most predictions range between 2028 and 2040. Meanwhile, the bulk of the shrinking global petroleum demand will center on the developing countries in southern latitudes, further away from the Arctic, jeopardizing the economy of new large-scale projects.

At the same time, the transition to low carbon also creates new opportunities for a sustainable Arctic development applying environmentally friendly technologies to boost "blue economy", tourism and renewables. Projects in onshore and offshore wind parks and blue and green hydrogen production are either implemented or discussed in Arctic regions around the world. They frequently involve outside investors and/or equipment suppliers, like the EU at the Berlevåg hydrogen project in Arctic Norway, Japanese companies in a wind park in Yakutia, or Italians in a Murmansk district wind park. Overall, the multifaceted adaptation of the Arctic to global warming is going to be one of the key drivers of change.

Third, comprehensive digitalization is already bringing unparalleled solutions to the Arctic, like broadband internet connections, delivering goods, and monitoring environment or infrastructure by unmanned drones, telemedicine, remote manufacturing, and so forth. There are two impressive projects for laying offshore fiber optic cable lines between Europe and East Asia over the Arctic Ocean, both along the NSR and via the American Arctic, with connections to key coastal settlements. The dramatically improved communications, safety of navigation, public security, opportunities for distance learning, and so on will, among others, attract new businesses to the Arctic (cold climate data processing centers are a prime example), making it an attractive place to live in, not just for the Indigenous Peoples. This may contribute much to achieving SDG 10 on reducing inequality within and among countries.

At the same time, contrary to SDG 8, new information and communication technologies will also change Arctic lifestyles themselves, including a loss of many jobs and a profound shakeup of traditional communities (Coates, 2020). Many non-native men and women living north of the Arctic Circle may feel forced to leave unless they find new reasons and means to stay there (Mineev, Dybtsyna and Mellemvik, 2020). The region will need reengineering and "reinventing", as it did a century ago, when industrial technologies created a completely new vision and image of the Arctic. Economic research also requires a new holistic paradigm, combining international economics, postcolonial development, and institutional and regional economy.

While these global trends are very likely to continue and gather strength up to 2035, the political reactions are much less predictable and more dynamic, and still very important for all Arctic economies.

In a way, the Arctic has moved away from the political focus since 2013–2014, overshadowed by the long series of acute "hard security" crises in southern latitudes and systemic shocks. Arctic policies today look more like a function of broader, overarching approaches than a relatively self-sufficient policy area. This being said, the politicians will always have a very powerful voice in Arctic development, including business. They can help the Arctic economies adapt to the changing global trends, seizing the opportunities and overcoming challenges as well as exacerbate their problems.

An international foresight workshop held in the Russian Arctic city of Naryan-Mar in November 2018 was a clear illustration of this. All three expert groups identified domestic policy as one of the two key drivers for their respective problem areas, while the second drivers varied (international relations, raw materials markets, and technological progress). The summing-up discussions helped chart four integral scenarios for the Russian Arctic development through 2050, by unifying the other drivers under generalized "external conditions":

- The *Harmonious Arctic* scenario, whereby the external conditions are favorable and the government policies encourage bottom-up innovation and creativity, among others, by maximizing the regional ripple effects of Arctic megaprojects.
- The *Self-Reliant Arctic* scenario, which implies a strong domestic policy aiming at developing the region despite unfavorable resource, international and/or technological conditions, largely by stimulating new businesses not related to resource extraction.

- The *Resource-Dependent Arctic* scenario, under which vibrant international cooperation, development of new technologies, and high global demand for Arctic materials are coupled with weak domestic policies, which turns the region into a net supplier of resources heavily dominated by large corporations and concentrating around extractive projects.
- The *Forgotten Arctic* scenario, whereby unfavorable external conditions are further exacerbated by weak domestic policies, resulting in a comprehensive stagnation and massive depopulation of the Arctic, which is largely left to survive on its own (Petrov et al., 2021).

A similar exercise in global Arctic forecasting by 2050, undertaken under the auspices of the Skolkovo School of Management in Moscow and Nord University in Bodø in 2020, produced similar scenarios. Although this study used a more market-minded terminology of "enabling environment", its essence was still largely dependent upon the governments' approaches and mutual coordination (Belostotskaya et al., 2020).

Key economic policy choice: Promoting development versus conservationism?

At least three fundamental political choices will shape medium- and long-term Arctic development. The first of these concerns the overall vision of the human role in the region.

No Arctic stakeholder today would declare the previous predatory approach to the region. Protecting its fragile environment, also highly appreciated as the Indigenous Peoples' habitat, is a universal priority (although recent industrial accidents in the North demonstrate that corporations still warrant close public supervision). However, there is no international consensus on the desired level of human activities in the Arctic. Contemporary research and policy papers demonstrate two fundamental attitudes.

One of these tends to absolutize the region's unique nature, the need to protect its vulnerable environment and the traditional cultures of northerners, who are perceived practically identical with Indigenous Peoples. Hence the conclusion that the non-Indigenous presence should be limited to the necessary minimum – which, paradoxically, would have the side effect of continued militarization of the Arctic, since the military presence enjoys a tacit status as an indispensable element of this minimum. This approach is popular among politicians and Indigenous Peoples in Sweden and Canada, in the US Democratic Party (US-Canada Joint Statement, 2016), and among some corporations and environmental organizations like Greenpeace.

The other approach relies on the belief that due protection of Arctic nature is compatible with sustainable economic development. The introduction of modern technologies, stringent environmental monitoring, and awareness raising are among the key success factors (Dmitrieva and Romasheva, 2020). The four consecutive recent and upcoming AC chairs, Finland, Iceland, Russia, and Norway, promote this view. Finland's Arctic Strategy stresses that "[b]y emphasizing and promoting sustainable economic activities, Finland can demonstrate that economic welfare, business development and environmental protection can be mutually supportive" (Government of Finland, 2021).

In terms of SDGs, the core issue is about reconciling economic development (Goals 7–9, 14) and Goal 13 on combating climate change. This debate is likely to gain new impetus in future, both within the framework of global warming mitigation and as a possible competitive tool, whereby countries that lag behind in Arctic technologies and investments may invoke environmentally motivated bans on economic activities to eliminate the handicap.

The Arctic oil and gas industry, viewed to be among the most promising businesses merely a decade ago, is especially vulnerable. At present, only Norway and Russia declare a strong political commitment to continued development. Still, inside Norway the issue remains highly controversial, which has, among others, caused the well-known unsuccessful "climate lawsuit" (Supreme Court of Norway, 2020). Meanwhile, most petroleum companies have abandoned Arctic Canada, Greenland, and Iceland because of discouraging exploration results and/or vehement resistance by politicians and Indigenous communities. The refusal by key US financial institutions, even under President Donald Trump, to supply funding to Arctic oil and gas projects has had a major impact on Alaska (Drill music, 2020; Tsafos, 2020). President Biden made this approach a coherent element of official US energy policy (President of the USA, 2021). The coming two decades may witness unprecedented political pressure exerted on the petroleum business, causing producer nations major foreign policy challenges.

Key governance issue: Arctic for the Arctic Five or global commons?

The central international contradiction around the Arctic, of direct relevance to the SDGs, lies in the evident fact that countries are not equal in their relation to the region. There are the eight Arctic nations with territories north of the Arctic Circle. There is a narrower "Arctic Five" (A5) group of Canada, Denmark/Greenland, Norway, Russia, and the US, the only countries which have jurisdiction over the Arctic Ocean coastline, its economic zones, and continental shelf areas. The A5 nations made it clear in their 2008 Ilulissat Declaration that they would resolve any Arctic issues themselves according to international law.

Meanwhile, there is a growing heterogeneous group of non-Arctic, even tropical states, and also international organizations, which are concerned about the global environmental impacts of Arctic processes and express an interest in the region's business and research opportunities. European and Asian companies are involved in extractive or green energy projects in virtually all Arctic countries. International tourists streamed into the region, halted only by the COVID-19 pandemic.

The structure of the AC clearly reflects this division. The eight Arctic nations have the status of permanent members and the Indigenous Peoples' organizations also have a strong say as Permanent Participants. Meanwhile, non-Arctic states may only be admitted as Permanent Observers subject to their compliance with a specific set of criteria, and there have already been cases of unsuccessful applications.

It remains to be seen, however, whether or not outside nations and institutions will demand a more prominent role in future. For example, China's Arctic Policy adopted in early 2018 sets a clear goal "to participate in the governance of the Arctic, so as to safeguard the common interests of all countries and the international community in the Arctic", including notably in the security domain (China's Arctic Policy 2018). Norwegian researchers have also questioned whether the A5 alone may ensure a true sustainability in the Arctic (Kristoffersen and Langhelle, 2017, p. 38).

Non-Arctic actors have already attempted to formulate international regulations or project their domestic rules as regards Arctic business operations. These include the Polar Code adopted by the International Maritime Organization, the WEF's Arctic Investment Protocol, or the Chinese national regulations for Arctic tourism, to mention a few. Most of these regulations are of limited scope and/or of an advisory nature, but the trend may eventually lead toward a search for supranational solutions profiling climate change, minorities' rights, or freedom of the seas and viewing the Arctic as "global commons".

The US quest for freedom of navigation in the Arctic (leveled at the Russian NSR and the Canadian North-West Passage) is of a rather dual nature in this respect. While the US seeks primarily to profile itself as an Arctic nation, establishing a visible maritime presence at high latitudes, its efforts equally help "open the door" to China, which also promotes freedom of the seas and closely scrutinizes Arctic shipping routes.

Key geopolitical issue: the Arctic as a region of cooperation or competition rivalry?

A similar yet very distinct problem concerns a stakeholder's general target image of the Arctic: whether it appears as an arena for cooperation or geopolitical rivalry.

The past few years have seen many signs of a return to the old-time neorealist rhetoric in public statements by politicians, admirals, and researchers. Starting with the 2019 Rovaniemi speech by US Secretary of State Mike Pompeo (Johnson and Wroughton, 2019), this confrontational approach has entered official foreign policy language, not only among the great powers. The current Swedish Arctic Strategy, for example, states that "as in the Cold War, the Arctic is a dividing line between western countries and Russia" (Government of Sweden, 2020, p. 23) – an assessment which is much more categorical than those of either Russia or Norway, a NATO member.

Arctic economic cooperation is already suffering from politically motivated interventions, like the expressed opposition to Chinese investments in several countries or the Western sanctions on supply of Arctic and deep water offshore drilling technologies to Russia. The Research Service of the US Congress has also suggested moving to suspend China's observer status in the AC as a cost-imposing action in respect to Chinese behavior in the South and East China Seas (Congressional Research Service, 2021).

However, there are also strong imperatives for enhanced transboundary cooperation, both to increase value creation and the well-being of the northerners and to address the universal problems of ultimately achieving the SDGs. Despite political tensions, the past few years have produced tangible results like the moratorium on central Arctic fishing, the establishment of the Coast Guards Forum, or the AC agreement on scientific cooperation.

There is both a risk of "renationalization" of the Arctic by the regional (primarily littoral) nations with a restored Cold War rivalry, much exacerbated by the US-China confrontation, and an opportunity to establish an unprecedented and more efficient and fair collaborative mechanisms. A confrontation and lack of coordination would obviously complicate achieving the SDGs in the pan-Arctic scale.

The North Atlantic, a key naval theater of the Cold War era centered round the Greenland-Iceland-UK (GIUK) gap, which is now regaining its strategic importance, is likely to stay at the very center of this development. Greenland is going to play a key role, pursuing its path toward independence and economic self-reliance under intensified political rivalry among China, the US, and Denmark (Mingming Shi and Lanteigne, 2019; Poppel, 2019).

We have proposed three potential long-term scenarios in this regard (Krivorotov, 2020, p. 178):

- Greenland remains politically united with Denmark but enjoys growing autonomy. US military presence is enhanced under tripartite agreements. Denmark, wary of both American and Chinese interest in the island, increases public spending there, while Greenland continues its quest for financial self-reliance.
- The US, although it does not "purchase" Greenland, still leverages its military presence and financial aid programs to have the island formally and informally more closely associated with the US, among others, by investing in military installations and potentially in rare earth elements production.
- Greenland obtains full statehood, pursues nonalignment and strives to get foreign investments on competitive terms in order to accelerate its economic development and well-being of the population. The emergence of a first-ever indigenously governed Arctic state, with a territory qualifying it for the top-12 worldwide, would certainly mean a profound stress test for Arctic development as a whole, unleashing a new round of competition and rule-making and reinforcing separatist moods around the globe.

It is highly relevant to note the recent major moves toward a reinforced cooperation inside the North Atlantic region, that is, Iceland, Greenland, the Faroe

Islands, and coastal Norway, all being territories with relatively sparse and scattered populations but immensely important geographic positions (Greenland Committee, 2020; Nordatlantisk utviklingsstrategi – NAUST, 2019). These optimistic and visionary ideas of North-to-North cooperation, somehow similar to the South-to-South collective self-reliance of the 20th century, may help create a new Arctic subregion, representing a positive alternative to geopolitical rivalry.

Conclusions and forecasts

We have outlined a rather lengthy but in no way exhaustive list of macro-trends framing the future Arctic economy. The scope of changes and the variety of policy options open up a broad spectrum of future development paths.

Besides the fundamental ideological choices, various Arctic actors and stakeholders also differ in their operational approaches, which are to a certain extent ambivalent to policy goals:

- Preserving the Arctic "as is" or making it subject to a Shumpeterian "gale of creative destruction" (which may equally well imply its accelerated development or abandonment).
- Addressing the Arctic as a unique policy object, which requires tailored technological and business solutions or treating it equally with other regions.
- Applying predominantly national or multilateral mechanisms. It is fairly obvious that a multitude of regulatory regimes would make the Arctic economic area increasingly heterogeneous.

Summing up, the Arctic will face serious challenges and is very likely to transform in many respects. In terms of international economic cooperation and governance, we may anticipate the following rough "scenarios":

- *Common Arctic market*, with international coordination mechanisms (eventually developed under AC auspices) for among others trade and investment cooperation, transport and communications, common information space, and so forth, with special responsibilities vested in the Arctic states.
- *Arctic as common heritage*, whereby it will be put under global control with a powerful role for non-Arctic actors like the EU, China, and Japan, with intensified international competition, especially in the investment area.
- *Split Arctic*, whereby it will develop heterogeneously within the respective national sectors, with heavy military components, low international cooperation, and potentially far-reaching restrictions on the involvement of non-Arctic actors.
- *Deserted Arctic*, a specific subscenario where it undergoes a decline either within closed national sectors due to lack of development incentives or as a result of creating supranational control mechanisms essentially prohibitive of economic activities in the North, thus turning it into a global nature reserve.

The positive cooperation experiences of the past three decades, be they the activities of the AC, joint investment megaprojects, or grassroots people-to-people contacts, constitute a strong and valuable legacy. But to succeed further, this cooperation needs to develop in a creative mode, addressing new issues and eventually designing new modalities and platforms. This may confront future business people and politicians with large and important missions.

December 2021

Ex-post reflections

I deeply regret needing to write this as continued battles rage in Ukraine. The Arctic, although far away, is strongly affected by the Russia-West crisis, with halted investments, people-to-people contacts, the AC, and Barents Region activities. Visions of a broad pan-Arctic cooperation seem nearly implausible, producing major negative effects for among other research, climate, and Indigenous interests. A militarization (especially if Sweden and Finland join NATO) looks more likely than collaborative rule-making, except for individual pragmatic issues like fish stocks management.

Regardless of eventual changes in Russian policies, most sanctions against Russia will stretch far beyond 2035. The country will undergo a painful yet timely revision of its "extract and export to the West" approach to the Arctic, targeting primarily national and Asian markets for investments, technologies, trade flows, and infrastructure.

Western attempts to cut hydrocarbon imports from Russia open up opportunities for domestic extractive industries and for climate policy adjustments. Yet, in the longer run, the isolation of Russia, the key proponent of sustainable economic development in the Arctic, will put Norwegian or Finnish projects under increased environmentally motivated pressure.

Together, these trends may jeopardize global Arctic economic development and cooperation, while potentially enhancing the influence of non-regional players like NATO, EU, and China.

In broader terms, the Ukrainian crisis may indicate that profound global transformations outlined in this chapter will unfold more dramatically, bringing about fundamental changes by 2035. This will impose even harsher requirements on decision-makers, to ensure that global Arctic interests are not sacrificed to seemingly more urgent policy considerations.

April 25, 2022

References

AHDR (Arctic Human Development Report) (2004). Einarsson, N., Larsen, J.N., Nilsson, A. and Young, O.R. (eds.). Akureyri: Stefansson Arctic Institute.

Belostotskaya, A., Kalinin, A., Lazariva, A., Middleton, A. and Nilssen, F. (2020). *Arctic 2050. Mapping the Future of the Arctic*. Skolkovo: Institute for Emerging Market Studies (IEMS).

Brady, A.-M. (2017). *China as a Polar Great Power*. Cambridge: Woodrow Wilson Center Press and Cambridge University Press.

Brox, O. (1984). *Nord-Norge: fra almenning til koloni* [Northern Norway: From Common Domain to Colony]. Tromsø: Universitetsforlaget.
China's Arctic Policy (2018). Available at: http://www.xinhuanet.com/english/2018-01/26/c_136926498.htm (Accessed: 27 May 2021)
Christensen, P. (2009). *Den norsk-arktiske torsken og verden. Torskefiskets historie* [North-Norwegian Cod and the World. Cod Fishing History]. Oslo: Sjøfartsmuseet.
Churkin, V.I. (2005). В семье арктических стран: экономика, жизнь, безопасность [In the Arctic Family: Economics, Living, Security]. *Severnye Prostory, 3–4*, pp. 6–8.
Coates, K. (1985). *Canada's Colonies: A History of the Yukon and Northwest Territories.* Toronto: Jack Lorimer & Co.
Coates, K. (2020). "The Future of Work in the Arctic" in Coates, K. and Holroyd, C. (eds.) *The Palgrave Handbook of Arctic Policy and Politics*. Cham: Palgrave Macmillan, pp.175–192.
Congressional Research Service (2021). *U.S.-China Strategic Competition in South and East China Seas: Background and Issues for Congress*. Washington, DC: CRS.
Dmitrieva, D. and Romasheva, N. (2020). Sustainable Development of Oil and Gas Potential of the Arctic and Its Shelf Zone: The Role of Innovations. *Journal of Marine Science and Engineering*, 8(12), pp. 1–18.
Drill music (2020). *The Economist*. August 22, p. 34.
Fukuyama, F. (1992). *The End of History and the Last Man*. New York: The Free Press.
Glomsrød, S. and Aslaksen, I. (eds.) (2006). *The Economy of the North*. Oslo – Kongsvinger: Statistics Norway.
Glomsrød, S., Duhaime, G. and Aslaksen, I. (eds.) (2021). *The Economy of the North – ECONOR 2020*. Oslo – Kongsvinger: Arctic Council Secretariat and Statistics Norway.
Government of Finland (2021). *Finland's Strategy for Arctic Policy*. Publications of the Finnish Government 2021:55. Helsinki: Prime Minister's Office.
Government of Sweden (2020). *Sweden's Strategy for the Arctic Region*. Stockholm: Government Offices of Sweden.
Greenland Committee (2020). *Greenland and Iceland in the New Arctic*. Reykjavik: Ministry for Foreign Affairs.
Grytås, G. (2014). *Kvotebaronar* [Catch Quota Barons]. Oslo: Samlaget.
Gustafson, T. (2012). *Wheel of Fortune: The Battle for Oil and Power in Russia*. Cambridge: Harvard University Press.
Heininen, L., Everett, K., Padrtova, B. and Reissell, A. (2020). *Arctic Policies and Strategies — Analysis, Synthesis, and Trends*. Laxenburg: IIASA.
Holst, J.J. (1991). Samarbeid forsvar-næringsliv i Finnmark [Armed Forces-to-business cooperation in Finnmark]. *FD-Informasjon, 7/8*, pp. 6–8.
Johnson, S. and Wroughton, L. (2019). *Pompeo: Russia Is 'Aggressive' in Arctic, China's Work There Also Needs Watching* [Online]. Available at: https://www.arctictoday.com/pompeo-russia-is-aggressive-in-arctic-chinas-work-there-also-needs-watching (Accessed: 27 May 2019)
Käpylä, J. and Mikkola, H. (2019). "Contemporary Arctic Meets World Politics: Rethinking Arctic Exceptionalism in the Age of Uncertainty" in Finger, M. and Heininen, L. (eds.) *The Global Arctic Handbook*. Cham: Springer, pp. 153–169.
King Hubbert, M. (1956). *Nuclear Energy and the Fossil Fuels* [Online]. Available at: https://web.archive.org/web/20080527233843/http://www.hubbertpeak.com/hubbert/1956/1956.pdf (Accessed: 04 April 2021)
Körber, L.-A. and Volquardsen, E. (eds.) (2020). *The Postcolonial North Atlantic. Iceland, Greenland and the Faroe Islands*. Berlin: Nordeuropa-Institut der Humboldt-Universität.

Kristoffersen, B. and Langhelle, O. (2017). "Sustainable Development as a Global-Arctic Matter: Imaginaries and Controversies" in Keil, K. and Knecht, S. (eds.) *Governing Arctic Change*. London: Palgrave Macmillan, pp. 21–41.

Krivorotov, A. (2020). North Atlantic in China-U.S. Relations. *China in World and Regional Politics (History and Modernity), XXV*, pp. 168–183.

Kryukov, V.A. and Kryukov, Y.V. (2019). Арктическая экономика — можно ли обеспечить гармонию общего и особенного? [Arctic economy—Is it possible to Harmonize Common and Specific?] *Scientific Works of the Free Economic Society of Russia, 216*(2), pp. 26–53.

Larsen, J.N. and Petrov, A. (2015). "Human Development in the New Arctic" in Evengård, B., Larsen, J.N. and Paasche, Ø. (eds.) *The New Arctic*. Cham: Springer, pp. 133–146.

Larsen, J.N. and Petrov, A. (2020). "The Economy of the Arctic" in Coates, K. and Holroyd, C. (eds.) *The Palgrave Handbook of Arctic Policy and Politics*. Cham: Palgrave Macmillan, pp. 79–96.

Lasserre, F. (2021). Exploitation des resources naturelles dans l'Arctique. Une evolution contrastée dans les soubresauts du marché mondial [Exploitation of Arctic Natural Resources. An Evolution Against the Backdrop of World Market Turmoil]. Études du CQEG no. 3.

Logunov, V. (2001). Третий срок [Third Term]. *Nezavisimaya Gazeta*, 20 November.

Maximova, D.D. (2015). Опыт российско-канадского межрегионального взаимодействия в Арктике [Experience of Russian-Canadian Interregional Cooperation in the Arctic]. *Kanadskiy Ezhegodnik, 19*, pp. 201–218.

McPherson, S.S. (2015). *Arctic Thaw: Climate Change and the Global Race for Energy Resources*. Minneapolis: Twenty-First Century Books.

Menezes, D.R. and Chater, A. (eds.) (2018). *Proceedings of the Sustainable Development Goals (SDGs) in the Arctic. High-Level Dialogue Series, No.1 (Session 1: Arctic Circle Assembly 2017, Reykjavik, Iceland, 14 October 2017)*. London: Polar Research and Policy Initiative.

Mineev, A., Dybtsyna, E. and Mellemvik, F. (2020). "Circumpolar Business Development: The Paradox of Governance?" in Coates, K. and Holroyd, C. (eds.) *The Palgrave Handbook of Arctic Policy and Politics*. Cham: Palgrave Macmillan, pp. 175–192.

Mingming Shi and Lanteigne, M. (2019). *China's Central Role in Denmark's Arctic Security Policies* [Online]. Available at: https://thediplomat.com/2019/12/chinas-central-role-in-denmarks-arctic-security-policies (Accessed: 27 November 2019)

Mitrova, T. (2019). "Arctic Resource Development: Economics and Politics" in Correl, R.W.et al. (eds.) *The Arctic in World Affairs. 2019 North Pacific Arctic Conference Proceedings*. Busan: Korean Maritime Institute, pp. 205–224.

Nergård, J.I. (2011). Når slutter en koloniprosess? [When Is Colonization to End?] in Jentoft, S., Nergård, J.I. and Røvik, K.A. (eds.) *Hvor går Nord-Norge? Tidsbilder fra en landsdel i forandring* [Where does Northern Norway go? Time Pictures from a Changing Region]. Stamsund: Orkana, pp. 119–128.

Nordatlantisk utviklingsstrategi – NAUST (2019). [*North Atlantic Development Strategy*]. København: Nordisk ministerråd.

Pelyasov, A.N. and Zamyatina, N.Y. (2019). Суперорганизации в управлении Арктикой: исторический опыт и современная интерпретация [Super-Organizations in Arctic Governance: Historical Experience and Modern Interpretation] in Филин П.А., Емелина М.А., Савинов М.А. (ред.) Государственные и общественные организации в управлении Арктикой: прошлое, настоящее, будущее [Filin, P.A., Emelina, M.A. and Savinov, M.A. (eds.) Public Agencies and NGOs in Arctic Governance: Past, Present, Future]. Moscow: Paulsen, pp. 160–173.

Petrov, A. (2018). "Re-Tracing Development Paths: Exploring the Origins & Nature of the 20th Century's Northern Development Paradigms in Russia and Canada" in Heininen, L. and Exner-Pirot, H. (eds.) *Arctic Yearbook 2018*. Akureyri: Northern Research Forum, pp. 21–34.

Petrov, A., Rozanova Smith, M., Krivorotov, A., Klyuchnikova, M., Mikheev, V., Pelyasov, A. and Zamyatina, N. (2021). Russian Arctic by 2050: Developing Integrated Scenarios. *ARCTIC, 74*(3), pp. 306–322.

Poppel, B. (2019). "The Path Towards Independence" in Correl, R.W. et al. (eds.) *The Arctic in World Affairs. 2019 North Pacific Arctic Conference Proceedings*. Busan: Korean Maritime Institute, pp. 120–132.

President of the USA (2021). *Executive Order on Tackling the Climate Crisis at Home and Abroad*. [Online]. Available at: https://www.whitehouse.gov/briefing-room/presidential-actions/2021/01/27/executive-order-on-tackling-the-climate-crisis-at-home-and-abroad/ (Accessed: 27 January 2021)

Schreiber, M. and Rosen, Y. (2021). *Biden Halts Oil and Gas Development in Arctic Refuge Hours After Inauguration*. [Online]. Available at: https://www.arctictoday.com/biden-halts-oil-and-gas-development-in-arctic-refuge-hours-after-inauguration/

Schwab, K. (2020). *Time for a Great Reset*. [Online]. Available at: https://www.project-syndicate.org/commentary/great-reset-capitalism-covid19-crisis-by-klaus-schwab-2020-06

Silin, A. (1989). Тюменский Север – не колония! [North-western Siberia Is Not a Colony!] *ECO, 7*, pp. 69–76.

Sokolov, A. and Terentyev, I. (2015). 10 российских городов, потерявших больше всего населения за 25 лет [10 Russian Towns Which Lost Most People Over 25 years]. [Online]. Available at: https://www.rbc.ru/photoreport/22/01/2015/54b92c319a79473fc05f3057 (Accessed: 27 July 2021)

Standlea, D.M. (2006). *Oil, Globalization, and the War for the Arctic Refuge*. New York: State University of New York.

Supreme Court of Norway (2020). *Judgment in the "Climate Lawsuit"*. [Online]. Available at: https://www.domstol.no/en/Enkelt-domstol/supremecourt/nyhet/2020/judgment-in-the-climate-lawsuit/ (Accessed: 27 April 2021).

Suter, L., Streletskiy, D.A. and Shiklomanov, N. (2019). Assessment of the Cost of Climate Change Impacts on Critical Infrastructure in the Circumpolar Arctic. *Polar Geography, 42*(12), pp. 1–20.

Tsafos, N. (2020). *Will New Funding Rules Kill Alaska's Oil Boom?* [Online]. Available at: https://www.csis.org/analysis/will-new-funding-rules-kill-alaskas-oil-boom (Accessed: 12 May 2021).

U.S.-Canada Joint Statement on Climate, Energy, and Arctic Leadership (2016). [Online]. Available at: https://obamawhitehouse.archives.gov/the-press-office/2016/03/10/us-canada-joint-statement-climate-energy-and-arctic-leadership (Accessed: 09 October 2020).

Wilson Rowe, E. (2018). *Arctic Governance. Power in Cross-Border Cooperation*. Manchester: University Press.

World Bank (2009). *World Development Report 2009: Reshaping Economic Geography*. Washington, DC: World Bank.

Yadrintsev, N.M. (1882). Сибирь, как колония в географическом, этнологическом и историческом отношениях [Siberia as a Colony in Geographic, Ethnologic and Historic respects]. St.-Petersburg: I. Sibiryakov Publishers.

Zellen, B.S. (2009). *Arctic Doom, Arctic Boom. The Geopolitics of Climate Change in the Arctic*. Santa Barbara: ABC-CLIO.

Zubov, V. and Inozemtsev, V. (2013). Сибирское благословение [Siberian Blessing]. Moscow: Argamak-Media.

5 "Gas Bridges" and Geo-Economics of the Arctic

Petter Nore[1]

Introduction

Energy relations between international partners in the Arctic have changed dramatically over time. After the collapse of the Soviet Union, Russia initially turned to the West to try to find capital and technology to develop its energy sector. This chapter will first analyze how this trend played out in the offshore Arctic centered around the Barents Sea and offshore Western Siberia.

It will concentrate on the Norwegian/Russian gas relationship and why it ultimately collapsed. After that, we will analyze the foundations of Russia's "Pivot to the East" and how from 2014 Russia turned its attention toward China. Here, special emphasis is placed on how the commercial and geopolitical relationship between the two countries will be affected by a "Net Zero" world.

The theoretical foundation of the article is "geo-economics" defined as "The use of economic tools to advance geopolitical objectives" (Schneider-Petsinger, 2016).[2] We have chosen this approach because the Arctic is an example of how abundant energy resources are used to advance geopolitical objectives and vice versa. The role of governments, commercial companies, and other stakeholders in the area can only be understood in the interplay between economics and geopolitics.

This chapter is concerned throughout with natural gas. In the Barents Sea, Russian and Norwegian companies and government institutions engaged in close cooperation, here described as a "Gas Bridge". The concept of a "Gas Bridge" is inspired by Thane Gustafson (Gustafson, 2020) who used it to describe the historic gas relationship between Russia and Europe. The Norwegian-Russian "Gas Bridge" was built around multiple objectives. Russia wished to obtain offshore technology and organizational competence; Norway wished to gain access to gas reserves for its national companies and open markets for its supply industry. There was a common geopolitical interest in maintaining the Arctic as a conflict-free zone. The Shtokman project played a key part in this era of cooperation.

Following the Ukrainian crisis in 2014, but also expedited by other factors, Russia's Arctic "Gas Bridge" turned toward China. Several large gas pipelines and liquefied natural gas (LNG) projects have since been completed, are under construction, or planned.

DOI: 10.4324/9781003246015-7

The last part of this chapter will analyze the dilemma of the "Sino-Russian Gas Bridge" faced with China's move toward "Net Zero". Two different outcomes are outlined in the form of narratives. First, there is a narrative of how the "Sino-Russian Gas Bridge" by 2060 will have declined in importance and become largely irrelevant as the world's energy mix will be almost totally dominated by low carbon solutions. The second narrative is based on a future where gas consolidates its position in China's energy mix and where gas imports from Russia will continue to increase and new projects will be initiated.

The formation of the Russian-Norwegian "Gas Bridge"

Norway and Russia had different reasons for engaging in close energy cooperation in the Arctic. Even at the time of the collapse of the Soviet Union, Russian geologists knew that the Barents Sea contained large amounts of hydrocarbons. Explorative drilling began in 1982 and five deposits were identified. Among these was the giant gas/condensate field, Shtokman, located 650 km Northeast of Murmansk in the Barents Sea. The size of the field (3,800 bcm [billion cubic meters]) made it the world's second largest offshore gas/condensate field with excellent production characteristics.

Norway, on its side of the still undefined border line that separated the two countries in the north, had made two potential finds which were later to become part of the Snøhvit LNG project, the first Arctic development in Norwegian waters. During the Cold War, there had not been any cooperation between the two countries in the oil and gas field, apart from the 46.5 MW hydropower project in the Pasvik River. However, this situation started to change during the Yeltsin presidency. Vast resources were ready to be exploited and there was no longer any ideological reluctance to cooperate with Western companies. However, the Russian energy industry clearly lacked the technological or organizational resources.

Norwegian companies wanted primarily to gain access to huge energy resources, which were assessed as accounting respectively for 13% and 30% of the remaining global oil and gas reserves (US Geological Survey, 2009). The two major Norwegian oil companies, Hydro and Statoil (later Equinor), established a presence in Moscow in 1991 and 1993. Both companies eyed with interest a possible role in a future Shtokman development. However, despite the close contact between the Norwegian and Russian companies and the fact that well number 4 on Shtokman was drilled by four Western companies with Norsk Hydro as operator in 1991, the chaos of the Yeltsin presidency hindered the development of any concrete projects on this scale (the financial crisis of 1998 further impeded the launch of such a project). But toward the beginning of the 2000s, things started to change.

The Russian companies, for their part, sought access to technology and competence. Contrary to common belief, the Russians did not primarily want Western oil companies because of their advanced technologies, as it was possible to buy such technologies from other sources. What they lacked was the skills that

Western oil companies possessed to plan, build, and ultimately operate complicated offshore projects in the Arctic (Lie Hansen, 2021). They pursued their objectives both by trying to obtain licenses in the Norwegian Arctic and thereby cooperate with Western companies and also by seeking direct bilateral cooperation with Norwegian companies. However, the latter was difficult to achieve in practice. The main difficulty was the Russian tendency to operate in a different way, always tightly controlled from the very top. There was little information sharing tradition. This contrasted with the approach of the Norwegian partners, who emphasized decentralized decision-making. The Russian companies and research institutions sought cooperation with Norwegian partners in the field of research and development (R&D). Twenty-four Norwegian/Russian research projects in cooperation between the Russian Academy of Science (RAS) and the Norwegian Research Council had been completed at a cost of NOK 190 million in the field of joint resources optimization during the period 2005–2011 (The Research Council of Norway, 2011).

The Norwegian government in a 2011 Parliamentary Report on the High North outlined ambitious plans to extend and intensify Russian-Norwegian cooperation in the fields of technology, commercial activities, and institutional cooperation (Norwegian Parliamentary Report, 2011). What did the Norwegian government seek to achieve? It strongly supported all efforts to internationalize its petroleum industry, which meant that the government supported Hydro, Statoil, and the very strong supply industry in their efforts to establish a foothold in Russia. The Norwegian government encouraged by all means the cooperation between its own energy industry and their Russian counterparts in the Arctic.[3] During this process, the government took several initiatives that encouraged Norwegian companies in making contacts with top Russian politicians. However, the Norwegian government also had its own geopolitical objectives in mind. The Norwegian foreign minister described the Arctic as "Norway's most important area of strategic interest" (Støre, 2006).

Norway wanted to bring Russia into a mode of geopolitical cooperation that underlined its wish to maintain the Arctic as a conflict-free area characterized by close institutional cooperation. This was well expressed through a major four-year project on "Geopolitics in the High North" financed by the government through the Research Council of Norway. The main purpose of the work was "to identify the conditions for cooperation, stability and peace in the Arctic" (Tamnes and Offerdal, 2016). Energy cooperation (Claes and Moe, 2014) was one of several areas described in the work as examples of how to achieve a non-conflict-prone Arctic. Norway's position was also mirrored in Russia's geopolitical approach. During this period, "Russia approached the Arctic as an area of low tensions, where cooperation with other powers in addressing common challenges was desirable" (Rumer et al., 2021, p. 3).

For both sides, economic cooperation became a way to achieve geopolitical objectives. The close relationship between the Norwegian and Russian governments culminated in 2010 with the resolution of the border dispute in the

Arctic that had remained unsolved for more than 40 years. Numerous factors were thus pointing in the right direction, both in the general relationship between Norway and Russia in the Arctic, but also for the Russian-Western Gas Bridge. The negotiations between Statoil and Gazprom (Hydro was merged with Statoil in 2008) about the development of the Shtokman field continued. In 2010, a leading Norwegian observer described the relationship between Russia and Norway in the energy sector as "peaceful and cooperative" (Moe 2010, p. 225).

Weakening of the gas bridge

However, underneath this calm surface, a combination of internal and external factors threatened the future of Norwegian-Russian gas relations. Internally, the Shtokman negotiations met some "bumps on the road". Costs increased dramatically while Russian partners at times seemed to be more interested in "harvesting rent" from a future development at the expense of finding low-cost common solutions (Lie Hansen, 2021).

The Shtokman project was overtaken by global developments such as the emergence of shale gas production in the US, as Shtokman was initially perceived as a project to export LNG to the US market. However, the US soon moved from being the central market for LNG export to become a competitor for Russian gas suppliers, especially in the European market. The Shtokman project then turned its attention toward a piped gas solution to the European market, which would increase the costs dramatically as new pipelines had to be built. Following an internal struggle within the company, Gazprom decided to develop the onshore Yamal gas project as an alternative to offshore Shtokman. This followed a bitter struggle between the "old" and the "new" factions of Gazprom represented by Alexander Ananenkov on the Yamal side and Alexander Ryazanov on the Shtokman side (Gustafson, 2020, pp. 303–304). In hindsight, it was possible to see that the battle in favor of Shtokman was "lost" already in 2006, when Ryazaonov was removed from his key position. Only in 2012, when both the Norwegian and French partners withdrew from the project, was the project "de facto" terminated.

Finally, and probably most importantly, Norway decided to join the Western sanctions against Russia following the Russian annexation of Crimea in 2014. Western sanctions banned Norwegian companies from assisting Russian companies in oil and gas exploration in the Arctic or in deposits that could be described as shale gas formations (US Department of the Treasury, 2014). Norwegian companies started to withdraw from doing business in Russia and a number of contracts were discontinued from the Russian side. The Norwegian petroleum supply industry, which had played an important role in the Arctic, saw its role disappear overnight. Statoil had entered into a strategic alliance with Rosneft in 2012, but the alliance has had difficulties in maneuvering in the sanctions-dominated world.

Post 2014: Russia's gas "pivot to the east"

Russia's "pivot to the East" was long overdue. By sheer accident, the world's largest exporter of fossil fuels (Russia) is located next to the world's largest energy market (China). Russia had for many years tried to turn to China to sell oil and gas, but with little success. The historic energy relationship between the Soviet Union and China had not been without problems (Henderson and Mitrova 2016, p. 1). Very little progress had been achieved in the first years after the collapse of the Soviet Union, but by 2000, Gazprom realized that China was a potentially large and rapidly growing gas market and was keen to sell gas there. Gazprom initially overplayed its negotiating hand and asked for commercial terms that were very favorable to Russia compared with what the Chinese were willing to offer. As late as 2012, Gazprom was asking for a base price and indexation in their gas negotiations with China that would guarantee the company the same netback as was the case for gas sold to Europe (Weitz, 2014, p. 82), yet the Chinese were not interested.

By 2014, the Russians had changed their strategy and presented new terms to the Chinese gas buyers. Following meetings between Presidents Putin and Xi Jinping at the Brazil, Russia, India, China and South Africa (BRICS) summit in Shanghai in September that year, Putin returned with a signed deal for Power of Siberia I (POS I). The negotiations took place "under the watchful eyes" of the two presidents (Chow, 2021). Political intervention in the ongoing negotiations was a prerequisite for the first large gas pipeline deal between the two countries.

Four factors spoke for building a Russian-Chinese Gas Bridge. In the first instance, Russia had almost unlimited supplies of natural gas, even if some of it was in the "wrong place". The Russians wanted the gas to originate in Western Siberia in order to direct supplies in two directions: either to Europe or Asia. The East Siberia option was chosen, as the Chinese wanted the gas from there. Second, China wanted more gas imports because of a strong and rapidly growing Chinese gas market coupled with lower than anticipated domestic production. The country's wish for a strategic diversification of its energy supply also contributed to its interest in buying Russian gas. Total Chinese gas demand in 2005 was a modest 50 bcm/year. By 2020, demand had reached 322 bcm/year with the expectation of a continued further rapid growth. Third, the weakening and politization of the European gas markets caused Russia to look for new outlets for its gas. Europe was the most important market for Russian gas in 2014. However, the prospect of competition from US shale-based LNG and the threat of an increasingly politicized gas market increased Russia's uncertainty. The first discussions and emerging conflicts about the construction and operation of Nord Stream II had already begun in 2014/2015. The view that gas was a "political commodity" spread and the general outlook for gas as a fuel weakened correspondingly. President Putin summed up the situation on his return from Shanghai in 2014: "We have to admit that energy consumption in Europe is growing more slowly due to low economic growth rates, while political and regulatory risks are increasing". He added: "Given these circumstances our desire to open up new markets are both natural and understandable" (Weitz, 2014, p. 81). "Lasty" cooperation in the gas sector was a way to strengthen the rapidly improving geopolitical relationship between the two countries, which shared a deep skepticism of what they viewed as "the US dominance" in international relations. This relationship further strengthened after the US imposed sanctions on Russia following Russia's annexation of Crimea in 2014. The sanctions "elevated their (Russia and Chinas PN) strategic and economic partnership to a higher level" (Alexeeva and Lasserre, 2018). Overland and Kubayeva (2018) have given a good overview of the role played by geopolitics in the Russian "pivot to the East".

Present status and future plans of the Sino-Russian "Gas Bridge"

Three projects of the "gas bridge" are already in operation or under construction. Russian companies have in addition been drawing up plans for new projects that will almost double the existing capacity. The USD 55 billion and 3,000 kilometers POS I piped gas project is the largest energy project to be undertaken in Russia since the collapse of the Soviet Union. The gas originates in East Siberia and is delivered on the Northwestern Chinese border. The first volumes were delivered in 2019. The commercial agreement which underlies POS I will last for 30 years, and deliveries will reach 38 bcm/year by 2025. In comparison, by 2019, piped imports almost exclusively from Central Asia reached 48 bcm while LNG imports reached 85 bcm, mostly from Australia and Qatar. In 2020, Gazprom announced talks with China to further increase the volume in the project to 44 bcm/year by 2030. The details of the POS I contract have never been made public, but different sources suggest a contract where the price of gas follows that of competing products like light fuel oil with a nine-month lag. It is a "take or pay" contract, where Russia guarantees 85% of income irrespective of Chinese offtake (Lalyveld, 2020). In 2020, when the average oil price was around USD 40/bbl (a barrel of crude oil), an observer said that the contract "under current market conditions does not look commercially attractive" but that the deal also contained a "political premium" that had to be taken into account (quoted in Lalyveld [2020] by Edward Chow).

Two LNG projects located in the Yamal area 500 kilometers north of the Arctic Circle are also part of the Sino-Russian Gas Bridge. Yamal LNG 1 produces today at full capacity (19 bcm of gas equivalent/year). But only part of this gas is destined for the Chinese market. Arctic LNG 2 with a capacity of 28 bcm/year is now being built and will start operations in 2025. These two LNG projects are examples of the close relationship between Russia and China in the field of Arctic energy. The Russian private company Novatek is the majority owner and operator in both projects while Chinese companies own 30% of Yamal LNG 1 and 20% of Arctic LNG 2 respectively. In 2016, two Chinese state banks loaned USD 12 billion for Yamal LNG 1 because Western sanctions made it difficult for Novatek to raise long-term Western finance (Overland and Kubayeva, 2018, p. 107).

In May 2020, Gazprom launched a feasibility study for a new pipeline "The Power of Siberia II (POS II)" that plans to start operation in 2032. It would have a capacity of 50 bcm/year and bring gas from Western Siberia through Mongolia to the greater Beijing area (the Altai connection). The advantage for Russia is that the gas will originate from existing sources in Yamal and production costs will be negligible. The production location will make it easier for Gazprom to balance the Chinese and European markets against one another. The disadvantage of the project is the high cost of transporting the gas: around 2,600 kilometers. So far, the Chinese authorities have not given any firm response to the Russian plans, thus keeping all their options open.

By 2019, the foundations of the Sino-Russian Gas Bridge seemed to be solid. There was no reason to rewrite the 2017 scenarios for gas imports that showed imports by 2030 ranging between 170 and 340 bcm/year compared with imports of 120 bcm/year in 2017 (Henderson and Moe, 2019 p. 112, 116). There was also an expectation of further growth over a longer time horizon in line with growing domestic gas demand. The strong geopolitical ties between the two countries further strengthened an optimistic view of the future gas trade. A "political premium" was built into the calculations of POS I and there is a belief that similar considerations would, if necessary, also be applied to future projects.

"Net Zero" and three uncertainties that will affect the Sino-Russian Gas Bridge

In September 2020, President Xi Jinping made a one-sentence statement in his video statement to the UN General Assembly. He said: "We aim to have CO_2 emissions peak before 2030 and achieve carbon neutrality before 2060". The world was greatly surprised and in large parts warmly applauded, but the Russians were stunned. They had planned for a long-term energy relationship with China that would guarantee increasing exports of fossil fuels and a steady stream of revenue for many decades to come. The Russians had not been informed beforehand and almost immediately started to worry about the future of their fossil fuel exports to China. In October, President Putin said: "I don't think it is realistic if everyone wants to be competitive to abandon hydrocarbons in the near future" (Valdai, 2020). In the future, Russia will need China as a market for its resources more than China needs Russia as an oil and gas supplier.

In May 2021, the International Energy Agency (IEA) presented a detailed analysis of what "Net Zero" (IEA, 2021a) would imply for the global energy system, followed in September 2021 by a special report on the consequences of "Net Zero" for China (IEA, 2021). Three factors will have great yet uncertain influence on the future of the Sino-Russian Gas Bridge in a "Net Zero" world.

Uncertainty 1: The future of China's gas demand

In a "net zero" scenario, China's gas will primarily be used for two purposes: gas as an intermittent fuel in power production and gas as basis for blue hydrogen for

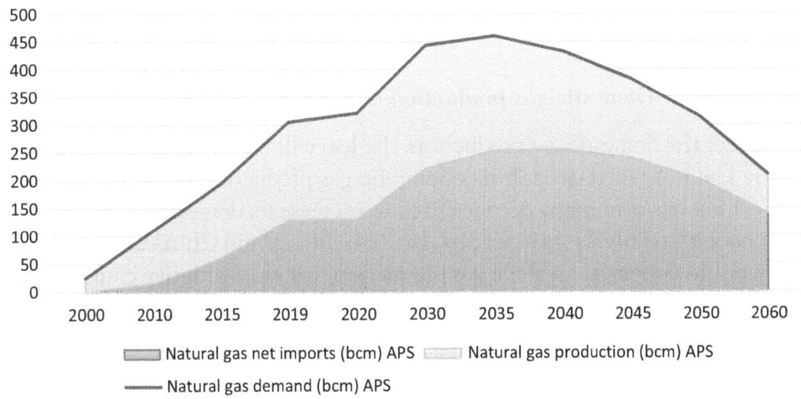

Figure 5.1 China's total gas demand – domestic production and imports (in bcm) 2000–2060.

Source: IEA, 2021b.

use in industry and transport. Gas for power today constitutes a very modest 3% of the total Chinese-generating capacity and this percentage is not expected to increase. Gas will not be the substitute for coal in power generation; renewable energy will take on that role. The full dominance of renewables in China's energy mix will require a very significant increase in the demand for intermittent power. The IEA assumes in its "Net Zero" scenario that natural gas with CCUS (Carbon Capture, Utilization and Storage) will constitute 15 % of total flexibility capacity by 2050 (IEA, 2021b; figure 3.6). Furthermore 20% of all hydrogen demand will at the same time be satisfied with blue hydrogen (IEA, 2021b, p. 96). BP (2020) estimates that blue hydrogen will constitute as a much as 47% of total future Chinese gas demand. In a "Net Zero" scenario, China's gas demand is expected to peak in 2035. By 2050 total Chinese gas demand is estimated to sink to a level of 310 bcm, marginally below today's level. This is in line with Kejun et al. (2021) who predict a small decline over the same period. During the decade 2050–60 gas demand is estimated to decline by a further 30%. There is downside in IEAs estimates if cheaper and better ways than gas are found to solve the intermittency problem in renewable power production. Green hydrogen can also over time outcompete blue hydrogen.

Should, however, the energy transition proceed more slowly than assumed in the "Net Zero" scenario and more in line with a "Business as Usual" type of scenario, the annual growth rate of gas demand toward 2050 will accelerate markedly. Yet, estimates vary: BP (2020 China) estimates a demand of 630 bcm in 2050 which corresponds to a yearly growth rate for gas of 2.6% annually, while the IEA (in its STEPS [Stated Policies Scenario]) sees an increase between 2020 and 2050 of around 200 bcm/year to 510 bcm, which corresponds to a growth rate of around 2%. Gas will in both scenarios retain an important role in the Chinese energy mix.

Uncertainty 2: Domestic gas production and imports in China

The higher the domestic gas production, the less will be the need for gas imports, be they LNG or piped gas. China's domestic gas production reached 191 bcm in 2020. China has over many decades tried to increase its domestic gas production, but so far with relatively little success. In 2014, the US and China entered into an agreement to cooperate in shale gas production, but this initiative came to nothing. Analysts have made different estimates regarding Chinese gas production. BP (2020) expects domestic production to increase by 75% by 2050, while the IEA expects production in "Net Zero" to drop to 110 bcm.[4] Domestic gas production is difficult to estimate. Geology has historically been distinctly unpredictable with even the most ambitious exploration programs. In China, part of the gas portfolio consists of "technologically immature" resources, such as shale gas and coal-based methane (Liab, 2020), while there will arguably be fewer resources devoted to the exploration of a commodity that will be viewed as being both in less demand while also being less "strategic".

Political factors are likely to play some role. Efforts to increase domestic gas production have historically been linked to a traditional geopolitical objective of increasing "security of supply" for strategic important commodities. Recently, this thinking has been reinforced by the introduction of the concept of "dual circulation", which was first presented by President Xi Jinping in May 2020 (*China Daily*). He wanted to tie China's development more closely to "internal circulation" – meaning increasing the domestic element of production, distribution, and consumption. Such a policy approach would mean a corresponding reduction in China's reliance on "external circulation". As one seasoned observer wrote: "The fallout of such a shift may be dramatic for countries depending on fossil fuel exports to China. China's new economic policy and new emphasis on energy security mean ever greater efforts to replace imports with domestic supply" (Myllyvirta, 2020).

The expected future import of gas to China is a result of the two variables addressed above: total anticipated demand and anticipated local production (see Figure 5.1). It shows a steadily downward trend from 2035 to 2060. In the business as usual (STEPS) scenario, the situation is opposite: expected imports are likely to steadily increase.

Uncertainty 3: Geopolitical relations between China and Russia

The third factor that will influence the future Sino-Russian Gas Bridge, and which so far has not played any major part of the analysis, is the geopolitical relationship between Russia and China. This will ultimately play an important role when countries are chosen as sources of gas imports and whether imports will take place via piped gas or LNG. Will geopolitical reasons cause Russian gas to be chosen over gas, for instance, from Central Asia or LNG? China has in the past let its geopolitical preferences influence its gas import policies, most notably in 2019 when the government decided to increase LNG imports from the US to settle a trade dispute between the two countries (GECF, 2020). It is therefore possible to argue that China will be willing to give Russian gas imports some kind of preference if there are strong enough geopolitical reasons for doing so. Both China and Russia have on numerous occasions underlined how energy cooperation was part of a much wider cooperation. The annual threat assessment of the US Intelligence Community for 2021 observed "a growing strategic cooperation between Russia and China" (Director of National Intelligence, 2021). Never since the mid-1950s has the overall relationship between China and Russia been stronger. We are no longer characterizing the situation as "a marriage of convenience", which in opportunistic ways commits to a strategic cooperation (Lo, 2008). We are now faced with a much more permanent cooperation (Gabuev, 2021).

Energy policies have so far played an important part in building this cooperation. Control over fossil fuels was at the time of the inception of the "Gas Bridge" a crucial element, both in defining geopolitical interests and the economic means to achieve these (Yergin, 2012). But the link between geopolitics and energy (and hence the basis for geo-economics) is changing irrespective of the scenario. Interests and conflicts between nations will move from being built around resource revenues for fossil fuels to being increasingly built around revenues from energy systems dominated by renewables and low carbon solutions (see IRENA, 2019). This change in China's geopolitical position is likely to express itself along two axes.

Russian gas will not in the long run get any "special treatment", for the simple reason that gas will be viewed as a commodity like any other. Long before 2050, gas will cease to be of key importance in defining the geopolitical relationship between the China and Russia. The Chinese will still be interested in "security of supply", but this interest will not be directed toward natural gas and other fossil fuels passing through the South China Sea or the Northern Sea Route. It will be directed more toward control over the rare minerals necessary to produce renewable energy. China today controls 85% of such "rare earth" minerals (CSIS, 2021), and there are many reasons to believe that China will pursue a policy in this field based on perceived national interest.

Two narratives for the future of the Sino-Russian Gas Bridge

We will present two "narratives" of the future based on different combinations of the three uncertainties analyzed above. The first narrative describes the *gradual decline and irrelevance* of the "Sino-Russian Gas Bridge" and rests on the following "building blocks":

- A rapid decarbonization rate in the wake of "Net Zero" in China, which, in turn, will decrease the demand for gas.
- A significant increase in China's domestic gas production. The resources, finance, and technology are at the disposal of the Chinese government and companies, so this objective is likely to be fulfilled if geological conditions are positive, reinforced by a political emphasis on "dual circulation".
- A weakening in the overall geopolitical relationship between China and Russia, possibly in combination with the narrative of how geopolitical interests change as the world moves toward a low carbon energy system. Russia in such a situation will not receive any positive treatment as gas exporter compared with its competitors in Central Asia and the LNG market.

In such a world, it is highly unlikely that any new major gas investments linking Russia and China will be undertaken. The plans for POS II and the expansion of existing projects will be put on hold. No new gas fields will be developed in East Siberia. The further expansion of Arctic LNG production will also be in doubt, even if this also depends upon demand for LNG in other parts of the world.

By 2035, the Gas Bridge will still provide a significant interim energy solution. Fundamental energy changes take time, and gas will continue to play a not insignificant part of the energy mix. However, from then on, "Sino-Russian Gas Bridge" will slowly start its downward move to oblivion. Gas will play a constantly declining role in the energy mix as the world turns its attention squarely toward low carbon.

The second narrative tells the story of the *continued expansion* of the Sino-Russian Gas Bridge.

- First, in this case gas demand will increase because the rate of China's energy decarbonization will slow down significantly. This lower rate may be due to a number of reasons, one being that the required energy "turnaround" will simply be too complex to be realized within the 2060 timeline. There will not be sufficient industrial and bureaucratic capacity to succeed, even for China. Other explanations may center around possible crises in the country as the direct result of the rapid energy transition. What will be the reactions of the 2.6 million coalminers, as their industry rapidly disappears? There may be regional and industrial elites that will find their positions threatened by the coming changes and do their utmost to maintain the status quo.
- Second, domestic Chinese gas production may not increase, a factor that will further increase the demand for gas imports. The geological conditions to increase production will simply not be in place, even if all other conditions, including a strong political commitment for more production, are in place.
- Finally, and despite the emergence of "low carbon geopolitics", other and more traditional geopolitical forces will continue to dominate the relationship between Russia and China. The Western powers may not manage to split the very close relationship between China and Russia, which continue to stand united in opposition to what they perceive to be the West's continued

attempts at domination. China will in such a situation continue to feel comfortable with relying on Russia as the main stable and politically secure supplier of piped gas to meet the anticipated increase in demand. In this expanding market, China will also continue to import gas from Central Asia. LNG (including deliveries from the rapidly expanding Russian LNG industry) will continue to be the "marginal supplier", giving China the needed gas importing flexibility. In such a world, the expansion of the Sino-Russian Gas Bridge will continue as planned. By 2035, huge new investments like POS II and additional LNG plants will have been built. New fields will have been developed in Russia to supply the Chinese market while the two countries will jointly try to develop and protect the transit through the Northern Sea Route.

Better health and lower emissions

Irrespective of the fate of the "Sino-Russian Gas Bridge", future developments will bring great improvements in health, thus partly fulfilling Sustainable Development Goal (SDG) 3 "Good Health and well-being". Increased use of gas in the short to medium term and an increase in renewables in particular will have a positive effect on the situation in northern China. Here, air pollution from the burning of fossil fuels, principally coal, is causing people to die on average 5.5 years sooner than they otherwise might (Flannery, 2015). Similarly, both narratives imply significant decreases in Chinese emissions and will contribute toward the achievement of SDG 13 "Climate Action". By 2040, China's energy-related yearly emissions in "Net Zero" will have dropped to 3.5 Gt (gigatons) compared to today's level of 10 Gt and reach "Net Zero" emissions by 2060. Even the less ambitious STEPS will see an emission drop to 6 Gt CO2/year by 2060 (IEA, 2021b, Figure 2.1).

Concluding comments

This chapter addresses two major issues. First, how Russia's gas relationship in the Arctic around 2015 pivoted from cooperation with Western companies and partners (exemplified by the Russian-Norwegian Gas Bridge) and entered into a closer relationship with China (exemplified by the Sino-Russian Gas Bridge). This pivot is often explained primarily as a result of geopolitics following Russia's annexation of Crimea in 2014. But our work shows that there were a number of other factors that also contributed toward explaining what happened: ranging from shale gas developments in the US and its market consequences to lack of results in the cooperation with Western companies.

Second, this chapter discusses consequences of China's aim to become carbon neutral by 2060 for the "Sino-Russian Gas Bridge". We have developed two narratives to analyze what may happen. Each narrative is built around three variables: the projected future demand for gas in China; the future level of domestic gas production; and the geopolitical relationship between China and Russia. The first narrative describes how the "Sino-Russian Gas Bridge" declines in importance and becomes largely irrelevant as China decisively moves in the direction of "Net Zero". Arctic will, in this narrative, change dramatically as oil and gas activities will be phased out. Russia will, in such a narrative, wake up to the fact that it has to "go green". It may start to loosen its links with China and seek for new partners like the EU. Links between Russia and Norway can be reestablished in the field of low carbon cooperation. The two countries share the challenge of moving out of their respective reliance on fossil fuels. The conditions are ripe for a much closer cooperation between the two countries in areas like wind hydrogen and ammonia. The next narrative describes how gas strengthens its place in China's energy mix and gas imports from Russia continue to increase as the energy transition in China moves more slower than expected in the "Net Zero" direction. In this case, the gas industry in the Russian Arctic continues its expansion as the Sino-Russian connection will gain rather than lose strength. Russia remains embedded in its fossil fuel economic and political structure. In the short to medium term, Russia can live with such a situation. But in the long run, this will be a disaster as markets eventually dry up and the world moves toward a low carbon future, albeit at a slower rate than in the "Low Carbon" case.

December 2021

Ex-post reflections

One of the most important geo-economic consequences of the Russian-Ukrainian war is likely to be dramatically decreased imports of Russian gas to EU. Within three to four years, a combination of accelerated LNG imports, increased pipeline production by non-Russian sources, and an intensified energy decarbonization will cut Russia's 155 bcm/year exports of gas to EU to almost zero. Russia may therefore lose up to 10 % of state income.

China is an obvious alternative market for Russian gas, especially given the good geopolitical relationship between the two countries. A new POS II pipeline with a capacity of 50–60 bcm/year from Yamal to Beijing via Mongolia has been planned for a number of years, even if no final decision has been taken.

It is unsure whether China needs the gas, especially if a "Net Zero" scenario is realized. It is also unsure whether China wants to increase its dependence on gas imports from one source (Russia) and it is uncertain whether China will support Russia so openly in its confrontation with the West and then be in danger of incurring Western sanctions.

Building a new pipeline to China is likely to take between eight and ten years. Thus, irrespective of whether there is demand for the gas, Russia is likely to suffer significant loss of income as there is no alternative market for the Russian pipeline gas.

Finally, any future cooperation in the field of renewables between Norway and Russia is likely to be many years away.

May 1, 2022

Notes

1 The author was from 2004 to 2007 President of Norsk Hydro Russia and has, in parts of this chapter, drawn on his experience from that period. I want to thank Thane Gustafson, Bengt Lie Hansen, and Leiv Lunde for inspiration in writing this paper. They have no responsibility for the content of the chapter.

2 The formation of geo-economics as a branch of geopolitics is often attributed to Edward Luttwak, an American economist and consultant, and to Pascal Lorot, a French economist and political scientist. This approach to geopolitics is being increasingly used where the interaction between economic and geopolitical issues are of special interest; cf. "Today's rising powers are increasingly drawn to economic instruments as their primary means of projecting influence and conducting geopolitical combat in the twenty-first century" (Blackwill and Harris, 2016, pp. 27–31).

3 In the winter of 2006, the Norwegian Prime Minister sent a letter instigated by the chief executive officers (CEOs) of Statoil and Hydro to the Russian government informing them that Statoil and Hydro would cooperate closely in their dealings with their Russian counterparts. The aim was to increase the competitiveness of Norwegian companies in their quest to gain an ownership share in the Shtokman license. The existence of this letter has never been officially acknowledged by any of the partners, but the author of this article has seen a copy of it.

4 IEA has not published an estimate for domestic gas production in 2060. In Figure 5.1, we assume that the proportion between domestic gas production and total energy demand will be the same in 2060 as in 2050.

References

Note: All links were accessed on 13 December 2021.

Alexeeva, O., and Lasserre, F. (2018). An analysis on Sino-Russian cooperation in the arctic in the BRI era. *Advances in Polar Sciences*, 29(4), pp. 269–282.

Blackwill, R. D., and Harris, J. M. (2016). *War by other means: Geoeconomics and statecraft*. Cambridge, MA: Harvard University Press.

BP (2020). *Energy outlook 2020; China insights*. Accessed from: https://www.bp.com/en/global/corporate/energy-economics/energy-outlook/country-and-regional-insights/china-insights.html.

China Daily. "Understanding 'dual circulation' and what it means for the world" (updated 6.9.2020).

Chow, E. (2021). Sino-Russian energy relations: A match made in heaven? *Carnegie Endowment for International Peace*. Accessed from: https://carnegiemoscow.org/commentary/83757.

Claes, D. H., and Moe, A. (2014). Arctic petroleum resources in a regional and global perspective. In Tamnes, R., and Offerdal, K. (eds.) *Geopolitics and security in the Arctic. Regional dynamics in a global world* (pp. 97–120). New York: Routledge.

CSIS (2021). *Does China pose a threat to global rare earth supply chains?* Accessed from: https://chinapower.csis.org/china-rare-earths/.

Director of National Intelligence (2021). *Annual threat assessment of the US intelligence community*. Washington, DC: Report and Publications.

Flannery, T. (2015). *Atmosphere of hope searching for solutions to the climate crisis*. New York: Atlantic Monthly Press.

Gabuev, A. (2021). *China and Russia's converging interests*. Podcast Carnegie Moscow edited by Bonnie Glaser. Accessed from: https://www.csis.org/podcasts/chinapower/china-and-russias-converging-interests-conversation-alexander-gabuev.

GECF (2020). Expert commentary – The impact of phase 1 of the US-China trade deal on the LNG market 07 February 2020.

Gustafson, T. (2020). *The bridge: Natural gas in a redivided Europe*. Cambridge, MA: Harvard University Press.

Henderson, J., and Mitrova, T. (2016). *Energy relations between Russia and China: Playing chess with the dragon*. Oxford: Oxford Institute of Energy Studies.

Henderson, J., and Moe, A. (2019). *The globalization of Russian gas: Political and commercial catalysts*. Edward Elgar Publishing.

IEA (2021a). *Net zero by 2050*. Paris: IEA.

IEA (2021b). *An energy sector roadmap to carbon neutrality in China*. Paris: IEA.

IRENA (2019). *A new world: The geopolitics of the energy transformation*. Accessed from: https://www.irena.org/publications/2019/Jan/A-New-World-The-Geopolitics-of-the-Energy-Transformation.

Kejun, J., Chenmin, H., Weiyi, J., Sha, C., Chunyan, D., Jia, L., and Pianpian, X. (2021). Transition of the Chinese economy in the face of deep greenhouse gas emissions cuts in the future. *Asian Economic Policy Review*, 16(1), pp. 142–162.

Lalyveld, M. (2020). Russia-China gas deal proves costly for both sides. *RFA* 10 July 2020.

Liab, I. (2020). Natural gas industry in China: Development situation and prospect. *Natural Gas Industry*, 7(6), pp. 604–613.

Lie Hansen, B. (2021). *SVP Norsk Hydro and later Statoil with responsibility for the Shtokman project*. Interview September 2021.

Lo, B. (2008). *Axis of convenience: Moscow, Beijing, and the new geopolitics*. Washington, DC: The Brookings Institution.

Moe, A. (2010). Russian and Norwegian petroleum strategies in the Barents Sea. *Arctic Review on Law and Politics*, 1(2), pp. 225–248.

Myllyvirta, L. (2020). Influential academics reveal how China can achieve its 'carbon neutrality' goal. *Carbon Brief*. Accessed from: https://www.carbonbrief.org/influential-academics-reveal-how-china-can-achieve-its-carbon-neutrality-goal.

Norwegian Parliamentary Report no. 7 (2011–2012). Nordområdene. Visjon og Virkemidler.

Overland, I., and Kubayeva, G. (2018). Did China bankroll Russia's annexation of Crimea? The role of Sino-Russian energy relations. In Blakkisrud, H., and Wilson Rowe, E. (eds.) *Russia's turn to the east* (pp. 95–118). Cham: Palgrave Pivot.

Rumer, E., Sokolsky, R., and Stronski, P. (2021). *Russia in the Arctic: A critical examination*. Carnegie Endowment for International Peace. Accessed from: https://carnegieendowment.org/2021/03/29/russia-in-arctic-critical-examination-pub-84181.

Schneider-Petsinger, M. (2016). Geoeconomics explained. *Chatham House*. Accessed from: https://www.chathamhouse.org/2016/12/geoeconomics-explained.

Støre, J.G. (2006). *The high north – top of the world – top of the agenda*. Washington, DC: Speech at Centre for Strategic and International Studies.

Tamnes, R., and Offerdal, K. (eds.) (2016). *Geopolitics and security in the arctic: Regional dynamics in a global world*. London: Routledge.

The Research Council of Norway (2011). *Norway-Russia research and cooperation in the high north*. Oslo.

US Department of the Treasury (2014). Announcement of expanded treasury sanctions within the Russian financial services. *Energy and Defense or Related Materiel Sectors*. Accessed from: https://www.treasury.gov/press-center/press-releases/pages/jl2629.aspx.

US Geological Survey (2009). Assessment of undiscovered oil and gas in the Arctic. *Science*, 324(5931). Accessed from: https://www.science.org/doi/10.1126/science.1169467.

Valdai International Discussion Club (2020). *Introduction by president Putin plenary session XI*. Accessed from: http://en.kremlin.ru/events/president/news/46860.

Weitz, R. (2014). The Russia-China gas deal: Implications and ramifications. *World Affairs*, 177(3), pp. 80–86.

Yergin, D. (2012). *The quest: Energy, security and the remaking of the modern world*. London: Penguin.

6 Business Cooperation in the Arctic

Learning Points from the Russia-Based Oil and Gas Projects

Andrey Mineev and Elena Zhurova Sæther

Introduction

After the socioeconomic disorder in Russia in the 1990s, in 2000–2020 we witnessed firmer control by the state, tense economic and political relations with the West, and increased cooperation with China. These changes also affected Russian experience of international cooperation in the Arctic. The ability of the Russian state to cultivate relations with foreign investors has played a key role in the oil and gas projects in the Russian Arctic and continental shelf. Did the Russian state succeed in learning from the experience and continuously improve its relations with foreign investors? What forms of cooperation proved better than others? What kind of knowledge can be transferred to future projects? These questions determine the rationale of our chapter. This chapter builds upon three case studies of Russia-based Arctic oil and gas projects: Kharyaga, Shtokman, and Yamal.

Our data come from publicly available sources and the findings of earlier studies on the Russian oil and gas sector. The research object is the governance regimes developed in the case projects. By governance regime we assume a set of legal rules related to the ownership, financing, and taxation as well as the organization of the project. The projects studied progressed along with changes in the politics and economy of post-Soviet Russia in the period from the 1990s until the present day. Therefore, we focus on the legal regimes in a historical perspective and in the context of these changes. Each of the projects studied ended up with a different legal regime. Although the Shtokman project has not been implemented, it was very close to take-off and its partners developed a legal regime worthy of analysis.

Several groups of actors with their own strategic agendas participated in the projects studied: the Russian state and its associated corporations, private Russian companies more or less associated with the state, transnational oil companies from the West, and national petroleum companies representing the interests of their home countries (e.g., China, Norway). Based on the analysis of the cases, we argue that the question of interaction between the Russian state and foreign investors remains key in international Russia-based projects in the Arctic. In each of the three projects, the Russian state aimed to achieve a high degree of control,

DOI: 10.4324/9781003246015-8

but acted differently. In the case of Kharyaga, the state gradually increased its ownership share. In the case of Shtokman, a "superstructure" was introduced by the state-controlled company Gazprom at the outset. In the case of Yamal, the state stepped in on the financing side to counterbalance the involvement of foreign investors, especially from China. Each of these tactics is associated with its own remaining contradictions. Understanding these contradictions, as we discuss, is important with an eye to future prospects for international cooperation in the Arctic.

The Russian Arctic contains a lot of oil and gas yet to be developed due to growing global energy demand. International cooperation is required. In our discussion, we highlight potential challenges to such cooperation. Cooperation with China can be challenging for Russia. Limited opportunities to involve companies from the West due to the sanctions imposed on Russia also present a challenge.

The remainder of the chapter is structured as follows. First, we introduce the three cases of cooperation projects. Then we present our reflections on the legal forms of cooperation in the three projects. The chapter concludes with a discussion on the future prospects for international cooperation in the Russian Arctic.

Case studies

By presenting the three cases, we address three empirical questions:

a What forms of governance were established and how did these change over time?
b Who were the main actors?
c How the actors structurally positioned themselves in relation to each other.

The locations of our three case projects are indicated on the map below (Figure 6.1).

Kharyaga oil field development

Kharyaga oil field is located 60 kilometers north of the Polar Circle in the Timan-Pechora basin on the border of the Nenets Autonomous District and the Komi Republic. This field was discovered back in the early 1970s and started producing oil in 1986. In 1995, the Russian government signed a production sharing agreement (PSA) with Total via Total E&P Russie and the Nenets Oil Company (NOC) to develop the Northern section of the Kharyaga field. One year later, Norwegian Norsk Hydro joined the consortium. The Kharyaga PSA agreement came into effect in 1999, uniting a multinational consortium consisting of Total (50%), Norsk Hydro (40%), and the NOC (10%). At that time, Total managed the project and was responsible to the Russian state under the Kharyaga PSA.

PSA is a form of contract between the government and an operator company concerning compensation of costs to be borne by the field operator and payment for its services with a share of the natural resources extracted under the terms of

Figure 6.1 The three case projects on the map (made by the authors).

the contract (Khartukov, 2009, p. 166). In PSA contracts, the operator company has direct access to and control over extracted resources (equity participation). After the development costs are covered, the profit from the resources is shared by the operator and the state.

The consortium planned to invest $16.7 billion in Kharyaga during the lifetime of the PSA, including $9.4 billion for contracts with Russian firms (Krysiek, 2007). The NOC, controlled by the Nenets Autonomous District, was not responsible for financing its share of the project (Krysiek, 2007). The company was included in the Kharyaga PSA in exchange for the support of the local authorities, as the PSA legislation required approval from the regional administration for oil developments located in places of traditional homelands of indigenous peoples (according to Federal Law N 225/FZ from 1995). By February 2006, Kharyaga had reached the cost recovery stage of PSA.

The initial Kharyaga PSA agreement implied that a Russian partner would eventually have a stake in the project, whereas Total and Norsk Hydro were to each sell 10% of their shares to that Russian partner. According to the 2001 amendment to the Kharyaga PSA, it was up to the Russian government to select a domestic partner. In 2009, Total announced the transfer of a 10% interest in the Kharyaga oil field to the state-owned Zarubezhneft of Russia. Another partner in the field, Norway's Statoil, which by that time had merged with Hydro, also transferred a 10% interest to Zarubezhneft[1]. Under the agreement, Total continued to operate Kharyaga with a 40% interest, along with its partners Statoil (30%), Zarubezhneft (20%), and the NOC (10%).

Again, the consortium changed its structure in 2016. Total and Zarubezhneft adjusted their participation in the Kharyaga PSA, whereby Total transferred a 20% interest together with operatorship to Zarubezhneft. According to both Total and the Russian Ministry of Energy, the agreement was of a purely economic nature: the increase of the Zarubezhneft share promised higher project profitability, given its successful experience with other developments in the region.

During its operatorship, Total experienced several serious conflicts related to approval of budget costs (2001–2002), compliance with the PSA and environmental infractions (2007), or extracted associated gas losses due to their local equipment contractor (2012–2013). However, in all cases, Total managed to reach compromising agreements with the Russian authorities and settle controversial issues peacefully, in contrast to some other foreign investors involved in upstream oil and gas projects in Russia, namely BP, Exxon Mobil, and Shell.[2] Total currently retains a 20% interest in Kharyaga PSA and is far from losing interest in Russian assets in general. In their official releases, Total has always underlined the importance of Russian projects and connected Total's ambitions in Russia with partnerships with Novatek and the ongoing Yamal project Kharyaga PSA structure and its changes over time are shown on the Figure 6.2.

Today, the Kharyaga consortium consists of Zarubezhneft via Zarubezhneft–Dobycha Kharyaga Ltd (40%), Equinor – the former Statoil (30%), Total via Total E&P Russie (20%), and the NOC (10%). Under current agreements, the field will produce until 2031, with a possibility for extension. The project showed lower operating costs (from \$7.5/barrel in 2016 to \$4.1/barrel in 2019) and better financial results for the Russian state, as Zarubezhneft reports. The Nenets Autonomous District doubled its revenues from Kharyaga, which is especially important since the Kharyaga PSA constitutes about 50% of the local budget. Among other existing PSAs in Russia, the Kharyaga project has been the most profitable for the Russian state (Tikhonov, 2020).

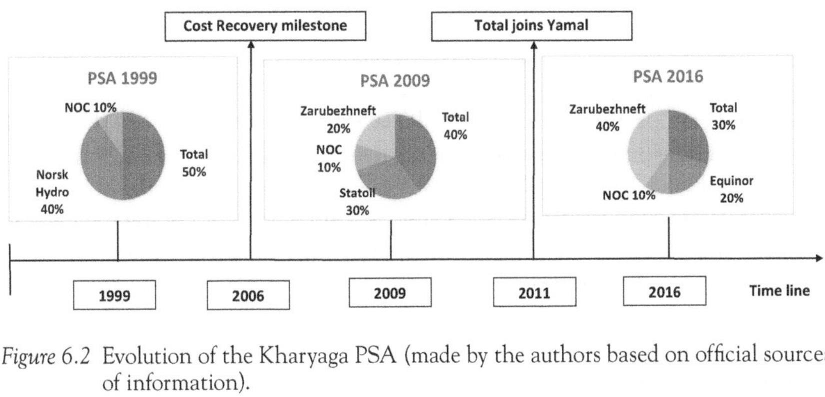

Figure 6.2 Evolution of the Kharyaga PSA (made by the authors based on official sources of information).

Since the beginning of the 2000s, the Russian state administration with President Vladimir Putin at the forefront has significantly raised its profile in the national energy sector. The government has endeavored to gain control over the Russian energy sector by seizing key assets, restricting foreign investments, and developing state oil and gas companies (Hanson, 2009). The Kharyaga oil field development aptly illustrates this trend. In the Kharyaga PSA, Russia steadily increased its share to 20% in 2009 and to 40% in 2016, taking over operatorship. State ownership has indeed become economically positive for the Kharyaga field, highlighting the controversial issue concerning the effectiveness of public companies in areas of high economic potential (Krivorotov, Finger, 2019). When it comes to Total, it seems that the company made the best of its situation. According to experts, in conflict situations with Russian authorities around Kharyaga, Total always managed to reach agreements in its favor (Krysiek, 2007). And even though Total had to renegotiate its share in Kharyaga, the company increased its assets in Russia in general via participation in the Yamal liquefied natural gas (LNG) field, where Total entered with almost 30% as the only Western partner (Bros and Mitkova, 2016).

Shtokman

The Shtokman field, one of the world's largest natural gas fields, lies in the Russian sector of the Barents Sea, 600 kilometers north of the Kola Peninsula. In the 1990s, "new" Russia was neither technologically nor financially prepared to take on such a project alone, and in August 1995 the Russian state-controlled company Gazprom and its subsidiary Rosshelf signed a letter of intent with Norsk Hydro, Conoco, Neste Oy, and Total SA to evaluate the possible joint development of Shtokman field (Victor, 2008, p.7). Although the Russian party had considered various technical and logistical solutions and potential partnerships, more concrete steps were only taken in the mid-2000s. At that time, many foreign companies actively competed to participate in the Shtokman project. In 2005, a shortlist of five foreign companies interested in Shtokman was formed, and bids from Statoil, Norsk Hydro, Total, Chevron, and ChevronTexaco were considered by Gazprom. The other companies interested in the project but not included in the shortlist were ExxonMobil, Mitsui, Sumitomo Corporation, and Royal Dutch Shell. Then, the initial idea of the government and Gazprom was to organize a project in the form of a PSA.

In October 2006, Gazprom rejected all five bids, saying that none of the foreign companies could offer assets with value corresponding to the volumes and quality of the Shtokman reserves (Walters, 2006). In addition, it was mentioned that foreign companies might be involved later as contractors but not as co-owners of the field (subsoil users). At the same time, Gazprom stated that priority would be given to pipeline gas transport to European markets instead of LNG shipment to the US market.

In parallel with Shtokman, the Russian government changed the rules of the game in another shelf project, Sakhalin 2, located in the Russian Far East. PSAs contracted with international operators in 1994 were revised in 2006.[3] In 2007, Gazprom entered Sakhalin 2 (after partial acquisition of shares from foreign partners Shell, Mitsubishi, and Mitsui) and became the major shareholder in Sakhalin Energy operating the project. Some experts and analysts assume that the decisions about Shtokman and Sakhalin 2 in 2006 were interlinked, merely signifying the changing attitude of the Russian government toward foreign companies (Krysiek, 2007; Veletminsky et al., 2006). After criticizing PSAs, the federal government decided to control strategic petroleum projects via national companies (Gazprom in the cases of Shtokman and Sakahlin 2, Rosneft in the case of Sakhalin 1), granting them both field licenses and majority shares in operating companies.

Later, in 2007, Gazprom decided to proceed with both an LNG plant with further shipping to the Atlantic basin markets and a pipeline to Europe. Foreign companies were invited to cooperate on the Shtokman project again, but now only as investors in field infrastructure, not as subsoil users. As a result of new negotiations, cooperation agreements with French Total and Norwegian StatoilHydro (Statoil since 2009) were signed in 2007. Total obtained 25%, StatoilHydro 24%, and Gazprom via its subsidiary JSC Sevmorneftegaz (SMNG) 51% in the special purpose company Shtokman Development AG (SDAG) registered in Switzerland. Gazprom (via SMNG) retained all rights to the market and to selling the gas produced. The agreement between the partners stipulated that SDAG would be the owner of the infrastructure of the field's first phase for 25 years after the field began production (Gazprom, 2008). Upon completion of phase one, Total and Statoil would transfer their shares in SDAG to Gazprom. In fact, Total and StatoilHydro entered the Shtokman project as service company partners.

Although the Shtokman project was not implemented, we can still learn a lot from this case about governance and the building of relations between the Russian party and foreign oil and gas companies. Organizationally established in 2007–2008, the project was ready to operate. Let us consider its organizational structure in 2008 (Figure 6.3).

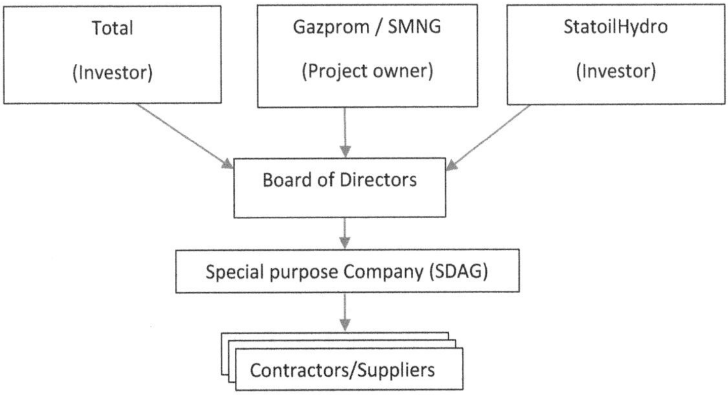

Figure 6.3 Organizational structure of the Shtokman project in 2008 (made by the authors based on information from SDAG)

According to the initial plans of SDAG, pipeline gas transportation to Europe and the production of LNG gas should have already started in 2013 and 2014, but at the beginning of 2010, it was decided to postpone the pipeline and the LNG until 2016 and 2017. The main reason for the postponement was the excess of global LNG supply and the shale gas revolution in the US, which had been presumed to be the target market for Shtokman LNG (Socor, 2012). Shtokman gas became too expensive to be extracted at that time. The shareholders' agreement expired in 2012 and then, as reported in the media, Statoil wrote off its investment in the project and handed its shares back to Gazprom. Total did the same in 2015. Gazprom declared its continued commitment to implementing the Shtokman project applying advanced technologies, but it has given to date no indication of the specific timing or of cooperation with foreign partners. Meanwhile, SDAG was liquidated in 2021 pursuant to the earlier resolution by Gazprom as its only shareholder.

Yamal LNG

Yamal LNG plant is located in the North-Eastern part of the Yamal peninsula in the Yamalo-Nenets Autonomous District. The project is based on the South-Tambeyskoye field discovered in 1974. The Yamal LNG project has developed through different phases.

Until 2005, the license belonged to JSC Tambeyneftegaz, where Novatek owned a 25.1% share. Later, the license was reassigned to Yamal LNG – a company established to involve foreign investors. Novatek, which is the largest non-state-owned gas producer in Russia, joined the project in 2009 by purchasing a 51% interest in JSC Yamal LNG. In October 2010, the Russian government approved a plan for the development of LNG production on the Yamal peninsula. The Yamal LNG project has been identified as a pilot project. In 2011, Novatek increased its equity interest in the Yamal LNG project to 100% from 51% by exercising call options through wholly owned subsidiary Novatek North West.

In March 2011, Christophe de Margerie, Chief Executive Officer (CEO) of Total S.A., and Leonid Mikhelson, Chairman of the Management Board of Novatek, signed a Memorandum of Cooperation at President Putin's residence in Novo-Ogaryovo. Then, Total bought 12% shares in Novatek and a 20% stake in Yamal LNG. By June 2013, Total's share had increased to 16%.

In September 2013, Yamal LNG acquired a third shareholder: China National Oil and Gas Exploration and Development Corporation (CNODC), a subsidiary of China's China National Petroleum Corporation (CNPC), purchased 20%. Inviting the Chinese company opened up a vast Chinese market for Yamal LNG, as Asia-Pacific countries account for the main growth in fuel demand. The main motive of the Chinese side in participating in the Yamal LNG project was interest in access to the Northern Sea Route as well as an economic interest connected to the growing demand for natural gas. In January 2014, the government of the Russian Federation and the government of the People's Republic of China signed a special agreement about favorable terms for the Yamal LNG project. For China, implementing a particular "pragmatic" approach while selecting investment projects in the Arctic (Petrovsky & Phillipova, 2018), Yamal LNG was part of their activities in the Arctic in the framework of their Arctic policy, officially formulated in 2018. Also, it is connected to the Belt and Road Initiative – the Chinese alternative to the global governance concept (Kopra, 2019) – an infrastructure development strategy adopted by the Chinese government in 2013 (Krivorotov, 2018). Initially, the Belt and Road Initiative was intended to strengthen infrastructure both on the westward land route from China through Central Asia and on the southerly maritime routes from China through Southeast Asia and on to South Asia, Africa, and Europe. In 2017, the Arctic area and the Northern Sea Route were added to the Belt and Road geographical scope. Increasing its role in the Arctic, China has shown interest in developing large-scale extractive industry and infrastructure projects such as Alaska LNG,[4] the Kirkenes-Rovaniemi railway, and the Northern Sea Route (Krivorotov, 2018).

In 2014, in response to the conflict in Ukraine, the US and the EU imposed a series of commercial and sectorial sanctions targeting Russia's financial services and energy sectors. It destroyed Novatek's plans to obtain Western funding. But the project survived because of Russian and Chinese support; in 2015, the Chinese State Silk Road Fund stepped in, acquiring 9.9% of Yamal LNG for €1 billion.

Figure 6.4 illustrates the existing consortium structure where the Chinese partners own 29.9% of the Yamal LNG (20% owned by the state-controlled Chinese National Petroleum Company and 9.9% by the Chinese Investment Fund). The company TotalEnergies (formerly Total), in addition to a 20% share in the Yamal LNG, also owns a 19.4% stake in Novatek. Novatek holds 50.1% of the shares in Yamal LNG.

As planned, the Yamal project commenced producing LNG in 2017. It reached full capacity in late 2018, a year ahead of schedule. Novatek's revenues for 2020

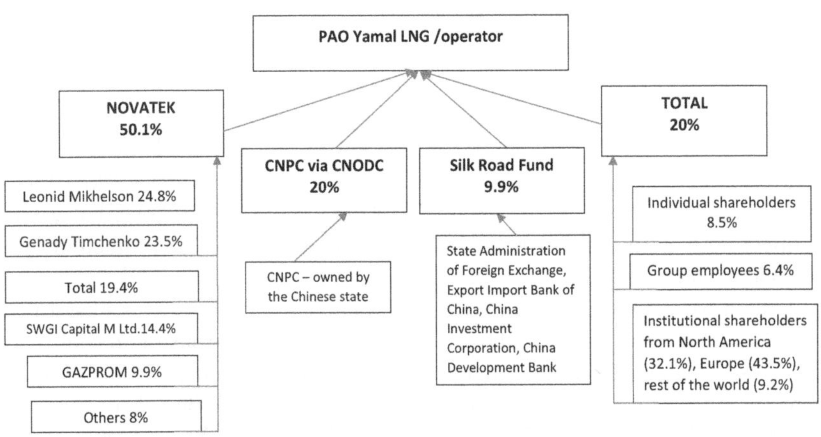

Figure 6.4 Extended ownership structure of Yamal LNG (made by the authors based on official information sources).

were lower than expected due to the fall in world prices for hydrocarbons. Still, the dividend payments increased by 10% compared to the previous year. It seems that the partner cooperation functions well, since in 2018 Total and Novatek signed an agreement about the joint work on the Arctic LNG-2 project, despite rather challenging conditions due to sanctions. Chinese CNPC and CNODC also entered the project. The project is planned to commence in 2023.

The total cost of Yamal LNG is $27 billion, consisting of Yamal LNG's capital and project financing of $19 billion, with 33.6% coming from the Russian side (the National Welfare Fund of Russia, Sberbank, Gazprom), and 58.6% from Chinese institutions. Along with it, the Russian state heavily supports the project with tax breaks, provisions of goods and services below market value, and finances lacking infrastructure in the area: construction and operation of Sabetta seaport (additional RUB 69.6 billion), ice-breaking fleet though Atomflot, LNG tanker fleet through Sovcomflot, and construction of Sabetta airport.

These enormous investments may seem reasonable in light of the political and geopolitical driving forces behind the project (Muradova, 2020). Yamal LNG reflects Russia's ambitions to develop its LNG industry and become one of the leading LNG producers globally (Henderson, Yermakov, 2019). The project will strengthen Russia's geopolitical influence in the Asia-Pacific region and assert the importance of the Northern Sea Route (Bros and Mitrova, 2016). The LNG project and the construction of the Sabetta seaport are key factors in the development of the Northern Sea Route, allowing a westbound winter route (from November until June) and an eastbound summer route for the rest of the year to European and Asian markets.

Given what we know about the endeavors of the Russian government aiming to gain control over the national energy sector, it would be natural to expect direct state participation in Yamal LNG as in the Shokman Project. However, Russian operator Novatek is a private company and the only state shareholder in Novatek is Gazprom PJSC with a 9.9% share (corresponding to 5% in Yamal LNG as such). So, how exactly do the Russian authorities control Yamal LNG and why safeguard Novatek's interests? The enormous state support confirms the great importance of Yamal LNG for Russia. Yet, rather than this, industry experts underline good personal connections between the major Novatek shareholders, Gennady Timchenko and Leonid Mikhelson, and the Russian authorities (Krutikhin, 2018; Yakovenko, 2018a), and even approach Gennady Timchenko as a "close friend of Vladimir Putin, possessing enormous lobbying resources" (Yakovenko, 2018b), connecting it with abilities to lobby for the project at the governmental level (Bros and Mitrova, 2016).

Yamal LNG provokes constant debates, where opinions differ. It is a flagship of the LNG industry in Russia and the largest Russian Arctic project in recent years. At the same time, economists note the low economic viability of Yamal LNG if it had not been for Russian state support (Belyakova, 2020; Bros, Mitrova, 2016; Lunden and Fjærtoft, 2014). The expected social benefits of Yamal LNG in terms of the government stake and supply contracts are not significant. In this light, the Yamal project is characterized as "subsidized by the state" (Lunden and Fjærtoft,

2014) and "unsustainable" (Belyakova, 2020), with expected low tax deductions and little direct benefit to society.

Observations from the case studies

The three cases demonstrate variations in the positions of the Russian state toward foreign investors. When it comes to Kharyaga, the evolution of its PSA deserves particular attention. PSA is known as a way to attract foreign direct investment for less technologically and economically advanced resource countries (as Russia was in the 1990s). However, from the literature, we know that resources-based countries tend to renationalize oil and gas developments over time (Goes, 2017; Stevens, 2008). Russia was no exception in the case of Kharyaga PSA. Yet, the share of the foreign partners gradually decreased in favor of the Russian state; today, this project looks economically viable for all concerned parties. An inherent disadvantage of PSA for a host country is that the foreign party gets an owner's share in the field resources. As for foreign companies, there is always a risk of renegotiated ownership by the host country. Yet this risk is lower than that of sudden changes in taxation or customs regulations without a PSA.

Some negative experiences from the PSA-driven projects (especially Sakhalin 2) may be a reason for the rejection of PSA as a legal regime for the Shtokman project. The case of Shtokman highlighted an excess of state intervention. By the mid-2000s, it was clear that the Russian state wanted to control its strategic oil and gas projects as much as possible. Here, the Russian party chose a corporate governance structure maximizing control by the Russian state and limiting the involvement of foreign partners.

The state-controlled company Gazprom had a majority holding in SDAG and retained all rights to the market and selling the gas produced. Further, the foreign companies Total and StatoilHydro were involved only as investors in the service company, not as owners of the resources to be extracted. This project structure was criticized as a "superstructure" or a system creating nonoptimal incentives (Osmundsen, 2010, 2011), because SDAG lacked commercial incentives as the company would have no access to the profits from the gas sold. For the Russian side, "superstructure" (nevertheless giving control) can be disadvantageous. It restricts commercial behavior (competition from foreign companies, commercial incentives of the service company). Further, it involves bearing high costs and economic responsibility.

The legal regime of Yamal LNG was set up nearly ten years after Shtokman. Organized as a joint venture, it looks like a golden mean between PSA (control by foreign companies) and "Superstructure" (control by the host country). At the same time, we can see that this governance regime is workable under Western sanctions on Russia. However, the project is heavily subsidized by the Russian state and its control may be too dependent on personal ties. Can future goals of the Russian state prosperity via a technologically developed Arctic be justified by the present-day practice of supporting a privately owned company (Novatek) with money from the National Welfare Fund and tax relief? Furthermore, while from

Table 6.1 Summary of the cases

	Kharyaga	*Shtokman*	*Yamal*
Position of the Russian state	Increase of share through state-owned company (after cost recovery point)	Imposed "superstructure"	Subsidizing Control through personal ties rather than state-owned company
Position of foreign partners	Equity (resource) partners	Service company partners	Operator company partners

the 1990s to the early 2000s, Russia was exposed to opportunistic behavior on the part of the Western transnational oil companies, nowadays it has favorable conditions for Chinese global expansion. The Chinese companies are deeply integrated into the project as implementing partners, investors in project infrastructure, technological partners, and buyers of the gas (Krivorotov, 2018). It seems that in times of Western sanctions, China tends to maximize its benefits in Russia. A summary of the cases is presented in Table 6.1.

Although cooperation with foreign companies through PSAs in the 2000s was not unproblematic, this experience can be used in improving the legislation. It seems that today's Russia, having had many years of experience with PSA, could benefit from such a form of governance and attract foreign investments. PSA must be then settled so as to prevent super profits for the foreign investor on the one hand and intervention by the host state on the other. However, this is hardly possible until Western sanctions on Russia are lifted. Foreign investors from the West would have difficulties financing their operations.

"Superstructure", as we have seen in the case of Shtokman, puts a lot of responsibility on the Russian partner and at the same time limits the incentives of foreign investors. When it comes to a "subsidized" form of project like Yamal LNG, today it appears to be the only workable solution. This form of project gives a lot of influence to the French and Chinese partners, not least through investments in strategic infrastructure. At the same time, the project is costly for the Russian state. Lacking money from the Western financing institutions, the Russian state seemed to be drawn into the project to counterbalance huge Chinese investments. If not counterbalanced by the Russian state, China might gain too much influence. Yet the Yamal LNG project remains "subsidized", which means that it may lack commercial incentives.

Can we assume that the development from Kharyaga to Yamal represented a learning curve from the 1990s until today? Can we say that Shtokman was better organized than the initial Kharyaga PSA, but that Yamal LNG was better organized than Shtokman? Not really. In an ideal situation for Russia, we can imagine the Yamal project to be a golden mean between PSA and "superstructure". This would mean still a majority share for a Russian company associated with the state, more financing from the West and less from China, and more incentives but fewer subsidies from the Russian state. Perhaps such a legal arrangement could be achieved if the Ukraine-related conflict between Russia and the West had not happened. However, in reality, it seems that the Russian state managed to achieve suboptimal equilibria in each project in the given circumstances.

Future prospects

The prospects for Arctic development in Russia are enormous: according to the US Department of Energy the Russian Arctic contains 60% of the world's Arctic reserves (National Petroleum Council, 2015). According to estimates by the National Ministry of Energy (Novak, 2019), the Russian Arctic zone contains more than 35 billion tons of oil and 210 trillion cubic meters of gas. In 2019, the Ministry of Natural Resources of Russia prepared a comprehensive plan, which included 118 projects aimed at developing and processing the Arctic resources and constructing the corresponding infrastructure. Along with Yamal LNG, this included a number of developments on different stages located in the Nenets, Yamalo-Nenets, and Komi Republics. Among prospective projects, there are Arctic LNG 2, Far Eastern LNG, Sakhalin-2&3 (2026), Vostok Oil, and others. The Shtokman project was officially postponed, but can yet be revitalized. The recent Russian Arctic Strategy confirms the state's firm intention to seize these opportunities. Since the extraction of the Arctic resources is costly, international cooperation to develop them would be inevitable. As the energy mix is improving toward the use of more environmentally friendly sources of energy, we assume that the growing world energy demand will continue to be a driving force for the development of oil and gas up to 2035.

What kind of legal regimes are to be expected in the future projects in the Russian Arctic? A dilemma for the Russian state would be between *control over* and *incentives for* foreign partners. As we have shown in our three case studies, the Russian state could act rather tactically to find workable yet suboptimal solutions in a given set of circumstances. A more strategic approach would be needed in the future. Since the late 1990s, Russia has learnt how to manage oil and gas projects involving Western partners. When it comes to strategic cooperation with China, Russia is still at the starting point of the learning curve. Chinese investment projects in Russia must be seen as a part of China's global strategy. This strategy is implemented, for example, through the Belt and Road Initiative. For Russia, it is important not to become heavily economically dependent on China. The policy of the Chinese Communist Party and the activities of Chinese businesses are well coordinated with one another. To be able to cooperate with such a strategic player as China, Russia and other countries need to be strategic too.[5]

Known for its pragmatic approach, China is expected to pursue its own economic and political interests. By investing in Russian oil and gas, China provides security to meet its own growing domestic energy demand. At the same time,

China actively invests in clean energy on other continents (see, e.g., Middlehurst, 2017). In other words, in making Russia more dependent on hydrocarbons, China emerges as a global leader in clean technologies.[6]

Involving petroleum companies from the West, and therefore counterbalancing the relationship with China, would be beneficial for the Russian strategy. West-based transnational corporations (such as BP, Shell, ExxonMobil, ConocoPhillips, Chevron, and TotalEnergies) seek their own interests and profits globally and represent neither their home countries nor their host nations but simply their own corporate selves (Miyoshi, 1993). Still, one can argue, they do enjoy very close and cordial relations with their respective governments. These corporations are owned by private institutional stakeholders mainly from North America and Europe. To some extent, their ability to cooperate with Russia is limited by continued sanctions imposed by the US and the EU. As follows from our cases, among these companies, TotalEnergies has so far been the most successful in Russia.

If further politics of the Russian state were more inclined toward Western liberal democracy (which is rather unlikely with the present government and in the situation of continued pressure on Russia from the West), then transnational corporations and financing bodies from the West would strengthen their positions in the country. However, having considered the experiences of PSAs from the 1990s, we cannot underestimate the importance of efficient regulation providing incentives for both sides and safeguarding against opportunistic behavior. The development of such regulations in practice can be challenging.

It is difficult to predict what forms of cooperation may evolve in future projects in the Russian Arctic. In addition to external factors such as geopolitical climate, world energy demand, technological trends, Russia's ability to deal with both China and the West would depend to a large extent on the internal situation in the country. Interplay between domestic politics, society, and economy in Russia presents a key uncertainty. Various outcomes of this interplay would also have implications for the future of international cooperation in the Arctic. Post-Soviet Russia has undergone dramatic changes during the 30-year period with a gradually strengthening role of the state. Thirty years is a whole generation. The individuals crucial to the development of the present-day national governance and economic system are those who grew up in the Soviet Union. They started their political or business careers in the 1980s and 1990s, and managed to quickly adapt to the new rules of the game after the dissolution of the Soviet Union since the early 1990s. A smooth transition of power to the upcoming generation of new Russian leaders[7] would depend on reforming the present *governance* and *economic* systems. The latter means diversification of the economy and reduced dependence on hydrocarbons and large companies (toward strengthening the manufacturing and service industries, increasing the role of small and medium-sized enterprises). The former means reduced dependence on key persons and their interpersonal relationships (toward an institutionalized governance system with distributed power and strong formal principles).

December 2021

Ex-post reflections

We write this text while tragic events are happening in Ukraine. We write in hope that peace to be established as soon as possible...

Our study, implemented under the previous geopolitical conditions, focused on three oil and gas projects developed in the Russian Arctic: Kharyaga, Shtokman, and Yamal LNG. One of the key conclusions was that the Russian state managed to achieve suboptimal equilibria in each project, every time dealing with a dilemma between *mounting control over* and *giving incentives to* the foreign partners. The cooperation was all the way influenced by changing geopolitical conditions. Today, due to the mounting tension between Russia and the West, Russia's access to partners, capitals, markets, and technologies from the West is rapidly diminishing. We already saw that the US and EU declared their goal to decrease dependence on Russia's oil and gas in the years to come. Russia is going to be more reliant on China. The Chinese investments in the Russia-based Arctic projects would probably assume the preferences claimed by China.

We stated that TotalEnergies was the most successful Western company in the Russian oil and gas business. This still seems to be true. According to the recent press releases, TotalEnergies remains a stakeholder in Kharyaga oil field PSA, Yamal LNG, and Novatek. In contrast, UK/US-based companies BP, Shell, and ExxonMobil have already announced their withdrawal from Russia and sales of their shares in Russian companies and projects. New buyers of these assets and their origins would indicate future geopolitical relations with Russia.

As we stated, the interplay between domestic politics, society, and the economy in Russia presents a key uncertainty. This means that Russia's ability to survive in the ongoing geopolitical shift depends to a large extent on the internal situation in the country. We hope that many people in Russian society will be able to reflect on the present-day crisis in a knowledgeable, responsible, and constructive manner.

May 2, 2022

Notes

1 Zarubezhneft JSC is a 100% state-owned Russian oil and gas company operating in Russia and abroad; the company counts its history from 1967 as a key external policy agent of USSR in the oil industry abroad.

2 As the operator and the main shareholder in Sakhalin-2 PSA, Royal Dutch Shell was accused of violating environmental regulations by Rosprirodnadzor in 2006, from which followed a public scandal. Later, in 2007, Shell sold 50% of its Sakhalin-2 share to Gazprom. Exxon mobile: annulation of Exxon Mobil licenses for Sakhalin-3 was in 2004, after the company had spent $60 million on development activity; the 2013 conflict between Gazprom and Exxon Mobil over Sakhalin-1 gas; the 2018 dispute between Rosneft and Exxon Mobil/Sakhalin-1 consortium ended up with an out-of-court settlement for $230 million paid by Exxon Mobil. BP: in 2008, BP was forced to suspend 148 staff in Russia in a dispute over visas a week after a police raid on the Moscow offices of BP and TNK-BP, 2008-2012 disputes around RNK-BP joint venture between Russian and British shareholders.

3 All designed in the 1990s, there were three PSA oil and gas projects in Russia: Sakhalin-1 and Sakhalin-2, both on the shelf in the Russian Far East, and Kharyaga – onshore oil project in the Russian Arctic. The conventional opinion of experts and analysts is that these PSAs were designed on conditions disadvantageous to Russia in favor of foreign companies. However, not only the opportunistic behavior of foreign companies but also the lack of efficient PSA legislation in Russia could be the two most viable reasons for revising the agreements (see, e.g., Mineev 2010, p. 316).
4 Despite the memorandum of understanding (MoU), Chinese companies withdrew from Alaska LNG later in the aftermath of the US-China trade war.
5 Cooperation with neighboring China is challenging by definition. Russia has a much larger territory and much more hydrocarbon resources. China has a high energy demand. Estimates vary, but one can say that the proven natural gas reserves of Russia are about ten times greater than China's. Russia's crude oil reserves are about three times greater than China's. The territory of Russia is nearly twice as big as China's. At the same time, the Chinese economy in terms of Gross Domestic Product (GDP) and population are both ten times larger than Russia's.
6 China stands for about 40% of global clean energy investment (according to estimates by Statista for 2019).
7 We assume that the upcoming generation of "new Russian leaders" in the perspective toward 2035 are people born around the 1990s and later. They definitely have a different mentality and knowledge base compared to the previous generation of leaders. By leaders, we do not necessarily mean the top political and business elite, but also key individuals at the level of regions, municipalities, businessmen, and directors of public sector organizations and nongovernment organizations (NGOs).

References

Note: All links were accessed on 13 December 2021.

Belyakova, M. (2020). Влияние системы лицензирования и налогообложения в нефтегазодобывающей отрасли на развитие национальной экономики: опыт Норвегии и России. [The impact of the licensing and taxation system in the oil and gas industry on the development of the national economy: The experience of Norway and Russia]. PhD thesis. MGIMO, Moscow.

Bros, A., and Mitkova, T. (2016). Yamal LNG: An economic project under political pressure. *Note De La FRS* no17/2016. Retrieved from: https://www.frstrategie.org/en/publications/notes/yamal-lng-economic-project-under-political-pressure-2016.

Gazprom. (2008, February 21). *Gazprom, Total and StatoilHydro Create Shtokman Development Company for Phase One on the Shtokman Gas Field* [Press release]. Retrieved from: https://www.gazprom.com/press/news/2008/february/article88266/.

Goes, S. (2017). Russian space meets western business practices: Understanding the law in the petroleum sector in Russia. Russian space meets western business practices. *Nordlit*, 39.

Hanson, P. (2009). The resistible rise of state control in the Russian oil industry. *Eurasian Geography and Economics*, 50(1), pp. 14–27.

Henderson, J., and Yermakov, V. (2019). *Russian LNG: Becoming a Global Force. OIES Paper: NG 154.* Oxford Institute for Energy Studies.

Khartukov, E. (2009). *Grand English-Russian Oil and Gas Business Glossary.* Moscow: Olimp-Business.

Kopra, S. (2019). "Climate Change and China's Rise to Great Power Status: Implications for the Global Arctic", in Finger, M., and Heininen, L. (eds.), *The Global Arctic Handbook* (pp. 125–139). Cham: Springer.

Krivorotov, A. (2018). China's polar policy: Gaining momentum, obtaining scope, multiplying vectors. *China in the Global and Regional Politics. History and Nowadays*, 23, pp. 183–199.
Krivorotov, A., and Finger, M. (2019). "State-Owned Enterprises in the Arctic", in Finger, M., and Heininen, L. (eds.), *The Global Arctic Handbook* (pp. 45–62). Cham: Springer.
Krutikhin, M. (2018). The owners of the Arctic: Why Putin and total believed in Novatek by Mikhelson and Timchenko. *Frobes.ru* article from May 28, 2018. Retrieved from: https://www.forbes.ru/milliardery/362219-hozyaeva-arktiki-pochemu-putin-i-total-poverili-v-novatek-mihelsona-i-timchenko.
Krysiek, T. F. (2007). *Agreements from Another Era: Production Sharing Agreements in Putin's Russia, 2000–2007.* Oxford Institute for Energy Studies.
Lunden, L., and Fjaertoft, D. (2014). *Government Support to Upstream Oil & Gas in Russia How Subsidies Influence the Yamal Lng and Prirazlomnoe Projects.* The International Institute for Sustainable Development/WWF.
Middlehurst, C. (2017). *China Emerges As Global Clean Energy Leader.* Retrieved from: https://dialogochino.net/en/climate-energy/8242-china-emerges-as-global-clean-energy-leader/?gclid=Cj0KCQiA-eeMBhCpARIsAAZfxZAbXQ0g5KZkMgpxlFnqU-qruewBc5E7RHPRksBdV5b9_VI9OCkaM6UaAsyxEALw_wcB.
Mineev, A. (2010). Development of Russian oil and gas regimes focusing on the Shtokman field: Institutional drivers. *Journal of East-West Business*, 16(4), pp. 303–339.
Miyoshi, M. (1993). A borderless world? From colonialism to transnationalism and the decline of the nation-state. *Critical Inquiry*, 19(4), pp. 726–751.
Muradova, A. (2020). Cooperation models for complex projects in the Russian arctic. Case study: Shtokman development AG and Yamal LNG. (Master's thesis in business economics). Nord University, Bodø.
National Petroleum Council. (2015). Arctic potential. Realising the promise of U.S. arctic oil and gas resources. *National Petroleum Council Report.* Retrieved from: https://www.npcarcticreport.org/.
Novak, A. (2019). The arctic - investing in the future of hydrocarbons. *Oil and Gas Vertical*, 3(4'19), pp. 10–11.
Osmundsen, P. (2010). "Chasing Reserves – Incentives and Ownership", in Bjørndal, E., Bjørndal, M., Pardalos, P. M., and Rönnqvist, M. (eds.), *Energy, Natural Resource and Environmental Economics*, Springer Verlag.
Osmundsen, P. (2011). Stability and efficiency. The importance of ownership. *University of Stavanger / NHH Presentation at the 2011 PETROSAM Conference Stavanger,* November 29, 2011.
Petrovsky, V., and Phillipova, L. (2018). The Chinese arctic strategy and the perspectives of Russia-China cooperation in the region. China in the global and regional politics. *History and Nowadays*, 23, pp. 171–182. https://elibrary.ru/item.asp?id=35729330.
Socor, V. (2012). Gazprom's Shtokman project: Relic of a past era. *Eurasia Daily Monitor,* 9(153). Retrieved from: https://jamestown.org/program/gazproms-shtokman-project-relic-of-a-past-era/.
Stevens, P. (2008). National oil companies and international oil companies in the Middle East: Under the shadow of government and the resource nationalism cycle. *Journal of World Energy Law and Business*, 1(1), pp. 5–30.
Tikhonov, S. (2020). In 2019 SPA projects brought 2.3 billion USD to the Russian budget. *Rossiskaya Gazeta,* article from November 10, 2020.
Veletminsky, I., Pakhlin, N., and Chichkin, A. (2006). [Every charter gets its Shtokman: Russia changes its energy priorities]. *Rossiyskaya Gazeta.* October 24. Retrieved from: http://www.rg.ru/2006/10/24/shtokman.html.

Victor, N. (2008). *Gazprom: Gas Giant Under Strain*. Program on Energy and Sustainable Development (PESD) at Stanford University. Working Paper.

Walters, G. (2006). Gazprom rejects foreign partners for Shtokman. *Dow Jones Newswires*, October 09, 2006.

Yakovenko, D. (2018a). Gas blaze: How Novatek built a plant at the edge of the earth. *Forbes.ru*, December 27, 2018.

Yakovenko, D. (2018b). Partnership from the God: Why Mikhelson and Timchenko are getting richer. *Forbes.ru*, September 19, 2018.

7 Arctic Innovation and the Potential for the Creation of a Circumpolar Innovation Ecosystem

Ken Coates and Carin Holroyd

Introduction

Countries and regions around the world are obsessed with the socioeconomic potential of scientific and technological innovation (Breznitz, 2021). Innovation refers to the commercialization of science and technology, to the ability of companies or other organizations to lift a concept or prototype out of the laboratory or a start-up company into broad circulation and use. Large-scale innovation requires the careful development of a tech-savvy commercial environment, based on access to development capital, technological expertise, business capacity, marketing reach, and retail acumen (Freeman, 1990). Innovation may start with scientific discovery and technological developments, but for an idea to become a commercial product requires a broad ecosystem, including entrepreneurs, companies, governments, universities, research laboratories, and venture capitalists. The remarkable success of Silicon Valley (Kenney, 2000) and the notable achievements of less famous but impressive places like Shannon (Ireland) and Waterloo (Canada) have government officials and politicians talking about commercial incubators, start-ups, and the other elements of the high-tech economy, contemplating a technologically enabled future that could address many contemporary problems (Erika and Watu, 2010). So it is also in the Arctic, where leaders are looking to technological innovation as the solution to the region's most significant challenges.

Technological innovation has been transformative for the North in the past, with the introduction of ocean-going ships, rifles, telescopes, oil-based energy, outboard motors, radios, Arctic-ready airplanes, television, the Internet, and dozens of other scientific and technological discoveries which brought radical, multidirectional changes to the people and communities of the Far North. For technology and innovation to be mobilized to improve northern life in the 21st century, the Arctic nations will need to work together. Beyond creating a market for northern-focused technologies, the North has a significant opportunity to rise and seize the science and technology opportunities that await by creating an Arctic innovation ecosystem. By coordinating research and commercialization activities, Arctic companies, research units, government agencies, and Indigenous and other northern communities could create

DOI: 10.4324/9781003246015-9

a circumpolar approach to innovation that could capitalize on scientific and technological discoveries to create a technology-enabled North (Hall, 2020). As part of imagining the Arctic in 2035, this chapter explores the potential for the creation of a circumpolar technology ecosystem. After a look at the state of Arctic innovation, this chapter discusses shared northern challenges (vast distances, extreme cold and darkness, high transportation costs, small populations, limited regional scientific expertise, suboptimal government attention, intense southern interest in natural resources, and the impact of global climate change) and notes that few emerging technological innovations have been developed specifically for Arctic purposes; most of those that are north-centric are related to the natural world (Hall and Vodden, 2019). Agreeing on the issues to be tackled and the technological fields which might help is the first step in the development of an Arctic innovation ecosystem. This is the subject of the rest of this chapter, which includes a discussion of the challenges which could hinder success and possible steps forward.

Few technologies have been developed for the circumpolar north because few entrepreneurs or the governments or venture capitalists who support them see the Arctic as having a population or commercial market large enough to warrant the investment. However, while the population of the northern regions of individual nations is not large, the circumpolar north collectively is home to almost four million people. In combination with much of the sub-Arctic, which shares many of the same northern challenges, there are over 15 million people and perhaps the same number again in other northern/cold weather and remote regions and among global consumers smitten with the cultural, geographic, environmental, and symbolic force of the circumpolar world. This is a large enough market to be of economic interest and could both support the development of northern entrepreneurs and turn the focus of southern businesses northward (Heininen and Southcott, 2010).

High-impact emerging technologies include mass digitization, artificial intelligence, robotics, alternative energy systems, multiple forms of biotechnology, and climate mitigation technologies. These technologies, individually and collectively, have vast potential, covering everything from the monitoring of ecological change; the future of work, basic retail, education, and medical care; building design; clothing; transportation; heating and electricity generation; and many other aspects of contemporary life (Diamandis and Kotler, 2020; Kaku, 2012).

Background – the Arctic environment

The economic, human, and geographic landscape is not uniform across the circumpolar world, but there are important commonalities. The population of the entire north is relatively small, particularly in relation to its size. The people of the Canadian North are sparsely settled and widely dispersed – 126,000 people spread over 3.5 million square kilometers with over 40% of that population living in three main cities. The Russian North is larger in both size (5.5 million square kilometers) and population (two million people). The climate of the Scandinavian north is much milder than that of Canada, Alaska, or Russia. The average temperature in Tromsø in January is −1 to −5 Celsius, while in Yellowknife it is −18 to −26 Celsius and in Yakutsk −36 to −42 Celsius (Nuttall, 2005).

Larsen and Huskey (2015) describe the Arctic economy as being composed of three main sectors: large-scale resource extraction, subsistence-based living (especially in Indigenous remote communities), and the public sector. The resources extracted are generally destined for international markets, but extraction and production costs in the Arctic are high due to "structural and persistent factors such as remoteness, lack of accessibility, long distance to markets and long supply lines, harsh climate and an often-inhospitable environment" (Larsen and Petrov, 2020, p. 81). The subsistence-based economy is culturally important and a valuable source of country food (the wild game and plants that are the traditional foods of Indigenous peoples) in some jurisdictions, but it is also declining in economic importance in sizeable parts of the North. In contrast, government is the major employer and a major contributor to the Gross Domestic Product (GDP) across much of the circumpolar world, particularly when defense expenditures are included. Even key "private sector" activities in areas like construction and medical care rely heavily on government contracts and subsidies. The major exception is the natural resource sector, where international firms serving global markets figure prominently.

Nonetheless, economic development outside these three main sectors in the Arctic is beginning to grow, especially in selected Arctic cities. Knowledge and creative industries, small-scale manufacturing, and professional work have been increasing. Governments around the circumpolar world are attempting to ensure the long-term viability of northern economies, generally by supporting the natural resource economy and seeking opportunities to promote "new economy" commercial development (Larsen and Petrov, 2020). There has been growth in industries outside the resource-extractive, transfer, or subsistence sectors such "as knowledge-based industries, arts and crafts, small-case custom manufacturing, professional and technical services, tourism/recreation and local retail trade" (Larsen and Petrov, 2020).

The Arctic's current innovation ecosystem

The challenge of creating an Arctic innovation ecosystem is formidable. Few Arctic or sub-Arctic regions have the research capability, highly skilled personnel, venture capital, accessible markets, and entrepreneurial drive necessary to compete in the global science and technology-based economy. Those countries with northern regions that have some of the above-listed building blocks of an innovation ecosystem could form the base of an Arctic innovation ecosystem. A range of northern communities in Sweden – Arjeplog (winter automobile testing), Kiruna (the commercialization of space exploration), Luleå (server farms and information technology services), Skellefteå (battery production and video game production), Umeå (environmental technologies) – have made significant efforts to build jobs and businesses beyond traditional northern economic activities (Coates and Holroyd, 2021). Tromsø in northern Norway is the self-styled "capital" of the Arctic, with a comparatively young but impressive university and a focus on science and technology for the circumpolar world. Bodø, a small Norwegian coastal city above the Arctic Circle, was recently named one of the European Capitals of Culture for 2024. The city was built on fish and later defense, but is building its future on being a smart city and a hub for investing in new sea-based industries such as algae and becoming the hub for maritime traffic through the northeast (Glomsrød and Aslaksen, 2008).

Northern Finland's cities of Oulu and Rovaniemi also demonstrate that creative and determined northern regions can compete successfully in the age of rapid technological transformation. Oulu (Finland), which was a core center for Nokia (the once globally influential wireless company), is a new economic powerhouse. Rovaniemi (Finland) is a northern logging town that has reinvented itself. The local university has helped but the establishment of the city as Santa's "official" residence was a masterstroke. Hundreds of thousands of visitors come each year to spend a few moments with Santa, fueling an economic revival that is rare in the North (Glomsrød and Aslaksen, 2008).

Russia's Far North has the largest share of the population in the Arctic, with millions of people in an area that comprises one-third of Russia's total land mass. The region has substantial natural resource wealth, including oil, gas, minerals, and forestry. The development of the North features prominently in Russia's national aspirations and remains one of Prime Minister Putin's top priorities. The development of the Northeast Passage has expanded the commercial potential of the Far North, much as has the country's willingness to expedite innovation projects (such as the floating nuclear power plant used to power mining operations in Chukotka), which in a democratic nation would take years to move forward. Russia has significant plans to expand its extraction of oil and gas in its Arctic regions and distribute them to global markets via the Northern Sea Route (Kryukov, 2019). Within the Russian Arctic, particularly in the Yamal and Gydan peninsulas, there are 35,700 billion cubic meters of natural gas and 2,300 million tons of oil and condensate (McGee, 2020).

Alaska, which built a prosperous state economy largely on North Slope oil and gas, also benefits from a large military presence, revitalized by increasing tensions between the USA, Russia, and China. The US military has invested heavily in Arctic science, providing an evidentiary background for the tactical and strategic resources marshaled in the defense of northwest North America. The presence of the armed forces has enhanced and sustained regional prosperity, but has not sparked the extension of the innovation economy. Given the scale and intensity of American military investments and the role of military research in expanding America's innovation economy, the comparative absence of flow-on economic development is somewhat surprising (Goldsmith, 2007).

The situation in other parts of the Far North is less promising. Greenland has focused largely on fishing, with some mining and tourism (Arnaut, 2021). Canada's research capacity is in southern and urban locations and only minimally in the Far North, although Whitehorse (Yukon) has made impressive progress promoting early-stage companies. The quality and reliability of regional infrastructure lags far behind that of more populated jurisdictions.

Building Arctic collaboration for an innovation ecosystem

How should the North approach the prospect of major technological change? The North could be completely passive, waiting for new technologies to emerge and responding, *or not*, as opportunities arise. Or the North could adopt a problem-based approach, identifying areas of intense need and driving science and business to respond to opportunities and challenges. Given that new technologies are rarely developed in and for the North, it is vital that the circumpolar world articulate its own requirements for future technological innovation.

It makes more sense to step back from technology and focus on the socioeconomic circumstances in the Arctic, identifying areas in need of urgent investment and region-wide commitment. Shared Arctic circumstances are obvious: cold weather, long and dark winters, vast distances, small and widely dispersed populations, the legalities and/or the realities of southern and colonial domination, marginalized Indigenous peoples, and the many issues related to peripheral economies and societies. The region's urgent needs are substantial and diverse, including language preservation, cultural transmission across generations, mitigation of climate change effects, and fresh food production in the North.

The need for economic development that creates reasonably paid and reliable jobs for regional residents is a key challenge for much of the North. This is the logic behind the promotion of resource development, tourism, or any other economic activity and remains a primary goal of all governments. One of the challenges of the "new" economy is that many new devices and technologies that can make companies more viable and profitable also eliminate jobs. While some of the eliminated work was difficult and dangerous or boring and repetitive, people without work leave communities (often heading for cities or larger urban centers), towns and villages decline, and an overall downward spiral in population and community development follows.

Innovation is a process more than a destination, with numerous roadblocks, failures, detours, and successes along a path that lacks signposts, let alone a detailed map. It is unclear which technologies will work in the Arctic and even harder to determine what impact these innovations will have on the Far North. So far, there are no substantial visioning processes underway that might show in an integrated way how science and technology could usher in a new era in the North. The Arctic nations would need to develop a strategy to bring large-scale and rapid innovation to the Arctic and sustain it in place.

The Arctic nations have long historical connections built through early European Arctic exploration, Cold War militarization, late 20th-century resource development, and an integrated Indigenous rights movement that has sustained one of the most successful intra-regional collaborations in the world, highlighted by the creation and operation of the Arctic Council (English, 2013). Yet, the lines of division and distinction are also strong, underscored by national and ideological differences, the sharp distinctions between Northern Scandinavia and the Russian and Canadian Norths, the extensive militarization of Alaska and Siberia, the unique political status of Greenland, largely unchecked resource development in the Russian North, and the special political and legal rights of Indigenous peoples in the Canadian North and Alaska (Coates and Holroyd, 2020).

Despite these divisions and cultural and linguistic differences, the circumpolar world has established an enviable track record for intra-regional collaboration. It has built a culture of cooperation, becoming one of the most politically and diplomatically integrated regions in the world. The Arctic Council established the foundation, reinforced by the creation of the University of the Arctic. The University of the Arctic sparked a surge in circumpolar scientific collaboration, sending out academic roots that will potentially bear fruit in future decades. These Arctic networks also reach well outside the region, with firm connections to Europe, southern North America, and East Asia.

Major and regular gatherings – High North Dialogue (Bodø), Arctic Frontiers (Tromsø), Arctic Circle (Reykjavik) – are key elements in ongoing conversations about present challenges and future possibilities. Growing scientific interest in the North, accelerated by heightened concerns about climate change, has drawn the EU and many non-Arctic nations into the development of collaborative approaches to northern development. The use of the UN Convention on the Law of the Sea to resolve long-standing disputes about Arctic boundaries and sovereignty claims is a solid and practical indication of the North's ability to address northern issues through legal and collaborative processes.

However, scientific and technological innovation cannot be addressed through geopolitical processes. Commerce is inherently competitive, with strong nationalist and regional elements. Inventors, scientists, and businesspeople do not sublimate their interests to nationalistic or regional considerations, particularly in Western nations. Innovation is driven by consumer demand and market forces; governments struggle to determine how best to support it. Furthermore, some of the common characteristics of the Far North, particularly small populations, high costs, and great distances, undercut commercial possibilities. Put simply, and putting aside firms working in natural resource extraction, it is hard to make money in the Arctic based on emerging technologies.

Global advertising and marketing bring southern and urban products into the Far North and, at least in theory, provide the opportunity for northern-based products and services to attract international customers. This has worked in selected areas – Santa Claus in Rovaniemi, aurora borealis viewing in other locations, adventure tourism across the Far North – but very few northern-based companies have found sustainable markets outside the North. Northern universities and colleges are trying, although their small incubators struggle to develop meaningful ties across the Arctic. Scientific researchers, in contrast, have made truly influential collaborative connections, particularly in the natural sciences and environmental domains. Large multinational resource companies have worked and collaborated across national boundaries and have borrowed technologies and made developments in new machinery, in minerals, and in forest products processing. Business groups, led by chambers of commerce, have developed regionalized ties, as between Yukon and Alaska, the NWT and Nunavut Chamber of Mines, and the Oulu-Luleå-Tromsø triangle in Northern Scandinavia, and have worked hard, with some success, to create northern commercial links. Individual entrepreneurs and business groups have been eager to explore opportunities in other northern regions – Yukon businesses made efforts to connect with Scandinavian companies, for example – but few lasting connections have emerged beyond the tourism and mining sectors. Much of this is connected to differences across the region. Norway's "blue economy" of ocean-based activities, for example, has only minor reflections in the Canadian North.

Russia is an outlier, a unique quasi-state commercial environment with very few formal connections with the rest of the circumpolar world. In the early years of *glasnost* in the former USSR, Alaskan businesspeople made an earnest, albeit self-congratulatory, effort to reach out to Eastern Siberia, believing that long-term commercial connections could be fostered. The campaign enjoyed a few successes, but the best of business intentions could not make permanent inroads against the rigid barriers between Russia and the US. A long-term effort to connect the Canadian and Russian chambers of commerce foundered over the years, with cross-border regulations and commercial restrictions impeding the local development of business contacts. Resource development in the Russian North has remained largely dominated by Russian companies.

The North, driven by intense and growing consumer awareness of international technological advances, is increasingly connected to the fast-changing worlds of innovation. Social media is as popular and commonplace in the North as in southern areas and indeed is much used by Indigenous communities. Emerging alternative energy sources, including wind and solar power, geothermal power, and run of the river hydroelectric systems, are being adapted to northern settings, although the economics of the Far North remain challenging. Internet-based services, including video streaming, electronic banking, e-education, and others, are available in many, even most, parts of the Arctic, although infrastructure deficiencies slow the complete rollout (Smith, 2010).

Well-financed resource projects are proceeding apace. While local firms have innovated in key areas – placer mining technologies in Yukon, permafrost technologies in Nunavut, deep sea drilling off Norway and Russia, Arctic Island construction to support offshore oil and gas development – much of the money and scientific expertise has come from outside the region. Local firms and residents have contributed substantially to the on-the-ground developments. Arctic mineral exploration includes the extensive use of drones, and some mines are capitalizing on autonomous trucks and remote mining operations. Norwegian offshore oil and gas developments are among the most technologically advanced in the world; while Russian systems appear less polished, they are effective. New sensors, satellite imaging, and other digital systems are improving wildlife and environmental monitoring, a particularly urgent requirement given the imperatives of climate change (Serreze, 2018). While rapid technological change remains more a southern and urban phenomenon than an Arctic one, some shifts are underway. The Arctic region is ripe for technological and scientific collaboration, but it is not clear if researchers and businesses can overcome the practical, political, cultural, and other barriers that stand in the way of collective problem-solving innovation.

Challenges

Politicians, government officials, post-secondary administrators, and business promoters speak enthusiastically about the potential for Arctic innovation, but they typically speak much more of potential than performance. Northern incubators produce small numbers of firms (in the North as elsewhere); they are much better at teaching employability skills to individuals than at creating export-ready regional businesses. Innovation is, moreover, a classic double-edged sword, with benefits in one area (e-health, for example) being offset by northern losses in other fields, such as the impact of global e-commerce on local businesses. It is exceptionally difficult for a small local company to compete with multinational corporations such as Amazon or Alibaba, to say nothing of thousands of specialized multinational firms. While optimism about the employment and commercial potential of innovation remains strong, there is a growing realization that the benefits shift largely to southern, urban environments with active commercial ecosystems and ready access to capital and advanced technologies. Few Arctic regions can claim, even in a limited way, to possess the combination of human, financial, and technological capabilities to compete globally. The infrastructure deficits in rural Alaska, Greenland, nonmetropolitan northern Russia, and across the Canadian North undercut efforts to innovate.

The state of innovation in the circumpolar world is far from encouraging. The region benefits from the general sharing of research, emerging technologies, and prototypes, and from the commercial ideas and regulatory developments in other nations. However, few innovations are being developed with northern needs specifically in mind. The global investment in addressing heat is much greater than the effort to respond to extreme cold. A vast international swimwear industry is many times larger than corresponding attempts to develop cold weather clothing. Certain high return areas, like Arctic oil and gas, attract large investments; low return areas, such as developing commercially viable products to overcome the health challenges associated with long periods of darkness, struggle to secure financial support.

To the degree that innovators and businesspeople wish to collaborate across the North, there are formidable barriers in the way. Long distances and high costs inhibit easy collaboration. It is much easier to travel from Whitehorse to Los Angeles or Bodø to Berlin than it is to make the journey from Rovaniemi to Iqaluit or Anchorage to Nuuk (Greenland). Practical considerations such as language and lifestyle differences and consumer preferences matter a great deal, although technology-based products are often less influenced by cultural considerations. Overall, northern-based firms struggle to find traction in the face of aggressive competition and international marketing by European, North American, and Asian companies. There are a few exceptions – the Yukon-based company Proskida sells digital monitoring devices to elite winter athletes across Scandinavia – but their rarity highlights the difficulties inherent in these market-building activities. The Far North remains the far end of the "long tail" of contemporary global business with only a small number of specialized companies able to sell innovation-based products and services internationally. This is hardly unique to the North and, in fact, is the norm for nonmetropolitan areas around the world.

There are many reasons for this. The extremely poor and inconsistent Internet services in the Canadian North and rural Alaska, in marked contrast to Northern Scandinavia, limit opportunities for the creation of digital innovations in the North American rural north (Delaunay, 2014). Myriad national regulations, commercial standards, business laws, and other requirements, particularly when working in Russia and Greenland, inhibit the easy migration of business and products across boundaries. Patent and licensing systems have become regularized internationally, but these processes do not protect north-centered innovations, nor do they block southern and global innovations from displacing northern businesses, as has happened internationally in recent years through e-commerce. At the same time, well-established global standards in such fields as Internet domain names and security standards, air safety regulations, pharmaceutical approvals, and the like have paved the way for the globalization of business, allowing products from around the world to make their way into Arctic stores and, at least theoretically, for Far Northern companies to sell their products and services internationally.

The commercial separation of Russia from the rest of the circumpolar world carries a significant cost. With Russia standing aside from the bulk of the circumpolar innovation effort, the Far North is deprived of one of the largest potential consumer markets, considerable resource development activity, and a vast area of wealth production. At present, however, there is little reason to anticipate that Russia will be an active participant in the circumpolar innovation system.

The story of Arctic innovation so far is more of missed and avoided opportunities than of technological achievement and experimentation. Three-dimensional (3D) printing/additive manufacturing, which has numerous potential applications of particular value in the Far North, remains largely underdeveloped (Coates and Holroyd, 2018). So too are e-education (although the school closures associated with the COVID-19 pandemic have likely accelerated developments in this field) and e-health. There are opportunities for remote surgery, technology-based mental health care, and remote monitoring of personal wellness and chronic ailments (like diabetes), but implementation has only occasionally gone beyond the pilot or test phase. Food factories or technology-based localized food production could be extremely beneficial in addressing small-town food needs, but the potential is largely unrealized. Smart cities, which seek to manage energy usage, traffic, and other city activities, have not yet emerged prominently in the Far North. Not even Svalbard (Norway) has capitalized on being one of the most digitally enabled sites on the planet.

In sum, despite some government efforts and great public enthusiasm, there has been limited commercialization of science and technology in the North and less than optimal experimentation with global innovative products and services. While there are pockets of north-centered innovation across the circumpolar world, few of these efforts extend beyond the natural resource sector. Many of the potential benefits of the age of technological transitions, such as the opportunity to work remotely, have not resulted in a major influx of new jobs and high incomes into most of the North. The global innovation system has demonstrated little interest in the Far North, leaving many Arctic needs unaddressed and opportunities substantially underdeveloped. There are enough successful locations (Oulu, Luleå, Tromsø, Bodø, and Whitehorse) however, it is clear that nordicity need not be an unsurmountable roadblock to the commercialization of science in the Arctic.

Steps forward

Several decades of massive global investment in emerging technologies have not yet produced substantial visions of a technologically enabled North. Many countries have extensive innovation strategies, although visions of socioeconomic development typically focus on major urban environments. There are, likewise, few substantial visioning exercises focused on rural and small-town areas in Canada, the US, Australia, and New Zealand. In short, and despite being in the midst of the greatest technological revolution in world history, the unfolding of innovation societies has been left to the vicissitudes of international commerce, national policy, and urban regional economic development, enhanced in a few cases by a small amount of local experimentation.

It is possible to imagine a technology-enabled North. Ever-improving satellites could provide both reliable, inexpensive, high-speed Internet connectivity and advanced monitoring of the Northern landscapes. Innovative energy systems, supported by advanced technology in home and building construction, could dramatically reduce the cost of heating. Food factories could reduce the cost and improve the quality of fruit and vegetables in the North. Advanced health care systems could provide high-quality preventative and emergency medical care in the North, using artificial intelligence, remote surgery robots, and personalized digital health monitoring to provide residents of remote regions with a higher level of medical care than they currently have. Additive manufacturing could address the challenges of ready access to a wide variety of manufactured goods, producing high-quality items on demand, on site and at sharply reduced costs.

Fully developed e-commerce systems could enable participation in global supply chains, providing northerners with ready access to a remarkably wide range of goods, quickly delivered to the North, as well as a properly developed distribution system for northern-developed and produced artwork, professional work products, and specialized digital and manufactured goods. With reliable access to global markets, e-commerce could produce long-term secure jobs for northern residents and substantial profits for northern businesses. Digital and remote employment could provide steady incomes for thousands of northern workers, many working for companies owned and operated thousands of kilometers away. This stabilization of the workforce could underpin a steady growth in the northern population and an improvement in prosperity and quality of life in the North. Arctic communities could be protected by a web of digital sensors, networked across the North to provide high-level monitoring of the changing northern climate and world-class monitoring of all forms of wildlife.

All these innovations are currently available, some in commercial distribution and others as prototypes and laboratory technologies. With the commitment of trillions of dollars to scientific and technological discovery, additional developments (faster, cheaper, more effective, user-friendly, and widely distributed) are inevitable. Artificial intelligence portends dramatic shifts in the economic order, as does self-writing software, personalized medicines, embedded devices (placed inside the human body), advanced automation, sensors and security systems, and gamification (commercializing video game technologies). However, the global innovation landscape is riddled with the ruins of failed commercialized innovation dreams. There are no assurances that the available technologies will work in practice or become commercially viable or that emerging innovations will be applied successfully in northern settings or will sustain the kind of economies, societies, communities, and lifestyles that northerners desire. Technologies are value-free and can have a wide variety of outcomes, with no assurance of a specific or predictable impact. An innovative North is, by definition, a completely unpredictable place, just as the state of global innovation c. 2035–2050 remains a matter of conjecture and at best calculated estimation.

Working toward an innovation-inspired and sustained North, if that is indeed possible, will necessitate concerted action. Supporting the commercialization of science and technology across the region will require an unprecedented level of cooperation among businesses and business organizations throughout the Arctic. The essential first step would be the creation of a Circumpolar Chamber of Commerce or Innovation committed to building businesses and sectoral networks across the region, coordinating/regularizing government regulations, working toward an Arctic free trade zone (or comparable form of trade liberalization). Such an initiative would be expensive and would require a serious and long-term commitment. It is unlikely that Russia would join such a collaboration, at least in the short to medium term, which would leave a large and economically important part of the Arctic outside the commercial network. Yet close business cooperation would exert pressure on governments and international organizations and could serve as a high-profile lobby for closer economic and social ties across the circumpolar world.

A north-centered commercial network requires the development of pan-Arctic consumer sensibilities, based largely on the shared experiences of life in the Far North. For inventors, researchers, entrepreneurs, and investors to become truly interested in Arctic innovation possibilities, the circumpolar world has to appear to a much larger extent as a single consumer market. The opportunities here are obvious, from cold weather clothing to improved winter transportation devices, remote and small-town energy systems to sub-Arctic localized and all-weather food production facilities. At present, consumers are part of national consumer networks and business systems. Norwegian and Swedish consumers, for example, operate largely within national and Scandinavian commercial networks, sharing common brands and many of the same retail and service companies. Canada and Alaska, in contrast, share fewer business ties, although the tourist industries and the resource sectors intersect in important ways. Nunavut and Greenland, although culturally similar and with much in common, have limited commercial connections, despite having attempted several times to strengthen these ties. A concerted and systematic consumer cultivation effort, coordinated between North-Central businesses and capitalizing on newly created social media networks and advertising investments, could create a pan-Arctic market, linking consumers and Arctic businesses and gradually building circumpolar retail, service, and promotional operations that could result in northern companies serving northern consumers with technological innovations built in and for the North.

This effort would have to be matched with Arctic-wide e-commerce networks, complete with production and distribution systems targeted specifically at the diverse and culturally rich societies of the circumpolar world. E-commerce would overcome logistical and geographical barriers, making it possible for producers of traditional Inuit clothing, for example, to sell directly and easily to Scandinavian consumers and for country food producers in northern Finland to reach new consumers in Alaska and Greenland. E-commerce is complicated, risky, and far from assured of success. A circumpolar business would require operations in multiple languages, well-integrated financial systems, and complex supply chains and would likely struggle until the emergence of a large and sustainable consumer market, interested in and committed to northern retailers and suppliers.

A collaborative retail and commercial effort will require a constant supply of new products and services connected to and comparable with the best in global innovation. This needs the elaboration and expansion of the embryonic pan-Arctic innovation system that connects universities, colleges, research institutes, private sector laboratories (particularly in the resource sector), and entrepreneurs in a creative, mutually supportive network. At present, innovation clusters and systems are community and region-specific, although they connect with investors, financiers, government officials, and other participants around the world. A circumpolar innovation system would have to overcome numerous barriers – different commercial cultures, distinctive patterns of government support and engagement, languages, regulations, and the like – and the forces for successful integration are much less pronounced and less well developed. Other variations, such as a remote/extreme weather commercial ecosystem or a network of circumpolar Indigenous businesses, might have a better chance of being accomplished than a more comprehensive and inclusive innovation system. Even here, however, the barriers to entry and sustainability are considerable and substantial funding, commercial support, government assistance, and long-term commitment would be required.

Conclusion

Imagining the Arctic in 2035 challenges commentators to consider both the opportunities and the barriers to achieving a preferred vision. A technologically enabled Arctic might emerge, in part if not in whole, within the next two decades. With the right regional, national, and international commitment, with a combination of government investment and private sector entrepreneurship, and with a concerted effort by northern researchers to convert global technological discoveries into commercially viable products and services, the North could be converted into a dynamic, creative, profitable, and exciting region better able to respond to the geographic and climatic challenges of the region and with enhanced abilities to build on the social and human capital of the circumpolar world.

The 2015–2030 UN Sustainable Development Goals (SDGs) outline a "blueprint to achieve a better and more sustainable future for all" and include SDG 8, focused on the promotion of inclusive and sustained economic growth, and SDG 9, which encourages the fostering of innovation and sustainable industrialization (Degai and Petrov, 2021). Industrialization, innovation, and economic growth are key components of global development, including in the Arctic. Achieving these will entail cooperation. Innovation could draw the North together, creating a shared market and the development of a commercial economy of scale for north-focused companies. Producing goods and services that meet northern needs could develop a sense of common purpose and shared destiny.

That something is possible is no assurance of success. The barriers to commercial integration are more significant and powerful than the important but less powerful and influential forces of integration and collaboration. This is a somber conclusion, with the possibility of an innovation-based economy, balanced off against the practicalities and realities of the circumpolar world. The vision of a truly technologically enabled Far North is highly unlikely to materialize. Much of the North lags behind southern and urban environments in many aspects of the technological era. The Far North's existing and severe digital divide may be diminishing in Northern Scandinavia, but it remains large and is even growing in much of the circumpolar world. National governments assign a comparatively low priority to Arctic innovation, the region lacks ready access to investment capital, any regional commitment to a technology-enabled future is highly localized, and on a circumpolar level, at best tepid. The northern research capabilities are focused on the natural world and not commercial products and research, with only small commitments to retail and industrial endeavors. External capital, scientific knowledge and expertise underpin the investment and innovation that does take place. Most of the profits – the source of subsequent innovation investments – leave the North. With key exceptions – permafrost and atmospheric work in Alaska, ice and ocean research in northern Norway, and cold weather research in Russia – the key technological work is done outside the North.

Ultimately, a truly innovative society requires national or regional consensus, sustained collaboration between business and government, research systems devoted to commercial outcomes, and consumers with a vested interest in innovation developments targeted to their needs and interests. These forces have come together at times but they do not exist at present in the circumpolar world, nor is there a likelihood of their coming together in the near future. Unlike the nationalistic and demanding consumers of East Asia, circumpolar consumers are passive and hard to mobilize on a regional, let alone on a circumpolar, basis. The Far North does not have great potential for commercial innovation despite the impressive efforts of individual entrepreneurs, several northern business schools and research institutes, and some subnational governments. Other issues, particularly climate change, Arctic sovereignty, environmental protection and sustainability, and Indigenous rights, garner much more attention than innovation and the Far North's technological possibilities. In all Arctic nations save Russia, northern innovation remains a small element for government and the private sector, with the circumpolar world having consistent trouble penetrating the national and international consciousness on anything but a symbolic and emotional level. The kind of national consensus and intra-regional collaboration needed to pursue a technologically enabled future with determination, long-term commitment, and scientific intensity is lacking.

It is difficult when the vision of a technologically enriched economy and society is clear and scientifically obtainable to conclude that the achievement of this aspiration is highly unlikely. This is not an indictment of the circumpolar world. Most nations, regions, and communities will likewise fail to capitalize on the technological potential of the 21st century. Indeed, this era of technological innovation is littered more with commercial failures than stunning and sustainable achievements. At a time when visionary entrepreneurs like Elon Musk, Jeff Bezos, and Masayoshi Son commercialize space travel, satellite Internet, software-based transformations, and other such developments, the reality is that much technological potential remains largely unrealized and without a solid, sustainable commercial foundation.

December 2021

Ex-post reflections

The circumpolar innovation system, after undergoing promising changes in recent years, faces change in Arctic affairs following the Russian invasion of Ukraine. The innovation file is built around complex multinational collaborations: shared academic research projects, regional partnerships, and close engagement with subnational governments. In a very short space of time, most universities canceled or limited collaborations with Russian institutions; dozens of foreign companies withdrew from Russia. Russia, previously the largest investor in Arctic innovation, continues to militarize the North, but innovation will be stymied by foreign withdrawals and sharply reduced access to Western intellectual property and technological innovations. While it will not be a hard stop, Russian innovation investments will almost certainly slow dramatically. The rest of the circumpolar world will likely see a burst in military spending, which could well support long-term technological innovation in the Arctic. In short, the Russian attacks on Ukraine have reversed the promising and hard-won course of Arctic collaboration in general and innovation engagement in particular. The circumpolar world without Russia is likely headed for a time of greater collaboration, which could well spark additional innovation. The invasion of Ukraine has changed circumpolar activity, simultaneously making greater Arctic commercial cooperation and innovation both more important and less likely, especially as regards engagement with Russia.

April 12, 2022

References

Arnaut, J.L. (2021). "The Political Economy of Greenland: From Colonialism to a Mixed Economy". In Høgedahl, L. (ed.) *Greenland's Economy and Labour Markets*. Routledge, pp. 30–47.

Breznitz, D. (2021). *Innovation in Real Places: Strategies for Prosperity in an Uncertain World*. Oxford: University of Toronto Press.

Coates, K. and Holroyd, C. (2018). *Revolutionary Building for the North: 3D Printing Construction*. Conference Board of Canada.

Coates, K. and Holroyd, C. (eds.) (2020). *The Palgrave Handbook of Arctic Policy and Politics*. Palgrave Macmillan.

Coates, K. and Holroyd, C. (2021). Northern Sweden and economic development. *Journal of Northern Studies, 15*(1), pp. 7–24.

Degai, T.S. and Petrov, A.N. (2021). Rethinking arctic sustainable development agenda through indigenizing UN sustainable development goals. *International Journal of Sustainable Development & World Ecology, 28*(6), pp. 518–523.

Delaunay, M. (2014). "The Arctic: A New Internet Highway". In Heininen, L., Exner-Pirot, H., and Plouffe, J. (eds.) *Arctic Yearbook*. Akureyri, Iceland: Northern Research Forum, pp. 503–510.

Diamandis, P.H. and Kotler, S. (2020). *The Future Is Faster Than You Think: How Converging Technologies Are Transforming Business, Industries, and Our Lives*. Simon & Schuster.

English, J. (2013). *Ice and Water: Politics Peoples and the Artic Council*. Penguin Canada.

Erika, K.M. and Watu, W. (eds.) (2010). *Innovation and the Development Agenda*. IDRC.

Freeman, C. (1990). *The Economics of Innovation*. Edward Elgar Publishing.

Glomsrød, S. and Aslaksen, J. (2008). *The Economy of the North*. Statistics Norway.

Goldsmith, O.S. (2007). *The Remote Rural Economy of Alaska*. Anchorage, AK: Institute of Social and Economic Research, University of Alaska Anchorage.

Hall, H.M. (2020). "Innovation, New Technologies, and the Future of the Circumpolar North". In Coates, K. and Holroyd, C. (eds.) *The Palgrave Handbook of Arctic Policy and Politics*. Palgrave Macmillan, pp. 117–132.

Hall, H.M. and Vodden, K. (2019). "Learning, Knowledge Flows, and Innovation in Canadian Regions". In Vodden, K., Douglas, D., Markey, S., Minnes, S., and Reimer, B. (eds.) *The Theory, Practice, and Potential of Regional Development*. Routledge, pp. 183–211.

Heininen, L. and Southcott, C. (eds.) (2010). *Globalization and the Circumpolar North*. University of Alaska Press.
Kaku, M. (2012). *Physics of the Future: How Science Will Shape Human Destiny and Our Daily Lives by the Year 2100*. Anchor.
Kenney, M. (2000). *Understanding Silicon Valley: The Anatomy of an Entrepreneurial Region*. Stanford University Press.
Kryukov, V.A. (2019). Studying the economy of Siberia: Continuity and integrity. *Regional Research of Russia*, 9(2), pp. 107–117.
Larsen, J.N., and Huskey, L. (2015). "The Arctic Economy in a Global Context." In Evengård, B., Nymand Larsen, J. and Paasche, Ø. (eds.) *The New Arctic*. London: Springer, pp. 159–174.
Larsen, J.N., and Petrov, A. (2020). "The Economy of the Arctic." In Coates, K. and Holroyd, C. (eds.) *The Palgrave Handbook of Arctic Policy and Politics*. London: Palgrave McMillan, pp. 79–95.
McGee, R. (2020). Mapping Russia's arctic hydrocarbon development scheme. *The Arctic Institute*, Feb. 18, 2020.
Nuttall, M. (eds.) (2005). *Encyclopedia of the Arctic*. Routledge.
Serreze, M.C. (2018). *Brave New Arctic*. Princeton University Press.
Smith, L.C. (2010). *The World in 2050: Four Forces Shaping Civilization's Northern Future*. Penguin.

8 Smart City Dialogue in the Arctic

Opportunities and Challenges

Evgenii Aleksandrov, Elena Dybtsyna, Nadezda Nazarova and Igor Khodachek

Introduction

In the past decade, the smart city concept has enjoyed growing popularity with its promise of sustainable development for cities and communities (Mora and Deakin, 2019). Smart cities have even become a part of Sustainable Development Goal (SDG) 11, to "Make cities and human settlements inclusive, safe, resilient, and sustainable" (United Nations, 2017). Broadly defined, the smart city suggests urban technological development with "investments in human and social capital and traditional (transport) and modern (ICT) communication infrastructure [that] fuel sustainable economic growth and high quality of life, with wise management of natural resources, through participatory governance" (Caragliu et al., 2011, p. 70). By these means, one of the smart city's key agendas is urban development for local needs via dialogue formation between global-local actors and with citizens' collaboration (Bolívar, 2019; Calzada, 2018; Kitchin et al., 2019; Vanolo, 2016).

Many studies have investigated smart city dialogue formation by holding up individual city cases in different contexts (see Mora and Deakin, 2019, for an overview). Yet, there is still a lack of knowledge about the role of dialogue in the global-local interpretation of smart cities (Burns et al., 2021). Moreover, there is a call to understand the possible tensions between global and local smart city trends through comparative studies and by capturing power geometries in global-local dynamics (Miller et al., 2021).

In response to that call, this chapter investigates the development of smart city initiatives in the Arctic, where the smart city concept is proclaimed to have the unique potential to resolve the region's socioeconomic challenges and secure sustainability (Raspotnik et al., 2020). However, smart city development has emerged only recently in the Arctic, with few empirical cases stressing the importance of smart city dialogue (Dybtsyna and Aleksandrov, 2020; Sköld et al., 2018). At the same time, the Arctic is an example of growing governance complexity, with multiple state and non-state actors operating locally, nationally, and internationally (AMAP, 2017; Smith, 2011). Such complexity, in turn, requires constant dialogue (Wilson Rowe, 2018a).

DOI: 10.4324/9781003246015-10

Thus, the Arctic represents a critical example of smart city dialogue formation in the context of global-local dynamics and power geometries. While the interest in smart cities and the promise of dialogue are growing in the Arctic, understanding of dialogue formation across actors is mostly missing.

In this regard, this chapter focuses on the state-of-the-art smart city dialogue formation in the Norwegian and Russian Arctic. Both countries are interesting to examine due to their different long-lasting governance traditions (Bourmistrov and Mellemvik, 2002) while placing a similar emphasis on Arctic importance and smart city development in their governance strategies. We mainly reveal and compare how *smart city dialogue unfolds across governance actors in these two countries*. Conceptually, we apply dialogic literature that critically assesses smart city developments in the two countries concerning the so-called Arctic paradox (e.g., Holm, 2020), that is, Arctic development but with the essential decisions taken elsewhere. Empirically, we analyze qualitative data and publications/working papers from three international research projects on smart city development in the Arctic in the period 2018–2021.

Our chapter proceeds as follows. First, we review the smart city literature concerning the Arctic and the dialogue approach. Second, we describe the setting and our methods, before presenting our empirical findings on smart city dialogue development in the Norwegian and Russian Arctic. Finally, we discuss our results and conclude with future implications.

Smart cites in the Arctic: between globalizing and localizing trends

The smart city global agenda has been flourishing in the last decade with two key trends. First, certain scholars currently argue for the smart city globalizing trend (Burns et al., 2021; Miller et al., 2021), that is, developing cities within global flows of capital, people, and information that form one dominant urban view. Globalizing also refers to specific urban projects, governance regimes, and processes in smart city initiatives, as widely adopted in Europe, with six characteristics: smart mobility, smart economy, smart people, smart governance, smart living, and smart environment (Giffinger and Pichler-Milanović, 2007).

This trend has provoked considerable debate on creating a dominant vision of the 'good city' (Burns et al., 2021, Vanolo, 2014) that ignores contextual importance and the diversities of agendas for cities. In this regard, a growing number of studies reveals the consequences and even dangers of ignoring the local context. For example, Grossi and Pianezzi (2017) demonstrate that smart cities can become utopian ideas promoting neoliberal values and distracting from locally driven city development. Similarly, Hollands (2008) highlights the danger of urban development privatization under a smart city agenda. Moreover, many scholars stress the dominance of business elites and their lack of consideration for citizens in urban development (e.g., Karvonen et al., 2018; Kitchin et al., 2019; Marvin et al., 2015).

Altogether, this forms the second—localizing—trend, with increasing emphasis on the varied nature of smart cities and acknowledgment of local context (Burns et al., 2021; Miller et al., 2021). This trend mainstreams alternative forms of smart city development through in-depth case studies with local interpretations of smartness and local knowledge formation, which cast doubts on the globally dominant agenda and vision.

Therefore, smart city development has witnessed two contradictory trends: the popularization of a global vision, and more recently, a preference for local images. Miller et al. (2021) assert that the encounter of global and local creates productive and complementary tensions. In other words, if we want to understand smart city development with global-local distinctions, it is essential to study the relationship between globalizing and localizing trends and how these unfold within the existing power geometries (Burns et al., 2021). This might raise the question of dialogue formation for global-local interpretations of urban development (Bolívar, 2019; Calzada, 2018; Vanolo, 2016). Dialogue is suggested to be crucial to prevent the aforementioned dangers and deliver on the sustainability promises of smart cities (Grossi et al., 2020). Yet, how dialogue unfolds within smart city globalizing and localizing trends in practice remains unclear.

The Arctic is an exciting space in which to examine smart city dialogue formation, as its governance is extremely complex and marked by global versus local aspects of sustainability (AMAP, 2017; Smith, 2011). While the Arctic gains increasing attention as a region rich in natural resources (e.g., fish, oil, and gas), and thus as presenting opportunities to ensure global economic growth and sustainability, it also faces the so-called Arctic paradox, indicating tensions between local and global interests. Locally, Arctic governance increasingly addresses the values of inhabitants and their sustainability (e.g., Russian Strategy of Arctic Development, Norwegian High North Strategy). Globally, increasing attention is paid to the Arctic by central governments and other players outside the Arctic (AMAP, 2017; Bourmistrov et al., 2015). Balancing these global and local interests creates a paradox: while talking about locally driven governance, decisions on the opportunities presented by the Arctic are made without considering those who live there. This results in most of the potential benefits flowing from Arctic regions and to resource exploitation without creating local value (Holm, 2020; Tennberg et al., 2014), thereby calling into question the sustainability of the region. As a response, there is a call for innovative solutions with dialogue across governance levels to ensure local stakeholders' capacity to handle this complexity (AMAP, 2017).

The smart city concept recently became one such solution actively promoted in the Arctic, addressing the situation with proposed technological development for Arctic cities (Dybtsyna and Aleksandrov, 2020; Raspotnik et al., 2020). Specifically, in line with global promises of urban sustainability under SDG 11, the smart city concept in the Arctic suggests using ICT for city infrastructure development, attracting new investments and residents, innovations building, and environmental and local social growth (McMahon and Raspotnik, 2019). Yet the concept is one developed outside the Arctic and so requires careful consideration of the global pitfalls (Dybtsyna and Aleksandrov, 2020). This corresponds to a situation where decisions are made regardless of the region's complexity or the inclusion of local actors (Bourmistrov and Johansen, 2019).

Thus, while a smart city can potentially facilitate sustainability of the Arctic, it can also become stuck in a similar (smart) Arctic paradox. Some undesirable trends are already perceptible (e.g., AMAP, 2017; BIN, 2019), but have so far been underestimated or ignored in relation to dialogue in the Arctic.

A dialogic approach to smart city development in the Arctic

The dialogic literature on public administration and governance (Brown, 2009; Rajala et al., 2018) is particularly valuable for studying Arctic smart cities' dialogue formation. This literature asserts that dialogue should be supported by divergent Arctic voices (Makki, 2012). In other words, for fruitful smart city dialogue, the borders between global and local visions (Burns et al., 2021) should be blurred in favor of learning from each other (Rajala et al., 2018). In that way, different parties to the dialogue become coauthors of collective actions (Bakhtin, 2010), that is, in this case, smart city initiatives in the Arctic. In that sense, the key idea of dialogue is to resolve contradictions between different worldviews not by denying their differences and upholding one dominant worldview, but by identifying and supporting the commonalities of those views (Rajala et al., 2018).

In the dialogic approach, smart city development should be viewed as a complex, political process, acknowledging the pluralism, difference, ideological conflicts, and power dynamics on the global-local agenda (Brown, 2009). In that sense, a smart city reflects the 'voices' of divergent governance actors within the Arctic and outside it, that is, unpacking smart city globalizing and localizing relationships. Analytically, it is key to distinguishing between monologue and dialogue formation. A dialogue affords divergent voices equal attention and promotes consensus-building within the smart city agenda, considering local, national, and international visions as equal. A monologue, in turn, suggests that smart city development will be framed by the dominance of one voice across governance actors. Therefore, in relation to the Arctic's smart city globalizing and localizing interrelation, the dialogic approach argues for the inevitability of power and the divergent interests of different governance actors (Brown, 2009). In that sense, it can help to track the tensions and conflicting nature of smart city formation within global and local agendas in the Arctic, emphasizing power and imbalance across actors, which have been undermined in the earlier smart city literature.

Based on prior theorization of a dialogue's underlying principles and assumptions, we propose five aspects of smart city dialogue formation to be considered in the Arctic (see Table 8.1), along with critical elements to consider when urging dialogue or monologue formation. These aspects will frame our interpretation of the data and presentation of the research findings.

Table 8.1 Application of dialogic literature to analyze smart city development in the Arctic (based on Bakhtin, 2010; Brown, 2009; Makki, 2012; Rajala et al., 2018).

Five Key Aspects	*Application to Smart Cities in the Arctic*	*Criticality*
Topic(s) and goal(s)	What are the common goals, elements, or framework of smart city initiatives?	Domination of a particular topic or goal
Parties/actors (local, regional, national, international)	Who are the actors involved in smart city initiatives?	Excluded or marginalized actors
Voices involved, the weight of each, and their balance	Who has the power and expert knowledge to decide on smart city developments, and how are the power and knowledge exercised?	Power is held by one voice. Expert knowledge comes from one voice; these are not shared
Format/rules of interaction	What financial aspects, agreements, network collaborations, and citizen involvement initiatives are there?	Symbolic dialogue that is highly controlled with predetermined answers
Outcome(s)	What is the status of smart city development so far?	Dominant voice versus multiplicity of voices

Empirical setting and method

This chapter studies the development of smart city initiatives with regard to its forefront dialogic promises in Russian and Norwegian Arctics. Despite certain similarities between the two cases, there are also definite differences. In this respect, comparison of the two contexts provides a fruitful opportunity to shed light on the issue.

We find several intersections of these two contexts in relation to various 'histories', that is, smart history, democratic history, and an Arctic history. On the one hand, smart history is relatively immature in both countries. The smart city phenomenon in Norway is under development (Rambøll Management Consulting, 2018) and is on the agendas of one-third of municipalities (Dybtsyna and Aleksandrov, 2020). Similarly, the official 'smartification' of Russia only started in 2017, stressing local city aspects for consideration in the Arctic areas (Khodachek et al., 2020). Beyond this, the democratic histories of the two countries lack commonalities. Norway's history of representative democracy, local autonomy, and self-government has flourished for more than 150 years. In contrast, representative democracy in Russia was only introduced in the 1990s after the collapse of a long-lasting totalitarian regime; accordingly, democracy is still a relatively new phenomenon. In this way, the democratic structures in the two counties differ significantly, which, in turn, gives rise to differences in the plots of their smart city stories. Finally, both countries have put strategic priority on the 'sustainable Artic' (e.g., Norwegian High North White Paper, 2020, and Russian Arctic Strategy, 2030), viewing the Arctic as 'the territory of the dialogue' (www.forumarctica.ru) and sustaining a peaceful Russian-Norwegian Arctic history (Bourmistrov et al., 2015).

To study how a smart city dialogue unfolds in the Russian and Norwegian contexts, we use data from three research projects focusing on smart city development in the Arctic with both countries as the main partners. In particular, the research project 'SMARTNORTH: Sustainable development and MAnagement by paRTicipatory governance practices in the High NORTH' (2017–2020) reveals the scope of developing smart city initiatives, their driving forces, and potential effects on sustainable development in Arctic communities. The second project, University of the Arctic's (UArctic's) 'Smart Societies in the High North Thematic Network' (2018–ongoing), has identified different approaches to the smart concept across the Arctic and proposes its potential amplification through intensive education and research cooperation between partners. Finally, the research project 'EduSmart: Education and Knowledge Development for Smart City Governance and Performance Management in the High North' (2021–ongoing) intends to enhance international education and research in the Arctic by incorporating knowledge on smart city governance and performance. Altogether, these projects cover smart city development in the northern regions of Norway (Nordland, Troms, and Finnmark) and Russia (the north coast of northwest, Ural, Siberian, and far eastern regions).

For our research, we conducted documentary and media analysis of the smart city rhetoric and citizens' involvement mechanisms in the period 2016–2019. The data studied included official documents (e.g., policy documents, decrees, and city strategies), newspaper articles, and national and city-wide websites. The documentary data of the project were moreover supported by collective and individual reflections on project-related seminars, workshops, conferences, and meetings with practitioners in Norway and Russia. Finally, we used data retrieved from interventionist research when the authors participated in ongoing smart city development projects in Norway and Russia. Such an approach resulted in interviews with public officials, interactive case-study work with students and practitioners, and observations of internal administration and public meetings regarding smart city-related projects.

Empirical findings

Smart city monologue in the Russian Arctic: top-down smartification and national smart city standard

Initially, before 2018, the smart city agenda in Russia developed chaotically, sustaining openness in interpretations and stimulating creativity at the local level, including Arctic cities (Aleksandrov et al., 2021). There is evidence that international smart city frameworks and ideas influenced Russian smart city development (UN Habitat and similarities to European smart city framework). Nineteen Arctic cities were involved in smart city initiatives in 2018, where neither state and cities nor companies or academia dominated the discourse. The key idea for smart city development was 'sharing knowledge and experience of city development' and creating a 'common sustainable future' for Arctic cities via dialogue between international, national, regional, and local parties with different ideas and visions as well as considering companies' and citizens' perspectives (Khodachek et al., 2022). Local Arctic actors represented by city administrations and universities emphasized the importance of democratic and human-centric initiatives in smart city development, enthusiasm, and creativity (Trunova et al., 2022).

However, documentary analysis demonstrates that after 2018, Russia's federal authorities—namely the Ministry of Construction and Development (hereinafter Ministry)—intervened and monopolized the discourse, sustaining top-down smart city bureaucratization all over the country, including Arctic cities. In this vein, in 2018 the Ministry issued a smart city roadmap stipulating five key principles for smart cities: the human dimension, urban infrastructure, utility management and urban planning, comfortable and safe urban environment, and economic efficiency. These smart city ideas were further included in the so-called 'Smart City Standard' in 2019. The Standard covered activities in eight areas: urban management, 'smart' utilities, innovations for the urban environment, 'smart' urban transport, intelligent public and environmental safety systems, communications network infrastructure, tourism, and services. By these means, Russia framed its national way of defining smart city objectives and elements (Khodachek et al., 2022), with implications for Arctic cities.

Concerning the different parties in the smart city dialogue on the Russian Arctic, academic actors and city management practitioners from before 2018 were joined by federal authorities and corporations. More specifically, in 2018, the federal authorities became active and dominant participants in the fragmented smart city agenda by launching mass smart city initiatives through the Roadmap and the Standard. Expecting federal funding for smart city initiatives and public-private partnership funding schemes, state-owned enterprises and large national companies then began developing digital solutions to city-specific problems (smart bus stops, smart lightning, e-government services, citizen engagement apps).

Interestingly, besides partnerships with state corporations, the Ministry reflected the 'voices of regions and local governments' as 'being heard in defining smartness' (www.russiasmartcity.ru). Nevertheless, as Khodachek et al. (2022) demonstrate, despite claiming 'a comprehensive and systematic approach to "smart" cities' development' (Minstroi Decree #38 2018, 1), the focus was on framing the importance of national priorities in smart city development, and therefore on downsizing the consideration and voices of citizens in the Arctic. In particular, when creating the Standard, the federal authorities set goals, performance indicators, funding, and an implementation plan for the whole country, including Arctic cities. Moreover, the Standard became a tool to convey the Ministry's vision to the local level since municipalities, including those in the Arctic, are not entirely part of the state and possess a certain autonomy from central authorities.

When it comes to the format and rules of the smart city dialogue, in parallel with consolidating power and interpretations into one national voice, the Ministry introduced the 'pilot smart city' initiative: selected municipalities were added to a secret list and granted access to direct communication with the Ministry. By 2019, 79 pilot cities in 47 regions had voluntarily joined the smart city project, including several Arctic cities (www.russiasmartcity.ru). The pilot cities committed to exceeding the basic requirements of Standard and to implementing additional elements. In turn, the Ministry promised methodological and financial support for the pilots. However, during the smart cities' implementation, the pilots faced issues as the Ministry insisted on 'detailed descriptions of activities', 'complete programs', and regular reports on the progress toward and changes in development plans (Project Office Report, 2019). Moreover, while the Standard implied there would be plenty of initiatives, there was no specific federal budget for their implementation at the local level. Instead, some activities could be financed through other Ministry subsidies (e.g., within the Comfort Urban Environment Programme). Curiously enough, these financial support mechanisms had been unexpectedly transformed into a control mechanism (Khodachek et al., 2022). Thus, even within the restricted pilot cities group, the dialogue was replaced by bureaucratic reporting and coercive financial levers, turning the smart city projects from bottom-up initiatives to top-down compliance for the Arctic cities.

Reflecting on the overall picture of smart city dialogue formation in the Russian Arctic, cities were forced into financial dependency and the smartification lost its inspirational quality with the projects facing issues of survival. Unexpectedly, instead of following local priorities, the municipalities became 'smart no matter what', with projects focusing on smart bus stops but ignoring potholes and setting up ICT solutions without activating them. In addition, citizens' involvement in the smart city agenda in Arctic cities became another mechanism of control through which upper-level authorities could monitor local authorities' performance (Khodachek et al., 2022). Hence, the national voice limited the creativity and autonomy of local actors in their interpretations of smart cities in the Arctic, forcing a technocratic vision to dominate instead.

Smart city dialogue in the Norwegian Arctic: between the European framework and local smart city visions

In 2019, nine cities and municipalities were engaged in smart city initiatives in the Norwegian Arctic, recognizing that the Arctic environment brings both opportunities and challenges for smart development (Dybtsyna and Aleksandrov, 2020). Dybtsyna and Aleksandrov (2020) demonstrate that diverse frameworks and agendas were set locally, including some to enhance city attractiveness to investments, new residents, and companies through smart city development, although the development in the Norwegian Arctic was dominated by technological/*information technology* (IT) development. For example, some cities claimed to develop Smart Arctic City as a platform to test Arctic-related technological solutions (e.g., Longyearbyen) and support smart technologies, buildings, and mobility (e.g., Alta, Harstad, Narvik). At the same time, some cities stressed that 'smart people' were an essential part of smart city development (Dybtsyna and Aleksandrov, 2020) and offered opportunities to discuss the smart city initiatives via dialogue (e.g., Bodø and Tromsø). Thus, it is evident that national and European frameworks guided many Norwegian smart city ideas, such as those concerning smart mobility, smart people, and more (Caragliu et al., 2011). The European discourse is also noticeable in frameworks such as the Design and Architecture Norway (DOGA) roadmap[1] and at smart city national conferences (e.g., https://nordicedge.org/).

The smart city dialogue in the Norwegian Arctic involved multiple local actors, such as city administrative bodies, businesses, and universities (e.g., Longyearbyen, Mo i Rana, and Narvik). In other cases (e.g., Vardø, Alta, Harstad, and Bodø), city and municipal actors actively sought to join professional networks for smart city development, nationally and internationally (e.g., the Nordic Network of Sustainable and Smart Cities). In many cases in the Norwegian Arctic, smart city development presented an opportunity to develop cities through local public-private partnerships and by stressing citizen involvement, with citizens as key actors (e.g., Bodø, Tromsø). In addition, national authority in the form of the Ministry of Local Government and Modernisation (MLGM) became central to smart city development across Norway, setting the smart city agenda via the ministerial report 'Smart cities and municipalities in Norway: mapping' (MLGM, 2019). Finally, the EU Commission's involvement is also evident in the development of Norwegian Arctic cities via research and innovation programs with related funding mechanisms such as Horizon 2020.

Considering different actors' power and the values of different voices in the Norwegian Arctic dialogue, it can be argued that despite strong local voice formations, there has still been a limited balance between local, national, and European actors' voices. At the national level, smart city development is supported by central actors via large projects and national funding. These raise the critical aspect of the national authority's dominance in setting the urban agenda with its own vision (e.g., the Bodø case of a smart city and airport development[2]). In addition to such a power aspect, the Norwegian MLGM also actively promotes local smart city initiatives' linkages to the SDGs (MLGM, 2019). For cities and municipalities, this raises a question about the immunity of smart city initiatives to influence from international (e.g., EU Commission) and national actors, which may compromise local legitimacy. Moreover, EU funding (e.g., Horizon 2020) often limits the chances for local Arctic voices to be heard under the EU agenda as strict reporting and funding routines may distract from the fulfillment of local needs.

Similar critical aspects apply to the format of dialogue across actors in the Norwegian Arctic. Particularly on the local level, most city initiatives recognize the importance of the involvement of citizens and non-governmental organizations (NGOs), but many lack efficient instruments to facilitate such engagement. The most common opportunities for participation are public meetings, breakfast seminars, surveys, and so on. These are mainly upheld as citizen participation is enshrined in the Norwegian Planning and Building Act that concerns urban planning and development (Dybtsyna and Aleksandrov, 2020). More advanced mechanisms of engagement also come into play when cities (e.g., Bodø) attempt to involve local actors in smart city dialogue formation between citizens, the city administration, businesses, academia, and NGOs, for example, via a so-called 'City lab'. Nevertheless, such mechanisms are still under development, and few cities have efficient instruments at their disposal with which to facilitate weighted and balanced decision-making for smart city developments (Dybtsyna and Aleksandrov, 2020) where expert knowledge dominates and is rightfully assigned most power at the local level. Thus, in many cases, instead of this, the smart city dialogue is organized by business development associations, which arrange for external experts to share their knowledge, reflecting the general agenda of the EU Program on Research and Innovation.

Therefore, reflecting on the status of smart initiatives' dialogue in the Norwegian Arctic, it can be seen that diverse actors represent many voices on local, national, and international levels. Yet those actors make a limited contribution to the dialogue since national and international agendas prevail. In these conditions, there is only partial consideration of voices from Arctic cities and the locals needs that they attempt to communicate. Hence, no matter how the smart initiatives are considered locally, there are still few dialogue forms to facilitate citizens' involvement in smart city development decisions in the Norwegian Arctic.

Discussion and conclusion

This chapter reported on smart city development in two Arctic countries as a follow-up to global trends (Mora and Deakin, 2019). Despite smart city promises of sustainability, certain warning signals regarding the lack of necessary dialogue formation to succeed with smart city ideas continue to be received (Bolívar, 2019; Kitchin et al., 2019). This chapter studied the issue in the Arctic context, exploring in particular how the smart city dialogue unfolds across governance actors in the Norwegian and Russian Arctic.

Our findings demonstrate that the formation of a smart city dialogue was problematic in the two countries (see Table 8.2 as the summary). Despite common global smart city ideals, the goals for smart city development, processes, and outcomes may differ significantly. In the Russian Arctic, state and corporate actors have overtaken the promise of a 'common sustainable future' and the initially open smart city dialogue in the Russian Arctic. Using their power and ability to dictate the rules of the game, they have silenced other (local) voices, transforming a symbolic smart city dialogue into a monologue of national authorities that fails to acknowledge voices from Arctic cities and the locals' needs communicated by them. Meanwhile, the Norwegian smart city dialogue started with the active involvement of national voices (as in the Russian case) along with European ones. While building on the European smart city framework was probably a significant initial driver of smart city implementation, those external voices tend to slightly dominate the smart city dialogue in the Norwegian Arctic (even if unintentionally) as a result of providing significant financial support. Hence, as in the case of state-controlled smart city development in the Russian Arctic, the initially more dialogue-inclined Norwegian efforts toward smart city dialogue formation have come to favor top-down development, and global and local Arctic voices are far from balanced (Brown, 2009; Makki, 2012).

Overall, reflecting on the role of dialogue in globalizing and localizing smart cities (Burns et al. 2021), our findings show that while national/international smart city agendas have a broad and overarching focus on sustainability (Miller et al., 2021), local Arctic actors' requirements and expectations of a smart city can be more concrete and complex (Dybtsyna and Aleksandrov, 2020). Ideally, the differences between the two should be reconciled through dialogue, where local voices are heard and considered by national and/or international smart city actors.

Table 8.2 Smart city dialogues in the Russian and Norwegian Arctic

Aspects of Smart City Dialogue	*Russian Arctic*	*Norwegian Arctic*
Topic(s) and goal(s)	From open promises of a 'common sustainable future' toward a national framework and smart city standard	Promises to enhance cities' attractiveness to investment, new residents, and companies but aligned with the European smart city framework
Actors involved	From domination by local actors (academia and city administrations) to domination by national actors and state corporations	Local actors (city administrations, universities, citizens, companies), regional authorities, national authorities, and the EU Commission
Power and balance of voices	National players (e.g., Ministry of Construction) dominate power and expert knowledge, leaving little space for other voices	Arguable balance between local, national, and EU voices with slight dominance of national and international agendas
Dialogue format and rules	Symbolic dialogue with much control and predetermined answers via pilot initiatives; funding restrictions formed top-down compliance	Limited dialogue with citizens; consistent prioritization of large national projects and EU funding
Outcome(s)	Monologue of national authorities, failing to acknowledge local voices and needs in Arctic cities	Limited dialogue, with national authorities and international parties setting the agenda and only partially considering voices from Arctic cities

However, a situation where smart cities faithfully serve local Arctic interests seems to be somewhat utopian (Grossi and Pianezzi, 2017). Despite the promise of dialogue for smart city development (Bolívar, 2019; Calzada, 2018; Vanolo, 2016), the Arctic cases illustrate the risk of a monologue taking over where the localizing trend is marginalized (Burns et al., 2021; Miller et al., 2021). In that sense, smart city development aligns with the Arctic paradox (Bourmistrov and Johansen, 2019; Holm, 2020) since, instead of preparing the ground for a fruitful dialogue seeking the best outcomes for local interests, monologues may end up led by actors from outside the region with different priorities in mind.

In such conditions, dominant global and/or national voices will hardly weaken or make room for local voices, even under sustainability pressures. To address this issue, particular attention in both countries should be directed toward reviewing existing governance traditions (Russian centralization approach vs. Norwegian bottom-up and network approach) and national/international players (Russian state corporations and smart city national vision vs. Norwegian public-private partnerships and EU smart city framing). Moreover, to restore the balance between globalizing and localizing trends (Burns et al., 2021; Miller et al., 2021), local actors should make their voices heard. In this regard, while the opportunities for bottom-up initiatives may be limited, collective Arctic voices can still increase their weighting via horizontal dialogue across the Arctic.

In particular, horizontal dialogue suggests that local actors may express themselves through various formal and informal organizations. For example, centers of competence[3] in Russia demonstrate that the smart cities' dialogue-friendly dimension may not be lost entirely, even under a top-down smart city monologue (Khodachek et al., 2022). In line with this, establishing a Nordic dialogue arena (e.g., via the Nordic Network of Sustainable and Smart Cities) could be a step toward strengthening local actors' voices and power across the Arctic. Moreover, opportunity for horizontal dialogue lies in international research and education cooperation projects and networks, with the capacity to integrate different voices of local Arctic actors, including academics, practitioners, and policymakers (e.g., UArctic Thematic Network on Smart Societies in the High North). Learning from each other's experiences through dialogue gives hope for the amplification of the Arctic smart city agenda. If the local voice gradually becomes stronger at national and global levels, shared understanding of the specific needs of Arctic citizens may be achieved via a true dialogue.

Implications for theory, practice, and the future of the Arctic

This chapter presented several important insights. Regarding its theoretical implications, this chapter contributes to knowledge formation on the role of dialogue in the global-local interpretation of smart cities (Burns et al., 2021; Miller et al., 2021). In particular, based on the example of smart city developments in the Arctic, we have shown that dialogue formation between global and local agendas is not a panacea for the smart city's bright future (Bolívar, 2019; Calzada, 2018; Vanolo, 2016). Instead, it is a matter of careful concern regarding power geometries and the balance of voices, where all sides should be heard (Brown, 2009). This is not always the case, as the Arctic smart city development illustrates. This has broader critical implications for understanding smart cities as the best solution for a sustainable urban future worldwide (Mora and Deakin, 2019) and in the Arctic (AMAP, 2017; Holm, 2020; Smith, 2011). Regarding the latter, this chapter has asserted that Arctic sustainable development lies in the careful interpretation of the smart city concept through horizontal dialogue formation across Arctic states. If such horizontal dialogue is lacking, smart cities risk diminishing sustainable development instead of supporting it (Tennberg et al., 2014).

When it comes to implications for practitioners and policymakers, we encourage these actors to be more critical toward challenges connected to smart city dialogue formation between global and local actors. As our two Arctic cases demonstrate, without the proper involvement of divergent actors and a critical attitude toward initial smart city promises and framing at different levels, smart city development can fall under the control of a monologue, even despite the rhetoric of a better urban future. When it comes to local actors, this chapter has asserted that even with a smart city monologue, there is hope for local players' cooperation and learning across borders via horizontal dialogue formation.

When it comes to implications for the future of Arctic development and sustainability, this chapter has stressed several points. First, in the resource-rich and climate-fragile Arctic region, the smart city presents an ambitious solution to tackle the sustainability agenda (Haarstad and Wathne, 2018) by creating safe, resilient, and sustainable cities (SDG 11). In particular, looking ahead, it is certain that global and/or national players will continue to dominate visions of what is smart for the Artic cities and communities. Second, despite such a negative trend, we are certain that Russian-Norwegian cooperation in education and research will strengthen the local capacity to embrace smart development and new urban technologies in general by addressing global and local concerns in a horizontal way between Arctic states (Wilson Rowe, 2018b). Hence, we predict that the smart city agenda will partially facilitate urban sustainability in the Arctic but will not reverse major contextual challenges like depopulation and the Arctic paradox.

December 2021

Ex-post reflections

If we rephrase well-known wisdom, a *dialogue* appears to be the first casualty in the overarching deterioration of the relationships between Russia and the West that we currently observe. That said, Russian-Norwegian cooperation in the Arctic has brought much value (with the potential to bring even more) so that there is a strong hope that both sides will make significant efforts to secure it. Russia and Norway will keep on sharing common challenges and opportunities in the Arctic and the High North. Thus, we hope that the dialogue will also survive. Nevertheless, this dialogue is unlikely to flourish at national or supranational levels after February 24, 2022. Rather, we foresee it to rest on peer-to-peer communications between entrepreneurs, tourists, and individual researchers when institutional educational and research cooperation will be inhibited in the immediate future. Nevertheless, the smartification of northern cities will continue in both countries, even though in different ways for now.

April 25, 2022

Notes

1 https://doga.no/globalassets/pdf/smartby-veikart-19x23cm-eng-v1_delt.pdf
2 https://www.nrk.no/nordland/skjalg-fjellheim-i-nordlys-er-kritisk-til-at-bodo-far-milliarder-til-ny-flyplass-og-ny-by-1.15402059
3 Special purpose regional government-funded NGOs responsible for citizens' engagement in comfort urban environment projects.

References

Aleksandrov, E., Dybtsyna, E., Nazarova, N., Khodachek, I., Middleton, A., Raspotnik, A. and Grossi, G. (2021). *Smart Cities for (Sustainability in) High North?: Mapping Initiatives in Norway, Russia, Finland, and North America.* The 25th Nordic Academy of Management Conference, Vaasa, Finland, 22–24 August.

AMAP (2017). Adaptation Actions for a Changing Arctic: Perspectives from the Barents Area. Arctic Monitoring and Assessment Programme, Oslo, Norway, pp. 267.

Bakhtin, M. M. (2010). *The Dialogic Imagination: Four Essays.* University of texas Press.

BIN (2019). [Online] *Maritime Traffic and Transportation Infrastructure along the Northern Sea Route.* Available at: https://businessindexnorth.com/reports/?Article=67 (Accessed April 25, 2022).

Bolívar, M. P. R. (2019). "Public Value, Governance Models and Co-Creation in Smart Cities", In Bolívar, M.P.R. (ed.) *Setting Foundations for the Creation of Public Value in Smart Cities.* Springer, pp. 271–280.

Bourmistrov, A. and Johansen, S. (2019). Governance in the High North: Rhetoric and Reality in the Barents Region, *Barents Studies*, Forthcoming.

Bourmistrov, A. and Mellemvik, F. (2002). Exploring Accounting and Democratic Governance: A Study Comparing a Norwegian and a Russian county, *Financial Accountability & Management*, 18(4), pp. 331–353.

Bourmistrov, A., Mellemvik, F., Bambulyak, A., Gudmestad, O., Overland, I. and Zolotukhin, A. (2015). *International Arctic Petroleum Cooperation: Barents Sea Scenarios, Routledge Studies in Environmental Policy Series.* Routledge.

Brown, J. (2009). Democracy, Sustainability and Dialogic Accounting Technologies: Taking Pluralism Seriously, *Critical Perspectives on Accounting*, 20(3), pp. 313–342.

Burns, R., Fast, V., Levenda, A. and Miller, B. (2021). Smart Cities: Between Worlding and Provincialising, *Urban Studies*, 58(3), pp. 461–470.

Calzada, I. (2018). (Smart) Citizens from Data Providers to Decision-Makers? The Case Study of Barcelona, *Sustainability*, 10(9), pp. 3252.

Caragliu, A., Del Bo, C. and Nijkamp, P. (2011). Smart Cities in Europe, *Journal of Urban Technology*, 18(2), pp. 65–82.

Dybtsyna, E. and Aleksandrov, E. (2020). Smarte Byer i Nordomradene: Samhandling Med Innbyggerne, *Magma*, 05/20, pp. 86–96.

Giffinger, R. and Pichler-Milanović, N. (2007). *Smart Cities: Ranking of European Medium-Sized Cities*. Centre of Regional Science, Vienna University of Technology.

Grossi, G., Meijer, A. and Sargiacomo, M. (2020). *A Public Management Perspective on Smart Cities: 'Urban Auditing' for Management, Governance and Accountability*. Taylor & Francis.

Grossi, G. and Pianezzi, D. (2017). Smart Cities: Utopia or Neoliberal Ideology?, *Cities*, 69, pp. 79–85.

Haarstad, H. and Wathne, M. W. (2018). "Smart Cities as Strategic Actors: Insights from EU Lighthouse Projects in Stavanger, Stockholm and Nottingham". In Karvonen, A., Cugurullo, F., and Caprotti, F. (eds.) *Inside Smart Cities*. Routledge, pp. 102–116.

Hollands, R. G. (2008). Will the Real Smart City Please Stand Up? Intelligent, Progressive or Entrepreneurial?, *City*, 12(3), pp. 303–320.

Holm, A. (2020). The Arctic Paradox, *High North News*. https://www.highnorthnews.com/en/arctic-paradox

Karvonen, A., Cugurullo, F. and Caprotti, F. (2018). *Inside Smart Cities: Place, Politics and Urban Innovation*. Routledge.

Khodachek, I., Aleksandrov, E., Nazarova, N., Grossi, G. and Bourmistrov, A. (2022). Smartocracy: Context entanglement of the smart city idea and bureaucracy in Russia. *Organization Studies*, OnlineFirst.

Khodachek I., Delva K. and Galustov K. (2020). Smart Cities in the High North: A Comparative Analysis of Arkhangelsk, Bodø, Murmansk and Tromsø. *Urban Studies and Practices*, 5(1), pp. 57–79.

Kitchin, R., Cardullo, P. and Di Feliciantonio, C. (2019). *Citizenship, Justice, and the Right to the Smart City*. Emerald Publishing Limited.

Makki, M. (2012). Evaluating Arctic Dialogue: A Case Study of Stakeholder Relations for Sustainable Oil and Gas Development, *Journal of Sustainable Development*, 5(3), pp. 34.

Marvin, S., Luque-Ayala, A. and McFarlane, C. (2015). *Smart Urbanism: Utopian Vision or False Dawn?* Routledge.

McMahon, R. and Raspotnik, A. (2019). Can Smart Societies Support Innovation in the Arctic?, *The Arctic Institute*. https://www.thearcticinstitute.org/smart-societies-innovation-arctic/

Miller, B., Ward, K., Burns, R., Fast, V. and Levenda, A. (2021). Worlding and Provincialising Smart Cities: From Individual Case Studies to a Global Comparative Research Agenda, *Urban Studies*, 58(3), pp. 655–673.

Mora, L. and Deakin, M. (2019). *Untangling Smart Cities: From Utopian Dreams to Innovation Systems for a Technology-Enabled Urban Sustainability*. Elsevier.

Rajala, T., Laihonen, H. and Haapala, P. (2018). Why Is Dialogue on Performance Challenging in the Public Sector?, *Measuring Business Excellence*, 22(2), pp. 117–129.

Rambøll Management Consulting. (2018). [Online] IT i praksis, *Smarte Og Bærekraftige Byer*. [IT in practice, Smart and sustainable cities]. Available at: https://no.ramboll.com/-/media/files/rno/publikasjoner/it-i-praksis-2018--smarte-og-brekraftige-byer.pdf?la=no (Accessed April 25, 2022).

Raspotnik, A., Grønning, R. and Herrmann, V. (2020). A Tale of Three Cities: The Concept of Smart Sustainable Cities for the Arctic, *Polar Geography*, 43(1), pp. 64–87.

Sköld, P., Baer, K. C., Scheepstra, A., Latola, K., Biebow, N. and Sköld, P. (2018). The SDGs and the Arctic: The Need for Polar Indicators, *Arctic Observing Summit*, 2018, pp. 2018-4-23.

Smith, L. (2011). *The New North: The World in 2050*. Profile Books.

Tennberg, M., Vola, J., Espiritu, A. A., Fors, B. S., Ejdemo, T., Riabova, L., Korchak, E., Tonkova, E. and Nosova, T. (2014). Neoliberal Governance, Sustainable Development and Local Communities in the Barents Region.

Trunova, O., Khodachek, I. and Khodachek, A. (2022). Visualising and calculating the smart city: a dialogue perspective, *Journal of Public Budgeting, Accounting & Financial Management*, ahead-of-print (ahead-of-print).

United Nations (2017). New Urban Agenda 2030, *Habitat III Secretariat*, A/RES/71/256.

Vanolo, A. (2014). Smartmentality: The Smart City as Disciplinary Strategy, *Urban Studies*, 51(5), pp. 883–898.

Vanolo, A. (2016). Is there Anybody Out There? The Place and Role of Citizens in Tomorrow's Smart Cities, *Futures*, 82, pp. 26–36.

Wilson Rowe, E. (2018a). *Non-state Actors and the Quest for Authority in Arctic Governance*. Manchester University Press.

Wilson Rowe, E. (2018b). *Arctic Governance: Power in Cross-Border Cooperation*. Manchester University Press.

Part III

International Cooperation in Science, Education and Culture

9 Arctic Climate Change – Perspectives on International Scientific Cooperation

Jan-Gunnar Winther, Larry Hinzman and Kim Holmén

Arctic climate change receives widespread attention and concerns (1990–2010)

Major campaigns like the International Geophysical Year in 1957 (also referred to as the Third International Polar Year [IPY]) called for international cooperation, but until the 1990s, systematically organized international scientific collaboration on Arctic climate change was limited. Important changes were occurring during that period. President Mikhail Gorbachev's memorable speech in Murmansk in 1987 largely shaped Arctic policy in the following years (Åtland, 2008). The changes in the Soviet Union eventually led to a substantial strengthening of Arctic scientific cooperation through the active involvement of Russian scientists. During those years, the evidence of climate change became stronger and stronger, and science reflected this by focusing increasingly on climate change research. The climate change agenda and more recently the focus on sustainability and teleconnections have resulted in a wide variety of research initiatives. The establishment of the International Arctic Science Committee (IASC) in 1990, six years after the founding of the Arctic Council (AC), marks a milestone (Rogne, Rachold, Hacquebord, and Corell, 2015).

The AC was founded in 1996 to promote and facilitate cooperation among the Arctic states, the Arctic Indigenous Peoples, and other Arctic inhabitants. Much of the scientific work accomplished by the AC is conducted within six working groups: the Arctic Contaminants Action Program, the Arctic Monitoring and Assessment Programme (AMAP), Conservation of Arctic Flora and Fauna, Emergency Prevention, Preparedness and Response, Protection of the Arctic Marine Environment, and the Sustainable Development Working Group. These working groups have been highly successful in coordinating international collaboration to conduct assessments on urgent issues. Task forces are also occasionally assembled to conduct projects or assessments of specific issues. The AC also worked with Indigenous Peoples throughout the circumpolar Arctic to establish representative groups that could promote their interests. These include the Aleut International Association, the Arctic Athabaskan Council, Gwich'in Council International, the Inuit Circumpolar Council, the Russian Association of Indigenous Peoples of the North, and the Saami Council. This recognition

DOI: 10.4324/9781003246015-12

of the need to engage indigenous groups has enabled better consideration of the rights of Arctic indigenous people.

The Intergovernmental Panel on Climate Change (IPCC) was established in 1988 by the World Meteorological Organization and the UN Environment Programme (UNEP). IPCC has proven to be much appreciated and highly influential as governments strive to develop climate policies that will reduce greenhouse gas emissions. The rigorous process used by the IPCC to develop critical assessments of possible climate change scenarios while identifying the risks and implications of response options provide local, regional, and national governments with the best scientific information available.

Although the IPCC published its first reports in 1990, 1995, and 2001 (Houghton, Jenkins, and Ephraums, 1990; Houghton et al., 1996; IPCC, 2001), a detailed analysis of the Arctic region was still lacking. In 2005, the AC published the Arctic Climate Impact Assessment (ACIA) (ACIA, 2005), which became the starting point for global attention to climate changes in the Arctic region. A timeline of the most important milestones in Arctic exploration and science is shown in Table 9.1.

Arctic Climate Impact Assessment

The 2005 ACIA was launched in response to a request from the ministers of the AC and was a follow-up to the preliminary evaluation of climate change included in the AC 1997–1998 AMAP (AMAP, 1997).

Table 9.1 Important milestones in Arctic exploration and science

982	Eric the Red first sailed North of the Arctic Circle and colonized Southern Greenland. Exploration and colonization
1818–1845	The British navy attempted to find the Northwest passage. Exploration and commerce
1868	Carl Koldewey, heading Germany's first Arctic expedition, sailed to Spitsbergen. Exploration and science
1878	Finnish-Swedish scientist Adolf Erik Nordenskiöld navigated the Northeast Passage. Exploration and science
1882–1883	First International Polar Year. Science
1903–1906	Roald Amundsen navigated the Northwest passage. Exploration, fame, and science
1909	Robert Peary and rival Frederick Cook claimed to have reached the North Pole. Adventure, fame, and science
1932–1933	Second International Polar Year. Science
1957–1958	International geophysical year (third International Polar Year). Science
1968	Ralph Plaisted reached the North Pole by snowmobile. Adventure
1990	International Arctic Science Committee founded. Science
1991	Arctic Monitoring and Assessment Programme (AMAP). Science
1996	Arctic Council established. Science and policy
2005	Arctic climate impact assessment published. Science
2007–2008	Fourth International Polar Year. Science

The objective of the ACIA, as defined in the AC Ministers' 'Barrow Declaration', was 'to evaluate and synthesize knowledge on climate variability and change and increased ultraviolet radiation, and support policy-making processes and the work of the IPCC'. ACIA was also instructed to address 'environmental, human health, social, cultural, and economic impacts and consequences, including policy recommendations'.

The assessment was produced by AMAP in collaboration with the AC's Conservation of Arctic Flora and Fauna working group and the IASC and was coordinated by AMAP. More than 250 scientists and six circumpolar indigenous people's organizations participated in the ACIA.

ACIA was the first comprehensive multidisciplinary assessment of the impacts of climate change in the Arctic. As such, it represents a baseline for later Arctic assessments.

ACIA was also a milestone, in that it included social science as well as natural science components. It was the first assessment of the impacts of climate change on socioeconomic conditions in the Arctic. The results of the ACIA were included in the IPCC fourth assessment (IPCC, 2007), and ACIA raised the profile of Arctic climate change issues in the UN Framework Convention on Climate Change and later IPCC efforts.

Other initiatives by the Arctic Council

The first thematic focus area of the AC was pollution. In 1997, *Arctic Pollution Issues: A State of the Arctic Environment Report* was published (AMAP, 1997). In the following years, AMAP published *Arctic Pollution 2002* (AMAP, 2002), *Persistent Toxic Substances, Food Security, and Indigenous Peoples of the Russian North* (AMAP, 2004), *Arctic Pollution* (AMAP, 2006), and *Arctic Pollution* (AMAP, 2009). Important knowledge about pathways originating from outside the Arctic was gained. Further, these analyses revealed a connection between nature and people that turned out to be valuable for climate studies undertaken in the following years. Thus, we argue that this work raised our awareness and understanding of the need for studies of the Earth system to address complex issues such as climate change. Also, it became obvious that a multidisciplinary approach was key, including cooperation between natural and social sciences, to gain deeper insights in complex issues. Here, the AC Sustainable Development Working Group plays an important role. The regional cooperation in the Arctic also served as inspiration for others, such as the Hindu Kush Himalayan Monitoring and Assessment Programme (Sharma, Molden, Wester, and Meher Shrestha, 2016).

In 2009, the next dedicated report on climate change by the AC was published (*Update on Selected Climate Issues of Concern* [AMAP, 2009]). Here, we interpret the following recommendation as an important recognition and amplification of the need for circum-Arctic cooperation: 'Integrate and expand monitoring efforts to enhance understanding of cause-effect relationships and temporal and spatial variability driving regional scale climate'. Monitoring is primarily done by national programs, and integrating such efforts requires collaboration between nations, institutions, and scientists.

The role of the IASC

The IASC (Rogne et al., 2015) was an idea that came to fruition in the late 1980s, some 100 years after the first IPY of 1882–1883 had demonstrated the powerful benefit of international collaboration in polar and especially Arctic science. Numerous multinational programs preceded the creation of the IASC, along with the subsiding of the Cold War (1988–1991) and the recognition of common concerns among Arctic rim nations. The IASC was formally established through international agreement on the Founding Articles in August 1990. Although our understanding of the Arctic region has significantly advanced in the past 30 years, we must recognize that our progress has been greatly facilitated and accelerated through formal agreements between national governments and through partnerships between individual researchers and scientific institutes.

The role of ICARP

The first International Conference on Arctic Research Planning (ICARP I; convened in 1995 and hosted by Dartmouth College in New Hampshire, USA) presented an important opportunity to assess the status of Arctic science and to build partnerships and create collaborations. More importantly, it was an opportunity to be visionary and think ahead toward a pan-Arctic science planning effort. The IASC emphasized the need to address critical issues that would benefit from multinational collaboration and those scientific challenges that required substantial investment in planning and organization. In the two years preceding ICARP I, the IASC Executive Committee developed a scientific agenda of four broad themes made up of ten specific topics. The goal was to agree upon a strategic approach to guide international cooperation over the next decade in conducting the research identified in the IASC Science Agenda. Additionally, several new ideas not proposed in the IASC agenda but of equal importance were launched during the conference. It is noteworthy that, similar to today's priorities, these concepts involve (1) integrating the contributions of the natural sciences and the social sciences to address matters of public concern, (2) communicating across distinct knowledge traditions, and (3) building mutually beneficial partnerships with Arctic residents.

ICARP II (convened in 2005 and hosted by the Danish Polar Center in Copenhagen) took a distinctly different approach by considering a more integrated and multidimensional perspective on the Arctic as a system. This perspective incorporated the human dimension, indigenous insights, and a broad consideration of Arctic processes in the Earth system. ICARP II differed markedly from ICARP I, in that the organizers made no attempt to establish a small set of IASC projects to be completed in the following years, but instead formulated science plans that were essentially visionary and more aspirational with far-ranging time frames. Planning for ICARP II both contributed to and benefited from the organization of ACIA (ACIA, 2005) and the 2007–2008 IPY. The ICARP II Steering Committee selected 12 topics where increased knowledge and concerted investments would markedly advance Arctic science. These 12 science plans were intended to promote and guide international cooperation through the next decade. In some cases, the results of these plans were presented as completed projects in ICARP III. The international collaboration set up in ICARP II launched several of the IPY programs as that planning framework required international teams working on regional and larger scales.

Arctic climate change enters the global scene

During the Conference of the Parties (COP) to the UN Framework Convention on Climate Change in 2009 (COP-15), Arctic climate change received prominent attention. Former Vice President Al Gore of the USA and Foreign Minister Jonas Gahr Støre of Norway initiated through the AC an assessment of climate change and the cryosphere. This assessment was presented at the COP meeting in Copenhagen. This work received global attention and was in many ways an update of (the) ACIA.

A few years earlier, in 2007, the UN Environmental Programme (UNEP, 2007) presented the Global Outlook for Snow and Ice at the World Environmental Day in Tromsø. Prominent world leaders participated in this event, including UNEP's Director Achim Steiner and the Chairperson of IPCC Rajendra Pachauri. Again, the world's attention was on the role of Arctic climate change. The same year, the first Arctic Frontiers conference was held in Tromsø. These events played an important role in raising awareness, strengthening funding possibilities, and developing stronger cooperation in Arctic climate change science. Also, the impact of science-policy interaction during these years should not be underestimated.

The nature of collaboration during the 'early days'

With rapidly increasing awareness and understanding of Arctic climate change and its impacts, there was a clear trend toward establishing coordinated efforts. A series of assessments, strengthening of monitoring, and organization of scientific initiatives through bodies such as AC and IASC marks a prominent development during this period.

Yet, the focus is biased toward natural science, a truly multidisciplinary approach is lacking, the pan-Arctic dimension is relatively weak, and the geographical focus is mainly on regional effects rather than global ones (IASC, 2020). However, we see this as a natural part of early development that will eventually give way to a more comprehensive and integrated approach in the next phase.

Consolidation, development, or radical change? (2010–2020)

In this section, we ask how the international scientific community continued its work in the decade after the spotlights were suddenly turned on the Arctic. We consider how the scientific community managed the great responsibility in producing relevant knowledge on climate change processes and for use in adaptation strategies. We ask whether the balances between natural and social science was optimal; indigenous and local versus scientific knowledge; disciplinary versus an earth system approach; and the interconnectivity between regional and global climate changes. Also, we elaborate on whether scientific data and recommendations shaped Arctic and global policies.

In one sense, the Arctic research community has benefited from its relatively small size. Communication of ideas, discoveries, and accomplishments spreads rapidly through highly integrated and generally collaborative channels. This somewhat unusual degree of collaboration is quite likely a product of necessity. *In situ* observations are quite sparse in the marine and terrestrial realms of the Arctic, in essence forcing an interdependence and reliance upon the collective community. The clear benefit of such close interrelationships is more efficient discovery and adoption of scientific achievements. The obvious drawback is the tough competition for limited resources. These partnerships among scientists have been elevated to the scientific organizations as quite often the same researchers are involved in multiple coordination mechanisms. This again yields benefits of reducing overlap through established pathways of communications.

AC – *SWIPA*

The Snow, Water, Ice, and Permafrost in the Arctic (SWIPA) (AMAP, 2011) assessment was the third AMAP assessment addressing Arctic climate issues and was a direct follow-up to the ACIA (ACIA, 2005). The SWIPA assessment was conducted by an international group of over 200 scientists, experts, and knowledgeable members of the Arctic indigenous communities. Lead authors were selected by an open nomination process coordinated by AMAP, the IASC, the WCRP/CliC (World Climate Research Programme/Climate and Cryosphere), and several national and international organizations.

The 2017 SWIPA assessment was an update of AMAP's 2011 SWIPA report (AMAP, 2017a). More than 90 scientists contributed to the 2017 assessment. The assessment mainly covered the period 2011–2015, with updates including observations from 2016 and early 2017. It concluded that the Arctic is shifting – rapidly and in unexpected ways – into a new state. Further, it suggested that if current trends continued, they would profoundly impact human health and safety, industries and economies, and ecosystems around the world.

The Special Report on the Ocean and Cryosphere in a Changing Climate (SROCC) in 2019 was a response to an IPCC Panel decision in 2016 to prepare three special reports during the sixth assessment cycle (IPCC, 2019). The SROCC addresses the entire cryosphere, but obviously the Arctic figure prominently. More than a hundred scientists from over 30 countries assessed the latest scientific knowledge about the physical science and impacts of climate change on ocean, coastal, polar, and mountain ecosystems, and the human communities that depend on them. SROCC declares that:

> All people on Earth depend directly or indirectly on the ocean and cryosphere. The fundamental roles of the ocean and cryosphere in the Earth system include the uptake and redistribution of anthropogenic carbon dioxide and heat by the ocean, as well as their crucial involvement in [sic] the hydrological cycle. The cryosphere also amplifies climate changes through snow, ice, and permafrost feedbacks. Services provided to people by the ocean and/or cryosphere include food and freshwater, renewable energy, health and well-being, cultural values, trade, and transport. Sustainable development is at risk from emerging and intensifying ocean and cryosphere changes. Ocean and cryosphere changes interact with each of the United Nations SDGs. Progress on climate action (SDG 13) would reduce risks to aspects of sustainable development that are fundamentally linked to the ocean and cryosphere and the services they provide. Progress on achieving the SDGs can contribute to reducing the exposure or vulnerabilities of people and communities to the risks of ocean and cryosphere change.

The SROCC both powerfully cements the importance of climate change research in the Arctic and clearly marks the onset of placing Arctic research in a Sustainable Development Goal (SDG) context (see section 3.1 and AMAP, 2017b).

ICARP III

ICARP III was hosted by the Science Council of Japan in 2015 in Toyama, Japan. It took a markedly different approach from its predecessors. Criteria were established to guide submissions and to evaluate proposed contributions to ICARP III; thus it was the research community that identified the most urgent needs and the way forward. While ICARP I and II were designed with significant IASC oversight and direction, ICARP III invested great effort in engaging the international community by including representation of sister organizations on the International Steering Committee. In terms of the goals of ICARP III, it was clear from the beginning not to undertake the development of new science plans, but rather to build on the numerous existing comprehensive science plans and to try to complement these by identifying gaps that might need attention. The ICARP III engagement period occurred over an extended time frame which enhanced the robustness of the contributions and the outcomes. ICARP III engaged the broader Arctic research community in identifying the most urgent research needs with descriptions of the best approaches to address those needs.

White House Arctic Science Ministerial (ASM)

In 2016, science ministers from 25 governments and the EU gathered at the White House to discuss Arctic research priorities and sign a Joint Statement on increased international collaboration on Arctic science and the inclusion of

Indigenous Peoples in understanding and responding to changes in the Arctic (The White House, 2016).

The four themes of the Ministerial and the Joint Statement were:

- Arctic Science Challenges and Their Regional and Global Implications
- Strengthening and Integrating Arctic Observations and Data Sharing
- Applying Expanded Scientific Understanding of the Arctic to Build Regional Resilience and to Shape Global Responses
- Empowering Citizens through Science, Technology, Engineering, and Mathematics (STEM) Education Leveraging Arctic Science

Among several commitments made at the meeting, the EU announced a new five-year project (2016–2021) coordinated by Norway to develop an Integrated Arctic Observing System. The US affirmed their support for 'EyesNorth', a US National Science Foundation research coordination network of community-based observation initiatives to strengthen community preparedness and response to environmental change. Further, support was given to the work of Sustaining Arctic Observing Networks – a joint effort of the AC (through its AMAP) and the IASC. Finland and the US decided to organize an international Arctic STEM Education Summit during the Finnish chairmanship of the AC (2017–2019).

By initiating the ASM, the US administration expanded the Arctic discussion. In addition, since science was the focal point, international cooperation on climate change became a natural cornerstone. The second Arctic Science Ministerial was co-convened by Finland and Germany in 2018 with a third planned for 2021, cohosted by Iceland and Japan (postponed from 2020 due to COVID-19).

MOSAiC (2019–2020)

The MOSAiC (Multidisciplinary drifting Observatory for the Study of Arctic Climate) expedition builds on a tradition of drifting with the Arctic sea ice. The practice was begun by Fridtjof Nansen with the *Fram* expedition, and for a long time was carried out by the Soviet and Russian drifting ice stations, thereafter the Surface Heat Budget of the Arctic Ocean study and the N-ICE2015 expedition.

N-ICE2015 (January to June 2015) was planned under the auspices of the Norwegian Polar Institute utilizing the research vessel *Lance* and involving 69 scientists from ten countries. N-ICE2015 was focused on the influence of climate change on the sea ice. A defining theme emerged: The Arctic sea ice has changed rapidly and dramatically in recent years, but our understanding is still based largely on observations made when thicker and older ice dominated the icescape in the Arctic Ocean. The subsequent MOSAiC expedition benefited greatly from the experiences shared from the N-ICE2015 expedition.

That MOSAiC expedition, with the German research icebreaker *Polarstern*, was the largest polar expedition ever organized (IASC, 2016). Throughout the year of the expedition, 442 experts from 20 countries traveled to the Arctic and were rotated in phases. Many more scientists will be involved in the processing of the data gathered. The budget was €140 million, of which 50% was covered by the 'Bundesministerium für Bildung und Forschung'. *Polarstern* was supported by five additional icebreakers (Russian icebreakers *Akademik Federov* and *Kapitan Dranitsyn*, German research vessels *RV Sonne* and *RV Maria S. Merian*, and the Russian research icebreaker *Akademik Tryoshnikov*). Sweden and China were originally planned to support the expedition with vessels, but the COVID-19 pandemic forced a reduction in international logistic support and a rescheduling of the entire expedition, including the elimination of one personal rotation.

The *Polarstern* set sail from Tromsø, Norway, in September 2019 and returned to its home port of Bremerhaven, Germany, in October 2020, having drifted with the ice across the Arctic Basin. The science program was interdisciplinary, but the defining theme was to take the closest look ever at the Arctic as the epicenter of global warming and to gain fundamental insights key to better understanding global climate change. The expedition was under the auspices of the Alfred Wegener Institute, Helmholtz Centre for Polar and Marine Research. The climate change theme and the attractive platform drew scientists from around the world; the results of MOSAiC will be forthcoming for many years. Despite being hampered by COVID-19, this expedition is clearly already a scientific and international teamwork success of unprecedented acclaim.

What characterizes this period?

During the decade 2010–2020, international cooperation on Arctic climate change evolved in many ways. Increased political and global relevance was prominent during this period. In 2013, the first Arctic Circle Conference was held in Reykjavik, Iceland. The World Meteorological Organization also addressed Arctic climate through initiatives such as Global Atmospheric Watch and Global Cryosphere Watch.

In addition, strengthening the social dimension and its integration into the sustainability agenda stands out as a feature, although there is the potential for further refinement (Tennberg, Lempinen, and Pirnes, 2019). Finally, the recognition of an Earth system approach continued to evolve, materializing in stronger multidisciplinary programs and cooperation

The way forward (today until 2035)

Where are we heading after three decades of organized international cooperation on Arctic climate change? A starting point should be lessons learned, today's situation, and future societal needs. In this context, it seems natural to discuss how the UN SDGs could shape the Arctic climate agenda.

Although the focus of this chapter is on *scientific* collaboration, we will elaborate on the connection between science and business. Globally, economic growth is a major driver for change, and the Arctic is no exception. However, we ask whether businesses in the Arctic could aim at the highest standards of sustainability and thus serve the interests of the region as well as being an inspiration for solutions and collaborative approaches adopted in other parts of the world.

Finally, we outline a few scenarios for what the world might look like in 2035, including the most likely ones and those the authors most desire.

The impact of the SDGs on Arctic climate change cooperation

Research in the Arctic has gravitated toward environmental issues, and in recent decades toward climate change issues. In the AC 2017 Agreement on Enhancing International Arctic Scientific Cooperation, the only specific research area explicitly mentioned was climate change. The sincerity of the international communities' engagement in Arctic research is customarily built on arguments of how Arctic change influences their nations of origin. Statements conveying the message that 'what *happens in the Arctic does not stay in the Arctic*" as well as the fact that most of the change in the **Arctic** originates from actions outside the region are frequently voiced. These are generally correct assertions and particularly true for climate change.

Arctic climate change research is dominated by detection, attribution, and the effects of climate change from a natural science perspective. With the increase in the general scientific understanding, there has been a trend from exploratory and process discoveries at single locations and in single scientific fields toward more comprehensive regional and system studies. This has markedly influenced international cooperation – since the complexity and extent of such studies are well beyond what a single scientist, institution, or nation can undertake alone.

The SDGs have become an important guiding principle for what the knowledge society is asking the scientific community to create. Research must continue to investigate climate change in the Arctic, but it must also address how the full set of SDGs are influenced by human interaction with and in the Arctic system. The SDGs thus complicate the scope of study, but will yield a fuller understanding of the Arctic providing enhanced direct meaning of the science for the peoples of the Arctic.

The SDGs also support the development of resilient ecosystems, economies, and societies that can prevent the most serious effects of shocks like the recent COVID-19 pandemic. We argue that building resilience through a planetary approach is an effective protection against unforeseeable situations (Steffen et al., 2015). It is too early to give an insightful and quantitative evaluation of the effects of the pandemic. Even so, there is reason to believe that the long-term effects of climate change will unfortunately be much more serious.

The Arctic is home to a vast variety of human circumstances and experiences. Many populated areas of the Arctic have high scores on the human development index (Larsen and Fondahl, 2020). The social sciences and indigenous knowledge need to be given even stronger influence on the planning and priorities of international research projects to ensure that the knowledge sought is meaningful for reaching all SDGs within the entire Arctic. The Arctic is rich in SDG important non-transactional values and assets (e.g., climate, wilderness, biodiversity, and culture). At the same time, it is abundantly endowed with commodities with transactional potential (e.g., fossil fuels, minerals, and marine living resources). Transactional entities are often easily quantifiable, but the non-transactional ones require enduring agreements to be quantified predictably.

The unique mix of bountiful assets of all kinds should be converted into opportunities to develop the Arctic without overexploiting nature. This will require new thinking and approaches that can build on the Arctic climate research experiences. Pooling our intellectual resources, knowledge innovation, and existing knowledge to address these issues is of paramount importance.

Can the Arctic become a global driver for sustainable businesses?

It is generally accepted that the Arctic environment is vulnerable and that special care needs to be taken when developing business in this region. In this sense, the public expects businesses to develop sustainable solutions for the Arctic.

On the one hand, some argue that activities should be low or even halted, for example, little or no oil and gas extraction activities. On the other hand, there is a need for value creation to sustain and develop societies which are attractive. A key to reconciling these two views is to develop innovative and sustainable industries. Here, the Arctic can serve as a 'workshop' where the highest standards for solutions and operations can be achieved. This would not only be good for the region, but may also imply the export of know-how and thereby drive a sustainable development globally. The notion 'what works in the Arctic works everywhere' could become a relevant argument to promote Arctic solutions to the rest of the world.

Paradoxically, climate change increases accessibility and potential for the industries that caused warming of the Arctic in the first place. That said, governments around the world know about the negative domestic effects of Arctic climate change. In India, for example, authorities are concerned about changes in the monsoon, China worries about outbreaks of cold air from the Arctic, and Singapore is preparing its infrastructure for severe sea level rise. These examples suggest that countries outside the Arctic region will also work toward limiting the consequences of climate change in the Arctic.

Thus, we argue there is a common understanding of the widespread effects of Arctic climate change and that this is a forceful driver for developing sustainable, climate-friendly business solutions.

We vehemently argue that development of the Arctic entails striking the right balance between protection and production. With political will, knowledge-based management, and strong international cooperation as set out in SDG 17 (partnership in the goals), this is achievable. The world needs the rich resources of the Arctic (e.g., seafood and energy). At the same time, it is crucial that exploitation be environmentally friendly and have eternity as its timeline. Short-term profit must concede to long-term sustainability. SDG 16 (peace, justice, and strong institutions; e.g., AC) are well represented in the Arctic. The Arctic must consider all SDGs simultaneously, but it has attained a strong position and should continue to be a leading region showing the way forward into the SDG era. The special environment combined with the high environmental standards required to salvage the unique nature and ecosystem should serve as an inspiration for solutions to create a balanced development through international cooperation around the goals of SDG 7 (industry, innovation, infrastructure). The protection demands can thus be converted into an incentive and competitive advantage worldwide rather than being conveyed as factors hampering opportunities for development in the Arctic.

Likely scenarios for 2035

The real value of closely examining our history is that it provides useful insight into our future. The information presented in this retrospective account demonstrates that our scientific efforts and our achievements have grown gradually over the centuries, dramatically and continually accelerating in recent decades. Similarly, we may note that the Arctic environment was remarkably stable in the first eight millennia of the Anthropocene. However, in the last century, the Arctic has experienced consequential changes in environmental conditions, economy, demographics, and connectivity to the global system. How then does this understanding of our past shed light on our future? While we cannot accurately forecast an endpoint, we can clearly see trends and, barring some unexpected catastrophe (e.g., Covid-19) or some rational change in human behavior, it is quite reasonable to expect those trends to continue.

The Arctic and the planet will continue to warm. Permafrost will degrade. Sea ice, glaciers, and ice sheets will melt. Plant and animal species will migrate northward or become extinct if their habitat ceases to exist. These are unfortunately easy projections to envision. The more uncertain future concerns how people and societies will respond and acclimatize to these changes. Humans are remarkably adaptable, living in every environment found on Earth, creating engineered solutions to protect us from the heat, the cold, the wet, the drought, and the unstable. What scenarios should we expect and prepare to address?

Due to the predictability of Earth's orbit around the sun, we know that the Arctic will always have winter, although it will indeed be less harsh. Summers in the Arctic have for millennia been well within the comfort zone of humans, but we should expect warmer conditions, with more frequent and more intense extreme events. The geothermal heat flux from our planet's interior will remain about 50 mW/m2 (milliWatt per square meter). When the winters warm to the point that heat no longer escapes through the surface to space, the permafrost will slowly but inevitably thaw. This point will mark the beginning of vast spatial changes in our ecosystems and our social infrastructure. We can design engineered structures that capture heat energy in the permafrost and release it into the air, effectively maintaining frozen ground below high-value structures. But we cannot protect whole ecosystems, long coastlines, or wildlife dependent upon habitat that ceases to exist. While people and societies may protect our buildings, airports, bridges, and even roads, we cannot halt the degradation of the environment around us. At sea, we will experience similar fundamental changes such as ocean warming, ocean acidification, sea level rise, ecosystem alterations, and sea ice melt.

These are the trends in environmental conditions, and they have already dramatically changed the Arctic as we have known it. While the Arctic will not be undergoing severe environmental degradation and environmental disaster 15 years from now, it will continue upon this trajectory, and evidence of that fate will become ever more apparent.

Desired scenarios for 2035

The obvious scenario that we would all like to see in the Arctic region is an area of peace, with healthy communities and individuals, productive and sustainable economies, and a place where environmental change is slow or stalled. World peace, while the dominant force determining national and economic stability is beyond the scope of this contribution. There are, however, other forces that we can harness to produce actions that will yield healthy and sustainable communities. Here, the representatives of the Arctic, in the framework of international collaboration within the AC, play an important role. They should promote and adopt international standards of ecosystem- and knowledge-based management strategies that incorporate predictions of climate and habitat changes to ensure sustainable populations of fish, birds, and other wildlife are maintained.

While it will be very difficult for polar communities to become carbon neutral, there are many existing technologies that can markedly reduce energy consumption. Improved building techniques such as superinsulated homes with adequate vapor management can reduce energy consumption and greatly increase the lifespan of structures. Road and airport construction incorporating natural passive cooling can slow or stop permafrost degradation, greatly reducing maintenance and replacement costs.

Coastal communities built upon barrier islands or other materials susceptible to erosion must initiate planning efforts to relocate. While there are many factors that influence such decisions, perhaps chief among them are cost and reasonable relocation sites. Planning, construction, and complete execution of village relocation may well be a multi-decadal exercise. Advanced planning is essential to ensure community acceptance, adequate financing, and appropriate site selection.

Globally, climate change will drive the need for resources like food, energy, minerals, transportation, and recreation. The ocean areas of the Arctic may provide many of these. The key, as emphasized earlier, is a sustainable development where we ensure a healthy ocean environment to sustain a productive one.

The works of the IPCC and ACIA document disturbing climatic trends. Politicians and other stakeholders must make rational decisions and visionary plans based upon such insights. Adapting to a rapidly changing environment will be a challenge. Research to address the most urgent needs can help us make informed decisions. An international effort such as an IPY in 2033, dedicated to adaptation and mitigation science, can yield great insight to helping our communities thrive and make more effective use of our limited financial and natural resources.

Conclusions

Arctic science has increased its global relevance due to strong teleconnections between the Arctic region and areas outside. Along with this, the AC has strengthened its role as a policy-shaping body. Further, the AC has expanded its thematic scope and the number of observers that actively contribute to knowledge about the region. These factors combined have resulted in an increasing scientific collaboration in the Arctic.

Today, the Arctic is changing very fast. Climate projections indicate that the region will be seriously affected and fundamentally changed. Hence, we expect the Arctic to play a vital role in the development of the UN SDGs. The overarching goal would be to develop an Arctic region of peace, with healthy communities and individuals, productive and sustainable economies, and a place where environmental change is slow or stalled. The Arctic as a model for sustainable solutions that can be exported elsewhere is an attractive perspective.

To meet the challenges of tomorrow, we need to strengthen the scientific understanding of the Arctic. Thus, we call for the establishment of the next IPY in 2033.

December 2021

Ex-post reflections

International collaboration in the Arctic is key to tackling the ongoing environmental changes we have been observing and quite likely to be even increasing in rate. Today, collaboration with Russia is on hold, hampering relations between scientists and cooperation on data access and sharing. Disturbingly, this will limit our contributions to combating the climate crises. Reliable policy recommendations require updated information from the entire Arctic region.

Given the unpredictable situation, it is currently impossible to judge whether this will be a temporary setback or have more long-lasting, serious consequences.

April 21, 2022

References

ACIA (2005). *Arctic Climate Impact Assessment. ACIA Overview Report.* Cambridge University Press, p.1020.

AMAP (1997). *Arctic Pollution Issues: A State of the Arctic Environment Report.* Oslo, Norway: Arctic Monitoring and Assessment Programme (AMAP).

AMAP (2002). *Arctic Pollution 2002: Persistent Organic Pollutants, Heavy Metals, Radioactivity, Human Health, Changing Pathways.* Oslo, Norway: Arctic Monitoring and Assessment Programme (AMAP).

AMAP (2004). *Persistent Toxic Substances, Food Security and Indigenous Peoples of the Russian North. Final Report.* Oslo, Norway: Arctic Monitoring and Assessment Programme (AMAP), p. 192. AMAP Report 2004:2.

AMAP (2006). *Arctic Pollution 2006.* Oslo, Norway: Arctic Monitoring and Assessment Programme (AMAP).

AMAP (2009). *AMAP 2009 Update on Selected Climate Issues of Concern* (Observations, Short-Lived Climate Forcers, Arctic Carbon Cycle, Predictive Capability). Oslo, Norway: Arctic Monitoring and Assessment Programme (AMAP). ISBN: 978-82-7971-049-3.

AMAP (2011). *Snow, Water, Ice and Permafrost in the Arctic (SWIPA): Climate Change and the Cryosphere.* Oslo, Norway: Arctic Monitoring and Assessment Programme (AMAP).

AMAP (2017a). *Snow, Water, Ice and Permafrost in the Arctic (SWIPA) 2017.* Oslo, Norway: Arctic Monitoring and Assessment Programme (AMAP).

AMAP (2017b). *Adaptation Actions for a Changing Arctic: Perspectives from the Barents Area.* Oslo, Norway: Arctic Monitoring and Assessment Programme (AMAP).

Åtland, K. (2008). Mikhail Gorbachev, the Murmansk Initiative, and the Desecuritization of Interstate: Relations in the Arctic. *Cooperation and Conflict*, 43(3), 289–311. DOI: 10.1177/0010836708092838.

Houghton, J.T., Jenkins, G.J. and Ephraums, J.J. (1990). *IPCC First Assessment Report.* Geneva: WMO.

Houghton, J.T., Meira Filho, L.G., Callander, B.A., Harris, N., Kattenberg, N.A. and Maskell, K. (1996). *Climate Change 1995: The Science of Climate Change, Contribution of Working Group I to the Second Assessment Report of the Intergovernmental Panel on Climate Change.* New York: Cambridge University Press. ISBN 0-521-56433-6 (pb: 0-521-56436-0).

IASC (2016). MOSAiC – *Multidisciplinary Drifting Observatory for the Study of Arctic Climate Science Plan.* Accessed from: https://mosaic-expedition.org/wp-content/uploads/2020/12/mosaic_scienceplan.pdf. Accessed: 25 August 2021.

IASC (2020). *State of the Arctic Science Report.* Accessed from: https://iasc.info/about/publications-documents/state-of-arctic-science. Accessed: 13 August 2021.

IPCC (2001). "Climate Change 2001: The Scientific Basis". In: Houghton, J.T., Ding, Y., Griggs, D.J., Noguer, M., van der Linden, P.J., Dai, X., Maskell, K. and Johnsen, C.A. (eds.) *Contribution of Working Group I to the Third Assessment Report of the Intergovernmental Panel on Climate Change.* Cambridge and New York, NY: Cambridge University Press, 881 pp.

IPCC (2007). "Climate Change 2007: The Physical Science Basis". In: Solomon, S., Qin, D., Manning, M., Marquis, M., Averyt, K., Tignor, M.M.B., Miller, H.L. and Chen, Z. (eds.) *Contribution of Working Group I to the Fourth Assessment Report of the Intergovernmental Panel on Climate Change.* Cambridge and New York, NY: Cambridge University Press, 996 pp.

IPCC (2019). "Summary for Policymakers". In: Pörtner, H.-O., Roberts, D.C., Masson-Delmotte, V., Zhai, P., Tignor, M., Poloczanska, E., Mintenbeck, K., Alegría, A., Nicolai, M., Okem, A., Petzold, J., Rama, B. and Weyer, N.M. (eds.) *IPCC Special Report on the Ocean and Cryosphere in a Changing Climate.* Cambridge and New York, NY: Cambridge University Press, 755 pp.

Larsen, J.N. and Fondahl, G. (2020). *Arctic Human Development Report.* DOI: 10.6027/TN2014-567.

Rogne, O., Rachold, V., Hacquebord, L. and Corell, R. (2015). *IASC After 25 Years.* ISBN: 978-3-9813637-7-7, ISSN: 1654–7594, International Arctic Science Committee.

Sharma, E., Molden, D., Wester, P. and Meher Shrestha, R. (2016). The Hindu Kush Himalayan Monitoring and Assessment Programme: Action to Sustain a Global Asset. *Mountain Research and Development*, 36(2), 236–239. https://doi.org/10.1659/MRD-JOURNAL-D-16-00061.1.

Steffen, W., Richardson, K., Rockström, J., Cornell, S. E., Fetzer, I., Bennett, E. M. and Sörlin, S. (2015). Planetary Boundaries: Guiding Human Development on a Changing Planet. *Science*, 347(6223). DOI: 10.1126/science.1259855.

Tennberg, M., Lempinen, H. and Pirnes, S. (2019). *Resources, Social and Cultural Sustainabilities in the Arctic* (1st ed.). London and New York: Routledge, 204 pp.

The White House (2016). Accessed from: https://obamawhitehouse.archives.gov/the-press-office/2016/09/28/fact-sheet-united-states-hosts-first-ever-arctic-science-ministerial. Accessed: 01 September 2021.

UNEP (2007). *Global Outlook for Snow and Ice*. Accessed from: http://new.unep.org/publications/search/pub_details_s.asp?ID=3945. Accessed: 05 August 2021.

10 Internationalization of Higher Education in the High North

Purposes and Strategies

Marit Sundet, Leif-Kristian Monsen and Elena Dybtsyna

Introduction

Knowledge and competence will always be the most critical resources in every society. In a complex world characterized by change, development, and a continuous supply of new insights, the ability of states to educate their populations is becoming increasingly crucial. However, no state is self-sufficient in terms of competence, and therefore they are dependent on knowledge cooperation and exchange between countries. In particular, it is essential when talking about increasing the degree of internationalization in the form of closer cooperation on higher education (Sundet, Forstorp and Ørtenblad, 2017). But what actually characterizes internationalization, what different purposes does it have and what strategic measures do higher education institutions (hereinafter abbreviated to HEIs) use when international education programs are to be established? Does it have any peculiarities regarding internationalization in the regions of strategic (both political and economic) importance?

The professional discourse around the internationalization of higher education has considerable ambitions to explain global movements and processes, contributing to theories and models becoming dominant in the representations of the phenomenon. The following is a definition of internationalization provided by de Wit et al. (2015, p. 29):

> The intentional process of integrating an international, intercultural or global dimension into the purpose, functions, and delivery of post-secondary education, in order to enhance the quality of education and research for all students and staff, and *to make a meaningful contribution to society*.
>
> (our emphases)

However, suppose we are to find out what characterizes internationalization, its purposes, and what strategies are used. In that case, we must focus on the individual HEI and the actors involved. In other words, we intend to provide empirical descriptions and context-specific characteristics of what the actors in a selection of HEIs present as the internationalization of higher education at their institutions and faculties.

DOI: 10.4324/9781003246015-13

This approach is based on the simple recognition of purposes and strategies and will always reflect differences between nations in culture, social structure, economic conditions, and jurisdiction. At the same time, our studies of HEIs in different countries provide opportunities to find out what meaning the content has, what kind of trust, and what opportunities and possibilities characterize the processes that establish and maintain international collaborative projects across national borders and between HEIs.

For the purposes of this chapter, we use the definition above of the internationalization of higher education phenomenon (see de Wit et al., 2015) to frame our empirical findings and to explain the purposes and strategies of international education programs in HEIs in the High North.[1]

This chapter has three main parts. The next part presents a brief description of the studies and data collected at the HEIs in the High North. It reviews the most important findings from a study of which strategies generally seem to characterize the internationalization of higher education at four HEIs in the High North. The second part is based on a study of a master's program run through a collaborative project between three HEIs in Russia, China, and Norway. As will be shown, these two studies point to different purposes and strategies, which will be discussed in more detail in the third and final part of the chapter. It will also highlight which criteria seem appropriate to include as preconditions for future projections on what the internationalization of higher education in the High North means for international Arctic cooperation.

Data and methods

Our chapter is based on two studies. One was conducted using personal interviews with key actors related to international education programs at four different HEIs in different ways. These HEIs are NArFU (the Northern [Arctic] Federal University in Arkhangelsk, Russia); the University of Oulu in Finland; the Luleå University of Technology in Sweden; and Nord University in Bodø, Norway. Four to six informants were interviewed at each HEI, in total 18 persons, of whom nine were female and nine male. Half of these are academic staff holding positions as prorectors of education and deans at faculties of technology, natural sciences, economics, and social sciences. The other half of the informers have administrative positions as heads of the HEIs' international offices or other key roles there. The interviews were conducted in 2019 and 2020.

The second part of this chapter builds on a single case study of a partnership program developed by three HEIs: Nord University in Bodø, Norway; MGIMO University in Moscow, Russia, and ECNU (East China Normal University) in Shanghai, China. The study uses secondary data, such as curriculum, program, and course descriptions, analytical reports, and previous research. The single case study was chosen since it enables us to go in-depth into the internationalization process between partner institutions to analyze how the cooperation has been going on and what is essential for such international programs to succeed. The educational program is organized in four semesters. Students enrolled in the program follow mandatory courses at their "home" institutions during the first semester. During the second semester, students study together in Norway. In the third semester, students are located in Russia. During the fourth and final semesters, students stay in China, working on their master's theses.

Internationalization as emergent strategies

Regarding the first study, the legislation in the four countries where our selected HEIs are located is not uniform regarding how and to what extent internationalization of higher education is mentioned and regulated. The HEIs face different imperatives, and these imperatives are in many cases also characterized by their being so-called political control signals. In other contexts, they are absolute; for instance, Norwegian master's programs will not be accredited unless they include facilitated opportunities for long-term student stays at HEIs outside Scandinavia. However, none of our informants claim that their HEI is subject to strict state regulations when it comes to how to shape strategies for internationalization. At all HEIs, the informants refer to "national ideas and guidelines", albeit without being able to immediately elaborate on the contents any further or specifying what limitations these may impose on what internationalization may be about. Most of the informants stated that developing expedient strategies for their own goals, ambitions, opportunities, resources, and capacities is mainly up to the management of their respective institutions.

However, the informants were likewise unable to provide us with clear answers as to what strategies have been developed at their HEI's. At NArFU, reference was made to goals expressing an expectation that internationalization is a high priority in all academic areas. In addition, informants described how all new agreements on international cooperation must follow certain procedures for establishing such. It is mainly maintained by their international offices. In Oulu, reference was made to the fact that government will change the funding system for HEIs from 2021 and financial support for student exchange will be withdrawn. Partly because of this, existing plans for internationalization have not been sufficiently followed up. New ideas for what measures to take and what efforts to initiate were to be formed over time. In Luleå, an internationalization strategy was developed in 2012. However, according to our informants, this was never followed up, partly because it was not sufficiently specific and partially because it was soon outdated and no longer relevant in terms of how the international student market developed. The same was the case at Nord University, where the "Strategy 2020" document devotes only a minimum of attention to quite a few airy pages dealing with internationalization. Wherever the term is mentioned, there is a reference to good intentions, yet issues such as organizing, resources, priorities, and actions are not included in the document. Another common feature of these HEIs is the informants' conviction that a separate strategy for internationalization will soon cease to be desirable, but instead that it will be integrated into an action plan covering all activities that the respective institutions want to emphasize and develop further.

Despite the lack of overall concrete strategies, the informants at all HEIs stressed conspicuously that it was nevertheless the faculties' responsibility to develop individual and targeted strategies for what internationalization should be about. The two most important arguments for this were that each academic community would know best which foreign HEIs are attractive and amenable to establishing cooperation and that they thus have to develop plans for how and with whom to contact to achieve long-term agreements with themselves. Second, it was argued that learning and experience from the faculties' various forms of cooperation with foreign HEIs over time would also guide how new agreements were shaped. When asked if this had a conserving effect that might reduce the faculties' potential range of actions, the answer was that praxis would clarify what is functional in cooperation and that the strategic element lies in doing what all experience indicates would be expedient. In this way, criteria for success in establishing cooperation projects will form a pattern for shaping the next project.

In this context, there was also an underlying way of reasoning that might explain the lack of overall strategies: because practices at the various faculties are so dissimilar and diverse, an internationalization strategy could only ever serve as a guide. Such a strategy would be able to only a limited extent to regulate and aim toward actions. However, it could provide guidelines and, first and foremost, communicate ambitions without being too specific. It would also be hard to implement in other ways than allowing the aims of the strategy to be interpreted by those with insight into the various international arenas in their respective academic fields who know their opportunities and who can best measure their chances of success. Thus, the strategy is formed along the way. It is a well-known phenomenon and consistent with what Mintzberg and Waters (1985) describe as "emergent strategies" and refers to strategic choices emerging through espousing whatever appears or emerges through practices.

So far, we have argued that internationalization appears as a mandatory social mission lacking an authoritative definition and justification to guide the HEIs' actions. The governments' vague expectations thus must be interpreted by each HEI. In this interpretation, they are given content and meaning adjusted to the HEIs' established activities, capacities, ambitions, and other institutional characteristics and features, imposing conditions on what is desirable, possible, and necessary. This reasoning is in accordance with theories about how concepts are translated, transformed, and adjusted to a local context on their way into organizations (Røvik, 2011, 2016). It goes a long way in explaining why the rationale for and ideas of internationalization appear to be shaped according to what practices are established over time at the various faculties at the respective HEIs. It is also reflected in the responses received to the question about the reasons for engaging in cooperation with foreign universities. The responses often differ, and while some are simple and rational, others are more complex and often connected to values and feelings. These responses are gathered into seven non-mutually exclusive or ranked categories to avoid going into excessive detail.

The first refers to the dissemination and exchange of research-based knowledge. Another category of reasons starts from the point of a need to benchmark and imitate foreign HEIs' activities and ways of working. A third and closely related category has to do with international ranking. The fourth category of reasons for international cooperation refers to exchange students as an essential factor in further developing programs and teaching.

The fifth category is about how academic identity, in contrast to national identity, knows no borders and where the feeling of community grows stronger the narrower the mutual academic interests and research areas are. The reasons in category six are less related to academic areas and programs as such and more focused on internationalization as a form of people-to-people cooperation with significance for world peace and sound bilateral relations. These are arguments with high validity given the political climate characterizing the High North. The seventh and final category consists of reasons related to the informants' personal backgrounds and careers and their interest in other cultures, meetings with people, travel, and exciting experiences.

Altogether, these reasons for engaging in internationalization are about learning and acquiring knowledge that goes beyond the purely academic field. The battery of reasons that appears to be the driving force behind everyone's participation in the internationalization of higher education is absent in the strategies reflecting how the HEIs have organized their input factors to achieve defined goals. The next question thus must be: how does this work out in practice?

In the absence of strategic goals, good intentions become guidelines. The disadvantage is of course that intentions are like poems; they are interpreted and contextualized depending on the reader. The advantage is that it opens the way to creativity and thinking that may go way beyond the framework allowed by a clearly defined and formulated goal. Thus, the vague and approximate ideas about what internationalization should be invite opportunities for anyone who wishes to take the initiative.

Our data suggest that internationalization is thus about bottom-up processes that often start with an idea developed in meetings between professors at a seminar or a congress somewhere in the world. As the idea appears to have a potential for implementation being realized, the HEIs' international offices are included as professional resources with routines and procedures for quality assurance of cooperation agreements, decision-making processes providing legitimacy as well as administrative anchoring. In addition to their facilitation tasks, employees from HEIs' international offices attend fairs and congresses where HEIs from many countries present their international education programs on offer. They also visit HEIs that they, through investigating, have found to offer programs of study that may be suited to be one out of several offers to the HEI's own students when it comes to international exchange. Thus, a clear division of labor emerges in organizations; professors primarily take the initiative, while those affiliated with various international offices assume the role of facilitators and offer analysis. Leaders at institutional and faculty levels have a third role as the authority empowered to accept or reject the agreements. However, in the words of one informant, most proposals are indeed accepted.

Internationalization as a partnership program

Nord University celebrates its 30th anniversary of cooperation with Russian universities. One of the first faculties to initiate such collaboration with HEIs in Russia was the Business School (Handelshøgskolen, Nord University). This part of our chapter deals with the development of a partnership program as part of the internationalization of higher education between Nord University Business School and its partner universities. The following describes how this educational program has been developing in the last 15 years and discusses what the Arctic/High North context means for its internationalization.

2004–2014: Heyday

Norway and Russia have a long history of cooperation. The signing of a delimitation agreement in 2010 was an essential step toward even closer collaboration between the countries, not least when it comes to the sustainable management of Arctic energy resources (Bourmistrov, Mellemvik and Vasiliev, 2011). The Arctic had become a particularly appealing field of study when Norwegian enterprises were allowed to participate in projects like the Shtockman. At that time, universities in Russia and Norway were training professionals and managers in the oil-and-gas industry but not explicitly paying attention to the Arctic region. Inspired by such political and business developments, the Business School at Nord University initiated and established contact with Moscow State Institute of International Relations (MGIMO University). The parties signed an agreement in 2004 where one of the aims was to develop a joint degree program at the master's level, focusing on managing oil-and-gas resources in the High North (Bourmistrov et al., 2015). Both institutions were eager to initiate energy management programs focusing on the most promising areas for cooperation for both countries and universities related to the High North (Bourmistrov, Mellemvik and Westerberg, 2011). MGIMO and Nord University decided to combine their strengths and competencies by enhancing knowledge in the field of energy resources in the Arctic. Nord University has been cooperating with Russian institutions in the field of business administration since 1991 (see, e.g., Bourmistrov and Mellemvik, 2002). MGIMO has since 2000 been building units competencies in foreign relations, diplomacy, and also its close connections with Russian energy companies. At the same time, the Arctic and the High North have become an important geopolitical region for both countries. Bourmistrov et al. (2015) showed how favorable political initiatives toward Arctic energy resources in both countries and opportunities for the harmonization of education brought about by the Bologna process have made it possible to embark on cooperation between universities. They underline that energy companies and the authorities supported the development of these programs by either participating in lectures or program boards or by providing financial resources for program development. When it comes to curricula, Bourmistrov et al. (2015, p. 115) point out that while designing a curriculum, many compromises had to be made due to the characteristics of the energy field, the High North region, the international nature of the program, and cross-cultural exchange and learning. When the master's degree program was launched, the program description stated:

> The program aims to broaden students' understanding of global energy management issues and highlight the opportunities and challenges for developing energy resources in new emerging energy provinces, especially in the High North. With oil-and-gas exploration now well underway in the Barents High North, and with the prospect of more such exploration along the coast of northern Norway, there is an increasing need for skilled professionals in the energy field—people familiar with the peculiarities of the oil-and-gas industries, possessing strong regional knowledge and expertise in energy management, energy diplomacy, and geopolitics.

The design and agreement on the courses in the program, including those for the second semester (Table 10.1), were implemented by dedicated working groups of professionals from both sides with relevant formal competence in each other's education systems and previous positive formal and informal experience of Norwegian-Russian education cooperation (Bourmistrov et al., 2015, p. 114). Faculty collaboration and trust have flourished as a result of this collegiality. The program enrolled its first students in August 2005, and in 2007–2014, more than 200 students graduated from this program.

2015–2021: Toward green shift and the orient express

Conflict over Ukraine deteriorated Russian-EU relations and shook Norwegian-Russian relations on energy cooperation in the High North. The fall in oil-and-gas prices has had a significant impact on the oil-and-gas industry causing job loss. Furthermore, global climate efforts to reduce carbon emissions through international agreements have occasioned a change in how extractive companies look at sustainable development. As a result, the energy management program has experienced a decline and fall in the number of students, despite the well-established program, harmonized courses, and dedicated lecturers. The numbers of students choosing to study at this program have decreased by almost 50% on the Norwegian side. MGIMO has still been able to recruit some candidates (although a smaller number) due to their prestige and close links with energy companies. This program was developed in accordance with European Transfer Credit Systems (ECTS). However, with structural changes in the higher education sector in both countries, in particular with regard to the quality assurance system, the program needed to be overhauled to meet the requirements of the universities' quality systems.

To deal with these challenges, those responsible for the program in both universities started to introduce the changes in the program. The program description at this time includes the following text:

> The program aims to broaden your understanding of energy management and sustainability. New professionals need to understand the complexities of the efficient use of depletable and renewable natural resources such as oil, gas, and renewable energy sources. Energy policies, exhaustible and renewable resources, Arctic oil and gas exploration and extraction, cooperation between Norwegian and Russian oil and gas companies, electricity generation, climate policy, systems and tools of energy management, sustainability, and corporate social responsibility are some of the subjects that are studied as part of the master's program.

Simultaneously, it was decided in the courses forming the second semester (Table 10.1) to shift the focus from solely fossil resources to include ideas of sustainable development, to teach more about the green shift, namely to impart a holistic understanding of the energy sector and its transition in the specific context.

Despite the central-level political difficulties between the EU and Russia, the High North and the Arctic have retained their relevance, and Norway and Russia have worked together to address mutual concerns such as environmental protection, resource management, and regional growth and employment. Also, cross-border cooperation, particularly in traditional fields like education, research, and people-to-people, has been going strong (see, e.g., The Norwegian Barents Secretariat, 2021).

At the same time, the research partners on the program included the Asian or "orient" idea of the program in the renewed version of the curriculum. Such trends as global Asian growth, Russia's diversification of its energy market toward the Asia-Pacific region (see, e.g., Overland et al., 2015), and Chinese interest in the Arctic (China's Arctic Policy, 2018) have been a part of the political and economic relations between Norway, Russia, and China for a while. In 2016, with the support of MGIMO, Nord University therefore initiated cooperation with ECNU in Shanghai, China.

The political and economic interests of Norway, Russia, and China in the Arctic and High North made it possible for the partner universities to build on the accumulated experience and combine their renewed strengths and competencies by enriching the knowledge in the field of international governance and business with a focus on the energy resources in the Arctic. This combines Nord University's competencies in Arctic governance and business management, MGIMO's specific knowledge about the energy sector from an international perspective, and ECNU's in-depth knowledge of regional studies, geopolitics, and international relations. In addition, recognizing the significance of international cooperation in higher education and research and building up solid institutional collaboration between Norway, Russia, and China, the partners have been applying for national financial support. This partnership program has been included in the prestigious "Innovative Study Programs" list of China's Ministry of Education and Research. Together with MGIMO University, Nord University applied for and received funding from the Norwegian Ministry of Education and Research for reciprocal student and faculty exchange and joint educational activities.

The dedicated working group of professionals from universities with relevant formal competence in each other's education systems and previous positive formal and informal experience of international education cooperation cooperated on the curriculum. Compromises had to be made, again due to the formal workload at the program level, admission requirements, the research component, and cross-culture exchange and learning. This time, the collegiality among not only faculty but also the administrative staff fostered the collaboration. The program description in the current version is as follows:

> An exciting partnership program with MGIMO in Moscow, Russia, and East China Normal University (ECNU) in Shanghai, China where you as a student can have the opportunity to study one semester in each city, on an innovative and provident (sic!) partnership program. You will acquire distinctive experience and competence after graduation, which will make you attractive in the labor market both globally, nationally, and regionally. The second semester (when students from partner universities taking courses together) has a distinctive focus on energy management and geopolitics, with a further focus on Arctic governance and business management.

From experiences of the development of this international master's program, it is possible that the Arctic/High North context and focus on energy has indeed retained its importance for internationalization between universities over the last 15 years. This corroborates with earlier research (see, e.g., Arruda, 2020) showing that energy initiatives in different geographical areas (or nationally and globally) reveal that definitive characteristics of the energy system have implications for governance, political decision-making, and educational systems. Thus, the High North context is essential for such education programs focusing on energy. In addition, when speaking about internationalization, it is necessary to take the following into consideration. In an earlier study on program development and implementation during the heyday by Bourmistrov et al. (2015, p. 7), the authors conclude that university cooperation in energy management with a focus on the High North has been possible because of the "continuous search for synergies, respect for differences, experimentation and the involvement of dedicated individuals". The reflections about the "Green change and orient express" period of

Table 10.1 Overview of the courses during the second semester

2004–2014	*2015–2021*
The geopolitics of petroleum and natural gas	Geopolitics and energy
Energy management – Norwegian perspectives	Energy business and management
Research methods	Circular economy
Norwegian language/Russian language	International perspectives on business and governance in the High North

program development and implementation shows that university cooperation in international governance and business with a focus on energy in the context of the High North/Arctic has connected three universities by reason of their genuine interest in each other's countries (supported by the political and economic climate), dedicated individuals, and opening up for better recruitment. Developing competencies and skills on such a program fosters an understanding and appreciation of each country and teaching and research on global perspectives in energy, the High North, and internationalization in partner universities.

Are there success criteria for internationalization in the High North?

Two different forms of development have so far been presented within the internationalization of higher education – one as a result of uncoordinated incremental processes at the individual HEI and the other as a far more deliberate consequence of collaboration between three HEIs related to regional challenges. The following will address our studies and show both coinciding and conflicting findings.

Prerequisites for successful internationalization

Our descriptions of internationalization at four HEIs in the High North and the collaborative project between three HEIs on a partnership program tell us something about what may be required for someone to take the initiative and succeed in the internationalization of higher education. The comparative study refers to the HEIs' internationalization characteristics that can be said to be common and more general features of how such processes proceed. The study of the partnership program, in turn, shows the contextual preconditions that characterized this development process and contributed to the actors' success in their project.

The first common premise seems to be that the professionals have had few formal limitations and thus wider opportunities to take the initiative and develop and realize their ideas. One of the reasons is that the HEIs, for their part, cannot demonstrate overall, specific, and operative strategies for how to run their internationalization processes. The advantage is that it opens up to creativity and thinking that may go way beyond the framework allowed by a clearly defined and formulated goal. A good example of this is how political and economic relations between Norway and Russia, and later China's interest, have supported the climate for such cooperation. The energy focus has moreover been inspired by combining multidisciplinary perspectives from partner institutions to learn more about energy development and its implications for governance, political decision-making, and sustainable business development in the Arctic as a region of strategic importance to all partners.

The second common premise is that for the individual academic community, it is not enough to have opportunities and freedom to develop international educational programs. It also requires acceptance that there is the best knowledge about the individual environment's professional potential and which foreign HEIs are attractive and in a position to cooperate. The partnership between NORD and MGIMO is an excellent example in this field as well.

The third common premise can be related to the characteristics of the individuals and professional communities engaged in internationalization. It refers to why faculty members spend much of their time and energy making their educational programs available to international students and establishing agreements about student exchange with HEIs in many other countries.

In sum, there is a significant coincidence between the factors these two studies find that help to open up and stimulate commitment to and development of the internationalization of higher education at all the HEIs mentioned in this chapter. However, to shed light on the forms of internationalization and transnational cooperation on education that can be most appropriate for the High North, we will move on to the next question.

Should more focus be placed on strategic partnership programs in the High North?

For HEIs in the High North, three different preconditions may constitute insurmountable barriers to participating in various international education programs. First, it is about economic preconditions. The requirement that HEIs, to a greater degree, must fund their activities through increased earnings contributes to increased competition in the international student market. In this context, HEIs located in geographically more peripheral areas appear to lose in the competition for domestic and international students and to attract the best qualified academics for positions in teaching and research. Economic muscle having a large academic community rich in tradition and also a central location are obvious advantages of internationalization.

The other kinds of preconditions to note include how international conditions make it challenging for HEIs and academic communities to cooperate. Western countries' sanctions following Russia's annexation of Crimea have led to significant problems for Russian HEIs.[2] For the HEI's in the Nordic countries, there are certain problems related to accepting exchange students from so-called third countries, that is, students coming from states outside the EU and European Economic Area (EEA) and the western world in general.

Language is the third precondition, often referred to as the essential precondition for participation in international education programs. Language constitutes significant cultural capital for anyone to cooperate with foreign colleagues or teach exchange students.

Two further questions are often repeated in the debates about the future development in the High North: how should we best build up the necessary knowledge and competence in vital areas and how should we facilitate a more constructive people-to-people cooperation across national borders?

Considering these two questions, findings from our two studies suggest that the more targeted and task-oriented partnership program aimed at transnational challenges in the High North has its clear advantages. At the forefront, the more emergent strategies seem to result either in programs developed locally and offered in English for international students or in collaborative projects that are either entered into between parties who, for various reasons, know each other from previous meetings and arenas or between parties wishing to strengthen the faculties' earnings by strengthening their positions in the student market. Some of these educational offers focus on businesses and issues concerning the Arctic, but then usually limited to their own country. Very few of them are actually collaborative projects between two or more HEIs in different Arctic states.

The partnership program between NORD, MGIMO, and ECNU exemplifies in many ways the potential that can be extracted from a more targeted collaboration between HEIs in the High North. First, it illustrates how professional environments at these HEIs in three states contribute with complementary academic competencies and joint interests and how this creates educational and research synergies within the relevant field. Second, it is not unreasonable to expect that the HEIs in the High North will take particular responsibility for contributing with the expertise they may have in relation to challenges of a common nature in the Arctic states. In other words, it is implicit in their geographical location that HEIs' social mission must, to a certain extent, be geared toward their region's tasks, challenges, and issues.[3] As a third point, this type of transnational educational cooperation will contribute to the HEIs involved being far more visible and more clearly profiled in an international and national context. It applies not only in relation to the marketing of programs and a potential increase in research production but also as institutions acquire an active voice in issues on the Arctic agenda. The fourth point is that the professional communities that are part of such educational collaboration will constitute a competitive network in terms of requirements for education and research funding from national and international funding sources. The fifth point is that, given that the first four points we have highlighted here can be realized, the HEIs will also be far more attractive and competitive in the international student market, where academic quality, the purpose of education, and to some extent also uniqueness are decisive. The final point is that the type of partnership program run by educational and research communities from different states but belonging to the same region, having joint interests and priorities, and facing the same challenges will undoubtedly strengthen people-to-people relations between the countries involved.

The internationalization of higher education will usually take place through emerging strategies. Bottom-up processes driven by institutional entrepreneurs will always contribute to growth and development at the individual HEI. Even where they fail, learning and experience can be utilized in the next attempt. From a regional development perspective, however, the six points above indicate that more targeted transnational partnership programs are likely to be most beneficial for the HEIs with joint interests in the region.

Conclusion

This chapter has presented findings from two studies. One included data from four different HEIs in the High North. They are generally characterized by a lack of defined goals and authorized strategies with clear guidelines for what the internationalization of higher education should be about. The variation in flora of different educational programs at each HEI results from ideas and initiatives taken by individual academics or individual academic communities. The second study refers to how an educational collaboration between partner HEIs is created by the parties involved, both having a clear focus on what kind of competencies this specific education can impart to the students and how they can combine their strengths and competencies to provide the program with high-level content that they cannot provide alone. Also, it shows that when the educational situation has changed, their willingness to cooperate and favorable external conditions have helped extend the partnership and made it even more strategic.

We can conclude that there were several similarities between these two studies. One is that the professionals have had few formal limitations and thus wider opportunities to take the initiative and develop and realize their ideas. Another essential prerequisite is that the academics at the faculty level themselves are best placed to assess the potential of their professional community and which foreign HEIs are attractive and amenable to establishing cooperation. The third common premise can be related to the characteristics of the individuals and professional communities engaged in internationalization. Here, we highlighted social and cultural capital as descriptive keywords.

What strikes us is that while the partnership program between NORD, MGIMO, and ECNU has a clear profile geared toward key challenges in the High North, we find no similar Arctic-oriented collaborative projects among the range of international educational offerings at the other four HEIs. This raises the question whether the HEIs in the High North should have more significant responsibility for the meaningful contribution to society that follows from being located in this region.

Education is a key factor in Agenda 2030 for sustainable development. The SDGs can be viewed as a trigger, while internationalization can be seen as a response (Ramaswamy et al., 2021). SDG 4 (Quality of Education) is a stand-alone target and pillar that underpins the achievement of all 17 goals, and it encapsulates internationalization. In addition, international cooperation and knowledge partnerships are essential to achieve other SDGs, as expressed in SDG 17 "Partnerships for the goals" (Norwegian Agency for International Cooperation and Quality Enhancement in Higher Education, no date). Hence, the internationalization of higher education in the High North, including actors from across the Arctic countries, disciplines, and sectors, has a profound transformative role to play in addressing the challenges of sustainable development in the High North through collaboration, continuity, trust, and engagement.

However, this is the beginning of a debate. It involves everything from the HEIs' right to self-determination and the academic freedom of scientists to adopt political perspectives on state governance and what characterizes the international relations at any given time, particularly when it comes to future projections on what the internationalization of higher education in the High North means for international Arctic cooperation. Considering what could be to the benefit of the High North, we welcome such a debate.

December 2021

Ex-post reflections

As you read in the Chapter 1, Europe is now in a tragic situation. In international cooperation in education and research, mutual respect and understanding are always the foundation for collaboration. In our chapter, we have shown that when changes occur, the willingness to cooperate built on social and cultural capital can help develop and strengthen the partnership between educational institutions. However, in the current situation, when institutional cooperation in education and research with Russian HEIs is put on hold, one should be aware that maintaining mutual trust may be difficult since we no longer have insight into each other's academic daily lives in these new contextual settings. In the present geopolitical conditions, cooperation in the Arctic is problematic and essential. We still need to address climate change, pandemics, and other challenges that bring an extra level of complexity to the current international cooperation in education and research in the Arctic. When reading this chapter, we kindly request our readers to reflect on the cases discussed and analyzed as an example of how educational institutions in the High North still have a significant responsibility to contribute to this region no matter what. HEIs need to restructure and rebuild their relationships while –maintaining a focus on developing international cooperation in education and research based on shared interests through collaboration, continuity, trust, and engagement.

May 30, 2022

Notes

1 We acknowledge different definitions, where the Arctic is the sea and land between the North Pole and the Polar Circle (66°33'north) and the High North (Nordområdene) is a term used by the Norwegian Government being more political and is not limited by the Polar Circle (Norwegian Ministries, 2017). However, for the purposes of our chapter, we use the Arctic and the High North interchangeably

2 Given 300 days' waiting time and a fee of $160 for a visa to the USA and more or less the same in the UK, Canada, and other states, this is tantamount to excluding Russian academics from attending conferences and other professional cooperation with colleagues from the HEIs in these relevant countries. Even NArFU, far north in the Arctic, instead turns south-east toward China, Japan, and the old Soviet states. NArFU also states that it has major difficulties recruiting students from Scandinavia and the western world in general.

3 For example, the central authorities in Russia have given NArFU the explicit responsibility for conducting High North-related research and education.

References

Arruda, G. M. (2020). *Sustainable Energy Education in the Arctic: The Role of Higher Education*. Abingdon, Oxon: New York: Routledge.

Bourmistrov, A. and Mellemvik, F. (eds.) (2002). *Norwegian-Russian Cooperation in Business Education and Research: Experiences and Challenges* (HBO-report No 11). Bodø, Norway: Bodø University College/Bodø Graduate School of Business.

Bourmistrov, A., Mellemvik, F. and Vasiliev, S. (2011). *Perspectives on Norwegian-Russian Energy Cooperation*. Olso, Norway: Cappelen Akademisk Forlag.

Bourmistrov, A., Mellemvik, F. and Westerberg, T.V. (2011). *Experience of Introducing ECTS in Cooperative Education Programs Between University of Nordland in Norway and Russian Universities* (HBO-report). Bodø, Norway: Bodø University College.

Bourmistrov, A., Gudmestad, O.T., Salygin, V. and Zolotukhin, A. (2015). "Norwegian – Russian Cooperation on Oil and Gas Education" in Bourmistrov, A., Mellemvik, F., Bambulyak, A., Gudmestad, O., Overland, I. and Zolotukhin, A. (eds.) *International Arctic Petroleum Cooperation: Barents Sea Scenarios*. Abingdon, Oxon: New York: Routledge, pp. 111–122.

de Wit, H., Hunter, F., Howard, L. and Egron-Polak, E. (2015). *Internationalization of Higher Education*. Culture and Education, European Parliament [Online]. Available at: https://www.europarl.europa.eu/RegData/etudes/STUD/2015/540370/IPOL_STU(2015)540370_EN.pdf (Accessed 12 July 2021).

Mintzberg, H. and Waters, J.T. (1985). Of strategies, deliberate and emergent, *Strategic Management Journal*, 6 (3), pp. 257–272.

Norwegian Agency for International Cooperation and Quality Enhancement in Higher Education. (2021). *Utforsk* [Online]. Available at: https://diku.no/programmer/utforsk (Accessed 15 July 2021).

Norwegian Ministries. (2017). *Norway's Arctic Strategy - Between Geopolitics and Social Development*. Oslo, Norway: Norwegian Ministries.

Overland, I., Bambulyak, A., Bourmistrov, A., Gudmestad, O., Mellemvik, F. and Zolotukhin, A. (2015). "Barents Sea Oil and Gas 2025. International Arctic petroleum Cooperation: Barents Sea Scenarios" in Bourmistrov, A., Mellemvik, F., Bambulyak, A., Gudmestad, O., Overland, I. and Zolotukhin, A. (eds.) *International Arctic petroleum Cooperation: Barents Sea Scenarios*. Abingdon, Oxon: New York: Routledge, pp. 11–29.

People's Republic of China, The State Council. (2018). *China's Arctic Policy*. Beijing: People's Republic of China, The State Council.

Ramaswamy, M., Marciniuk, D. D., Csonka, V., Colò, L. and Saso, L. (2021). Reimagining internationalization in higher education through the United Nations sustainable development goals for the betterment of society, *Journal of Studies in International Education*, 25 (4), pp. 388–406.

Røvik, K. A. (2011). From fashion to virus: An alternative theory of organizations' handling of management ideas, *Organization Studies*, 32 (5), pp. 631–653.

Røvik, K.A. (2016). Knowledge transfer as translation: Review and elements of an instrumental theory, *International Journal of Management Reviews*, 8 (3), pp. 290–310.

Sundet, M., Forstorp, P-A. and Ørtenblad. A. (2017). *Higher Education in the High North. Academic Exchanges Between Norway and Russia*. London: Springer International Publishing AG.

The Norwegian Barents Secretariat. (2021). [Online]. Available at: https://barents.no/nb/prosjekt (Accessed 19 July 2021).

11 Challenging Boundaries from Below

Cross-Border Culture in the Arctic

Bjarge Schwenke Fors and Yngvar B. Steinholt

Introduction

Where two or more people meet, there is culture. Where people are separated, culture, for good or for ill, imagines human connections and explanations for their absence. By counteracting the development of negative narratives of the other, cross-border cultural cooperation is a means to ensure peaceful and constructive coexistence across national borders. In this chapter on Russo-Norwegian relations, we argue that culture itself and cultural initiatives from people in the borderland have been more prominent than commonly held. Culture, we believe, will remain a decisive factor in the future development of Russia's relationship with the West, locally and internationally.

The contributions to this anthology all refer to the UN Sustainable Development Goals (SDGs), 17 global goals for strengthening environmental, social and economic sustainability by 2030 (United Nations, n.d.). The SDGs have faced criticism for making culture "the missing pillar of sustainability" (British Council, 2020, p. 9). Although more advanced understandings of the role of culture in sustainable development are found elsewhere,[1] these are usually more instrumental than the qualitative definition above. Three of the existing SDGs relate to cultural cross-border cooperation, as discussed here. SDG 4 promotes "lifelong learning opportunities for all" (United Nations, n.d.). A primary form of lifelong learning, culture is our fundamental way of making sense of ourselves, others and the world. SDG 5 aims to "achieve gender equality and empower all women and girls" (ibid.). How patriarchal tradition and gender discrimination may be challenged without culture playing its part is hard to imagine. Moreover, what is the origin of *empowerment* and how is it expressed if not in culture? SDG 17 encompasses culture with its aim to strengthen "the means of implementation and revitalize the global partnership for sustainable development" (ibid.). Constructing such partnerships lies at the core of cross-border cultural cooperation. Despite their limited understanding of culture, we keep these SDGs as reference points. We hope this chapter can contribute a more profound understanding of the integral role of culture in sustainable development.

Internationally, cultural cooperation with the Russian Federation (and the former USSR) has helped maintain contact on civic, diplomatic and political levels,

DOI: 10.4324/9781003246015-14

even in times of strained East-West relations. Often a necessity in difficult times and in the first establishment of cultural exchange, cooperation *from above* (i.e., initiated by the respective political authorities) is what we choose to call *border affirming*. Modeled on the culture and traditions of diplomatic exchange, its logic reflects a system of official visits and return visits across the physical border. Culture is instrumentalized and formalized, reducible to the violin playing at the reception. The potential of culture to bring peoples together on a more profound level is not necessarily understood, explored or utilized.

When seen *from below*, from the perspective of a border region's inhabitants, cultural workers and creative idealists, borders become obstacles that beckon transgression, more than instruments for organizing the world. People living either side of a peaceful, closed border are, hopefully, united by a desire to learn about and interact with each other. Thus, initiatives of cultural cooperation from below relate to the border differently from macropolitical initiatives. We term such initiatives *border transgressive*; since they focus on shared local and regional experience and/or universal existential questions, they think around and beyond the physical border in various ways.

Our distinction between cultural cooperation from above and from below resembles Ang, Yudhishthir and Mar's (2015, p. 365) between *cultural diplomacy*, an "essentially interest-driven governmental practice" – an instrumental use of culture for national interests – and *cultural relations*, ideals-driven practices by (largely) non-state actors (ibid.). In our context, cultural relations are also linked to *public diplomacy*, "a form of intercultural dialogue based on mutuality and reciprocal listening [also involving] civil society and/or private sector stakeholders" (ibid., p. 368). In practice, border-affirming initiatives from above and border-transgressive initiatives from below coexist and interact even within the same cultural events. They do not always directly connect, but mutually enable, influence and sustain each other. The top-down perspective of political analysis often ignores initiatives from below and their influence on the macropolitical level, yet if we wish to say something about the future development of Russia's relation with the West, it is imperative to better understand their role in bilateral cultural cooperation.

Norway offers a promising site for studying these issues. In the Russo-Norwegian cultural cooperation, a continuous top-down state involvement has coexisted with equally continuous cultural initiatives from within the border region. Both levels endured the Cold War. Norway's relationship with Russia is dual: a central one, governed by national demands, international agreements and alliances; and a regional one, influenced by the proximity of the border and the people living on the other side of it. The distance of the region from the respective capitals has provided space for local ideas about the other that coexist with and sometimes challenge narratives of national and international politics.

This chapter investigates initiatives from below in the Norwegian-Russian cultural cooperation represented by two cases: a theater company based in Kirkenes since 1990 and an exhibition of contemporary Russian art in Tromsø in 2013. What can these cases tell us about cultural cooperation initiatives from below, their support, realization, impact and complex relationship with formal initiatives from above? Focusing on this microlevel of bilateral cooperation, we aim to introduce some contrasting perspectives on past, present and future East-West relations. Following a brief historical survey of Norwegian-Soviet/Russian cultural cooperation, we turn to our two cases, for which we conducted six semi-structured, qualitative interviews with organizers and participants.[2] Some are quoted in the text and all (interviews) have informed our discussion. Finally, we assess the current state and possible future of Norwegian cultural cooperation with the Russian Federation across their Arctic border.

Cultural cooperation during the Cold War

By the late 1950s, direct contact between Norway and the USSR, former ally and liberator of Eastern Finnmark from the Nazi occupation, had been broken for a decade. Although diplomatic initiatives and trade agreements were pursued by the respective governments (Moon, 1964), the two states belonged to opposing alliance systems. In such circumstances, successful initiatives for cultural cooperation from below seemed unlikely. Yet, in the summer of 1959, the fishing vessel *Yngvar* sailed from the coastal town of Vardø in Arctic Norway to Soviet Russia with a bold plan of performing a brass band concert in Murmansk (Finnmarken, 1959). The adventurous musicians failed to notify Norwegian or Soviet authorities of their quest, so *Yngvar* was boarded by the Soviet navy upon entering Russian waters. The brave attempt to reestablish cross-border cultural relations could have ended here, but intense diplomatic negotiations allowed the journey to continue into Murmansk harbor and a cordial welcome with the appropriate official pomp. Thousands attended the concerts, which were covered by local TV (ibid.).[3] This daring and haphazard initiative from below marked the beginning of a lasting cultural cooperation across the Norwegian-Soviet border (Tjelmeland, 2012, p. 176).

Although the mutual suspicion surrounding Norwegian-Soviet relations persisted, cultural exchange continued throughout the next three decades, even when geopolitical tensions ran high (Cuba, 1962; Prague, 1968). In the border municipalities of Sør-Varanger (Norway) and Pechenga (Russia), cooperation was particularly active. In 1973, a formal friendship agreement was signed, emphasizing cultural cooperation (Avtale om [...], 1973) involving creative amateurs, local bands, orchestras and choirs. Norwegian culture professionals and their institutions were less actively involved, whereas the Soviet organization of cultural life[4] meant that creative amateurs participating in sensitive cross-border cultural exchange would always be institutionally affiliated.

After 1959, in parallel with official cultural diplomacy, local actors continued to establish and maintain cross-border cultural partnerships. Little could be achieved without official support, but Oslo and Moscow, if wary of excessive fraternization across the border, also valued culture as a means to reduce tension. When based on the rituals of diplomatic exchange, cultural cooperation affirmed contacts on political levels, building mutual trust and improving public relations. Cultural cooperation served Soviet slogans of "friendship of the nations" as well as Norway's self-image as a peaceful regional mediator. In culture as in sport, the USSR competed with the West and showcased its great achievements.

When responding to initiatives from below, cultural cooperation projects on both sides relied on the authorities taking a leading role. Implementation from above entailed a deliberate instrumentalization of culture for the higher goals of foreign policy. Official representatives were always present. Meetings and performances were thoroughly planned, formal and accompanied by flags and national anthems. Visiting Pechenga in 1973, a participant from Sør-Varanger recalls "endless speeches [...] about the greatness of cross-border cooperation, peacemaking, friendship between neighbors and things like that" (Hojem, 2014, p. 286).

Barents cooperation

With *Perestroika*, Mikhail Gorbachev's famous Murmansk speech encouraged international cooperation in the North (Gorbachev, 1987). Norway embraced this opportunity to normalize its relations with the USSR and from 1991 – the Russian Federation. The 1993 establishment of the transnational Barents Euro-Arctic Region confirmed this. The region covers 13 provinces in the north of Norway, Finland, Sweden and north-western Russia (Stokke and Tunander, 1994).

The Barents Region marked a significant step forward, but Russo-Norwegian normalization in the High North was limited by the border itself. A Schengen border, it imposes control on all travelers and a strict visa regime; a NATO-Russia border, it is also highly militarized. These factors prevented the Barents Region from becoming a closely integrated Euro region. Instead, it remains a form of "dreamscape", an area of potential, limited mainly to initiatives of "people-to-people cooperation" (Hønneland, 2005). Thus, culture became a guarantee for the continued relevance of the Barents Region. A bilateral agreement between Norway and Russia in the fields of culture, education and research was signed in 1994 (Avtale mellom [...], 1994). This established a framework for bilateral cultural cooperation, forming a trajectory for the subsequent agreement on cultural cooperation in the High North (Norsk-russisk handlingsplan, 2009) and for the program for Russian-Norwegian cultural cooperation (Program [...], 2019). In official rhetoric, culture has thus become fundamental to the Barents cooperation. As in the past, the respective national authorities promote cultural cooperation in order to achieve more abstract aims of friendship, normalization, shared identity and peace.

Norwegian political will is confirmed by generous economic support for cultural cooperation projects. In 1993, the Norwegian Barents Secretariat was established. Funded by the Norwegian Ministry of Foreign Affairs, its primary function is to support Norwegian-Russian cooperation projects. More than 3,000 Norwegian-Russian projects have been funded since 1993, roughly half of them on cultural cooperation (MA). In 2007, the Barents Secretariat launched BarentsKult, a grant program specifically designed for cultural professionals and larger art and culture projects. Geographically restricted to projects within the Barents Region and including the Russian "Northern Capital" of St. Petersburg, BarentsKult is financed by the respective Norwegian Ministries of Foreign Affairs and Culture, the counties of Northern Norway and the Barents Secretariat. Since its establishment, it has, with an annual budget of NOK 9–11 million, become the primary source of funding for cultural actors seeking cooperation with Russian partners within the Barents Region (MA).

The foreign political aim of BarentsKult is defined as strengthening the identity-building processes of the Barents Region (Breit, Rowe and Skedsmo, 2014, pp. 8–9). The program is also intended to support "the democratization process in Northwest Russia through compulsory and active participation of Russian artists and cultural actors" (ibid., p. 9). Probably motivated by the contributions of North Norwegian counties, BarentsKult is intended to nurture cultural entrepreneurship and employment in Northern Norway while creating more "good and innovative" art projects (ibid., p. 8). Thus, the program also nurtures an ambition to create art for the sake of art itself.

Cultural cooperation today

Today, foreign political interest continues to legitimate and motivate Norwegian-Russian cultural cooperation. The respective national authorities still regard cultural cooperation as an instrument for achieving other goals, be they dialog, "democratization" or region-building. This top-down instrumentalization of culture inherited from Cold War times is partly explained by strained Norwegian-Russian relations following Russia's annexation of Crimea in 2014. Even so, national authorities' direct involvement in cultural projects is less prominent today, leaving actors more freedom to choose partners and artistic agendas.

In many ways a top-down initiative, Barents cultural cooperation still generates grassroots initiatives on the part of independent local actors pursuing their own goals. The Barents Region has made initiatives from below a stronger, if not the dominant, element in cultural cooperation. Its financial support has increased the number of actors involved in cross-border cooperation, and these are more professional and geographically widespread than before. The immediate border region houses the most extensive cooperation. Cultural organizations in Kirkenes remain at the forefront, in particular Pikene på broen (Girls on the Bridge), curators and organizers of the annual cross-border festival Barents Spektakel (Fors, 2019, pp. 126–164; Viken, 2014) and the Samovar Theater (see below). Meanwhile, cultural organizations elsewhere in Northern Norway have joined the fray, including major festivals such as TIFF (Tromsø International Film Festival), the Northern Lights Festival (Tromsø), the Melting Pot Music Festival (Mo i Rana), and the Riddu Riđđu festival of indigenous culture (Kåfjord). On the Russian side, the Barents cooperation has expanded cross-border activities into regions as remote as Arkhangelsk *oblast*, the Komi Republic and the Republic of Karelia.

Cross-border cooperation may have become easier, but numerous challenges, bureaucratic and practical, remain. Natural differences and antagonisms between the two sides are often amplified by the mere fact of the border's existence. This affects the quality of all forms of cross-border interaction. Nevertheless, a more profound exchange has developed. In the following we present two cases which stand out as practices aiming to achieve mutual understanding on a more profound level.

The Samovar Theater is a permanent institution in the town of Kirkenes. The theater stages plays on a regular basis and tours on both sides of the border. It has in recent years become a landmark cultural institution in Norwegian-Russian cultural cooperation (Fors, 2014, 2019). Re-Aligned Art (RAA) was a one-off event, showcasing contemporary Russian art in Tromsø, outside the border region. It included art exhibitions in three galleries, artists in residence, street art and a seminar involving artists and academics. BarentsKult covered more than half of the overall expenses (SP). The event took place in September–October 2013, when Russian contemporary art was characterized by a political edge and an activist stance soon to become suppressed.

Case 1: The Samovar Theater

The Samovar Theater, a professional theater employing six people and featuring an acting school for children, is based in Kirkenes, Norway. Established in 1990 by Bente S. Andersen, who remains its artistic director, the Samovar Theater has staged more than 60 performances over the years based mostly on freshly written material. From its very conception, the theater decided to work transnationally. Andersen envisaged it as an agent of change between Norway and Russia, an institution that could create new meeting points while simultaneously transcending old borders and producing new and interesting art (BSA). The transnational profile is reflected in the name: the Russian word *samovar* is understood in Norwegian, while signifying Russia[5] (Fors, 2014, p. 163).

During the 1990s, the Samovar Theater established lasting partnerships with, among others, the Murmansk Regional Drama Theater, the Murmansk Children's Theater, the Arkhangelsk Regional Drama Theater and the National Theater of Karelia in Petrozavodsk. Over the past two decades, it staged numerous performances with these partners and with Russian freelance composers, actors and dancers. Its slogan "Breaking boundaries" refers both to artistic and physical boundaries and relates to a wider artistic discourse originating in the 1980s art group BAW/TAF (Border Art Workshop/Taller de Arte Fronterizo) in the US-Mexican borderland (Prieto, 1999). "Mixing art and political activism, the BAW/TAF turned the US-Mexican border into a resistance scene, a site for protest – against the border itself" (see Fors, 2019, pp. 40–41). The leading figure of the group, performance artist Guillermo Gómez-Peña, has "consistently over many years [...] turned 'his' border and borderland (US – California) into a performance space, [inventing] an art of border performance that transgresses and translocates borders of every kind" (Copland, 2012, pp. 512–513). The influence of Gómez-Peña's border-transgressive art was confirmed by his visit to Kirkenes in 2004 (Fors, 2019, p. 129).

The Samovar Theater never fully embraced Gómez-Peña's political militancy, yet its interpretation of the "breaking of boundaries" is undoubtedly radical.

The theater aims not only to cross the border but to transgress, even annihilate it (Fors, 2019, pp. 145–146). This aim deviates from the official foreign political goals for cross-border cooperation where cultural exchange is expected to take place in controlled forms between two sides across a clearly marked division line. The theater does not necessarily challenge the official aim for cooperation, but its slogan overtly signals an ambition to go further. This border-transgressive vision manifests itself in the theater's relationship with its Russian partners and in the messages of its productions.

The interaction of the theater with Russian partners seeks to overcome distance and achieve actual, reciprocal exchange. It has succeeded in sustaining long-term relationships, including Russian partners in the entire creative process. The Samovar Theater creates an environment where everyone knows each other by building on established connections with the same pool of institutions (BSA). Spending so much time together can be challenging, Andersen states, but working intensely together over time creates mutual trust and enables the partners to bond beyond the polite and formal behavior of first-time encounters. In Andersen's words, the partners know each other well enough not to be afraid to insult each other. This enables an open discussion of otherwise problematic issues, such as politics. Whereas cooperation processes are rarely devoid of friction and discussions may sometimes be tough, as long as they are constructive, they are regarded as an asset, Andersen explains. Friction, she says, can be a catalyst in the creative process and testament to an exchange of ideas on a more profound level. According to her, working closely and intensively together makes it easier to move beyond the "us-them" dichotomy and reach a point where the two sides merge into *one* collective.

These border-transgressive ambitions also affect the Samovar Theater's performances. It has developed its own style, *border dramatics*, partly inspired by Gómez-Peña's border performances (Fors, 2019, p. 139). This is characterized by a deliberate mixing and matching of art forms (theater, dance, video, music) and genres (comedy, tragedy, realism, fantasy). Nationalities and languages blend on stage and Norwegian and Russian actors (and sometimes actors from other countries) perform in the same play in their own language. These meetings on stage between actors, languages, art forms and genres become symbolic of the continuous encounters taking place in the real Barents Region borderland. The theater "enacts" the borderland and stages its liminality. Ultimately, the outcome of all these meetings is a "scenic idiom where text, movement, music and multilingualism is merged into a whole" (Samovarteateret, n.d.-a). It becomes a form of *Gesamtkunstwerk* where all the borders involved are transcended in various ways.

The plots of the plays are typically set in the Barents Region and address muted aspects of borderland life, less heroic stories that rarely feature in official contexts. This includes tales of oppression, of persecution and of the harmful effects of border-making in general. The bilingual Norwegian-Russian performance *Vodka, Water and Glasnost – A Foxtrot for Three* (2013) illustrates this. A cooperation between the Murmansk Regional Drama Theater, the National Theater of Karelia, and the Samovar Theater, the performance builds on the experiences of three individuals from the borderland, victims of World War II and the Cold War. Similarly, *Arctic Voices* (2009) tells the tale of how the shared lands of the North were split up between the nations, often with tragic consequences for the inhabitants. The powerful message is that the border has been forced upon and served to oppress the local multiethnic population and that it ultimately should be erased. In the program of the play, the theater poses a rhetorical question: "what would you do if someone suddenly divided your living room in half, removing you from family and friends, explaining that you now live in different countries?" (Samovarteateret, n.d.-b). The play ends with the people of the borderland rebelling. The rulers' speeches are ridiculed and governmental treaties demonstratively discarded (Fors, 2019, p. 133). The overall message is that the region belongs to its people, that their lives and their welfare should be the main priority and that the interest of nation states are, if not altogether illegitimate, at least of secondary relevance.

Case 2: *Re-Aligned Art*

In September 2013, *RAA* became the largest-ever display of contemporary Russian art in Northern Norway. The event came about after Svein Pedersen, then intendant of Tromsø Kunstforening (Tromsø Art Society), fused his knowledge of Marita Muukkonen, a Finnish curator with extensive networks in the Russian art scene, with that of BarentsKult funding. The resulting project proposal *A New Non-Aligned Art* combined the exhibition of works by 14 young Russian artists from all across Russia with roundtables, workshops, artists in residence and a seminar with Russian and Norwegian artists, academics and cultural workers (SP). It soon snowballed to involve the Tromsø Art Academy, two minor art galleries, street art and outdoor projects. The Russian artists were invited to Tromsø for a week and many of the works exhibited were created on site. The exhibition later moved on to Helsinki (Kiasma) and further locations in Europe (RAA website).

Thus, around an exhibition giving a broad representation of contemporary Russian art, political and activist art included, the event became a meeting place for Russian and Norwegian artists and culture workers, creating lasting networks and connections. Muukkonen brought in cocurator Ivor Stodolsky and the art group Chto delat' from St. Petersburg led by filmmaker Dmitri Vilensky (SP). This opened up tensions between strong-headed characters with diverging ideas about the functions of art in general and the planned events in particular. While challenging for the organizers, Pedersen remembers, this added urgency to discussions on and off the program.

Unlike previous art exchange programs involving regional Russian art institutions presenting well-established and traditional local artists, RAA did not showcase pre-contextualized works, Pedersen explains. Its synergy effects were not carefully planned and predicted. RAA went beyond established notions of border, neighborhood and exchange. From focusing on affirming and developing existing images of the other, it turned toward challenging, debating and diversifying those images (SP). One explicitly border-transgressive contribution was created by the

Krasnodar art collective Zip Group, who built a washing machine for national flags. Flags hung on a washing line were, as the machine was set in motion, soaked in water and detergent, passed through various rolls, then returned to the original position to dry. As visitors repeated the process, the flags gradually turned grayish-brown, losing all individual features.

Most of the exhibition depicted Russian conditions but focused on Russia and its political everyday as a totality rather than on specific parts, on which some form of a Norwegian-Russian consensus had already formed (SP). It raised more questions than it answered, and its picture of Russia was, if rich and dynamic, not necessarily pretty. The material exhibited was different from Norwegian reality to the extent that the audience were forced to relegate previous knowledge to a secondary level. The overall effect fused shock and excitement. Pedersen recalls the reaction of a young Russian couple residing in Norway who, upset by the artists' lack of respect and national pride in their representations of contemporary Russia, broke down in tears in the main venue's offices.

Artists' critical views of Russia were balanced by, among others, Nikolai Oleinikov's *Life in the Wilderness – a short commentary to 22 July, media-memory, visual neo-conservatism and the oil-welfare state*. Part of the work, consisting of wall paintings, addressed the national trauma of the recent Utøya massacre. The artist exposed disturbing contrasts between Norway's self-image as a peace-loving nation and its deeply embedded cultures of violence, exemplified by wildlife hunting. Motifs inspired by the speculative imagery of glossy Norwegian hunting magazines were combined with quotes from the public debate in the wake of the terrorist attacks, raising pertinent questions about the dangers of ignoring the inherent violence in human nature. Provocative yet sensitive enough to tap into the ongoing Norwegian self-scrutiny, the work appealed to a seasoned Norwegian art audience prepared to engage in dialog with the artist outsider.

This simultaneous excitement, shock and frustration, in Pedersen's words "an exchange for real", continued during the Sunday seminar. Organized by the research group Border Aesthetics at Tromsø University (UiT), Tromsø Kunstforening and RAA, it featured guest speakers from Russia, Germany and the UK. Presentations focused on Russian political art and activism and the atmosphere soon became heated. Andrei Rogatchevski's talk *Subversion as Propaganda: The case of Igor' Cherchenko, a National Bolshevik Artist from Israel* was interrupted by a furious Dmitri Vilensky, who demanded the speaker stop because his material did not qualify as "art".

More controversy followed. Transgender activist and member of art activist collective Voiná (War), Seroe Fioletovoe (Gray Violet), was speaking on *The Russian "Opposition Movement" and Minorities*. He began by announcing a change of topic. What followed was an exercise in disruptive conceptualism, Russian style. While this stunt could be understood, even admired, in retrospect it made the next half hour a trying time for everyone present.[6] First, the speaker's (feigned?) inadequate proficiency in English failed to convey any intelligible message. Guest speaker Dr. Mischa Gabowitsch volunteered as interpreter. Thus, Grey Violet's controversial ideas were now literally voiced by the Western expert. The paper presented two cases not normally associated with art as forms of conceptualism.

First, the audience were invited to read the media appearances of a Chechen religious extremist as a form of conceptualist art, many finding the challenge stimulating. The second case upped the ante. Here, Grey Violet insisted on interpreting the actions of a south Russian schoolteacher as conceptualism. Said schoolteacher allegedly removed the bodies of recently buried children from the local graveyard, preserved them as mummies, dressed them and kept them in his home, reading stories to them. Now, audience members demanded an end to the talk, but a quick poll showed a majority in favor of letting the speaker proceed. The organizers, reluctant to send the crowd home at this delicate point, accommodated a last-minute contribution. During this, unseen by the audience but witnessed by the moderator and speakers, Grey Violet celebrated his disruptive success by silently dancing at the back of the auditorium. Interviewee SH recalls:

> I can remember the tension in the audience, people didn't know what to think. The uneasy atmosphere. It was not fun to be part of it, although it was fun to realize afterwards what it was all about. [...] Re-Aligned [Art] was inspiring even if it wasn't all pleasant.

Transgressing the border

What makes our two cases border transgressive? First, they are far enough removed from diplomatic and foreign policy agendas to remain independent. They are free to create permanent or temporary networks, establish contacts, cooperate and create art without being defined or instrumentalized by macropolitical agendas. They also fulfill a function assigned to culture by the Barents cooperation: to bring the region's inhabitants together and forge stronger connections between them. The cases demonstrate how cross-border cultural cooperation has *expanded* geographically and institutionally since the establishment of the Barents cooperation.

The Barents Region's inclusion of Northwest Russia and the three northernmost Norwegian counties has spread ways of experiencing the border, formerly limited to the immediate border region, into 13 provinces in four countries, into areas far from the physical border. With this experience comes the desire to transcend the border, directly or metaphorically, to freely share ideas and experiences across it, to come together to create, perform and display culture. Here, our first case, the Samovar Theater, taps straight into borderland experiences and involves Russian and Norwegian cultural actors and institutions directly in dealing with them. Our second case, enabled by established funding opportunities, takes the opportunity to connect artists from either side of the border, to showcase contemporary Russian art and explore the possibilities of a temporary shared environment for artistic creation, exchange and debate. In this manner, the event – as suggested by Zip Group's flag-washing machine – seeks to erase the significance of nations and their borders.

On the institutional level, BarentsKult has enabled projects to operate more freely in relation to the respective foreign ministries and diplomatic institutions, yet without ruling out cooperation with them. If they wish, cultural actors can tap into diplomatic rituals and official ceremonies and simultaneously question or comment on them. They have achieved more control and integrity to prevent the passive instrumentalization of cultural events by the authorities. Where culture was previously brought in as a feature in diplomatic exchanges, diplomats and government ministers are now more likely to come together on the occasion of cultural events, where content and messages are determined by the cultural projects, artists and organizers.

On the microlevel, three factors facilitate border-transgressive cooperation. The first is *location*. For an event such as RAA, this entails bringing artists from across Russia to Tromsø in the Barents Region to showcase contemporary Russian art and facilitate joint creative projects between visiting and local artists. For the Samovar Theater, location is more fundamentally related to the proximity to the physical border. In Kirkenes, cross-border cooperation is normalized to an extent unseen elsewhere in Northern Norway. The city of Murmansk is a mere three-hour drive away, the towns of Nikel and Zapolyarny less than an hour. As we have seen, the Barents cooperation has increased opportunities for – and interest in – cooperation throughout the region, but physical proximity to the border influences the forms cooperation takes. Second, successful cross-border cooperation demands *knowledge* about people, resources, infrastructure, bureaucracy, logistics and practical possibilities. Theater director Andersen regards such knowledge as an advantage – if not a necessity – for achieving meaningful cross-border cooperation. The Samovar Theater accumulated such knowledge through decades of trial and error, making it a hub for cross-border cultural cooperation. For RAA, as a one-off event Tromsø Kunstforening depended on connecting with a similar hub of knowledge provided by a Finnish curator with a substantial network in the contemporary Russian art scene.

A third facilitating factor is *motivation*. The Samovar Theater taps into the local population's desire to transcend the border from both sides. The theater channels this energy into practical cooperation and into performances about the border area experience. Funding opportunities are another motivating factor. For intendant Pedersen, who initially had little knowledge of the Russian art scene, it was his fusion of available funding for cross-border projects with a hub of knowledge that brought about RAA. The major inhibiting factor is the border. Visa regime and opening hours make border crossing nearly as strenuous today as during the Cold War and prevent spontaneous forms of cross-border interaction. For Tromsø Kunstforening, overcoming this was a labor-intensive process, although Norwegian-Russian relations were favorable at the time and few problems were encountered (SP). Even the experienced Samovar Theater sometimes finds the border regime discouraging (BSA).

Finally, *cultural differences* pervade all these macro and micro factors. The people of the North may be tuned to similar wavelengths, but differences run deeper and are more complex than they may first appear. Ingrained in tradition, they make themselves felt on many levels, from religion, philosophy and political preferences to value systems, esthetic ideals and definitions of public and private spheres. Cultural differences may constrain communication and inhibit mutual understanding. Language plays its part here; it is not common for Norwegians to understand Russian and vice versa, and English is not necessarily mastered by both sides. For short-term events such as RAA, the controversies created by cultural differences can also prove an asset. For a long-term venture like the Samovar Theater, much depends upon successfully navigating the cultural gap. Notably, this gap also responds to changes in the cultural and political climate between the two nations. Nonetheless, for those involved, these challenges are part of the excitement and attraction of cross-border cultural cooperation. They may generate frustration, but also the momentum of a mutual, positive curiosity.

On the macro level, the success of transgressive cultural cooperation depends on the existence of political will on the part of both governments manifested in continuing economic support. Such support will depend on the right balance between a normalization of Norwegian-Russian relations on the one hand and intensifying geopolitical antagonism on the other. The level of cultural difference between Norway and Russia will always put intercultural dialog in demand, but were Norwegian-Russian relations to become normalized to the level of Norwegian-Finnish or Norwegian-Swedish relations, the political will to support cross-border cultural cooperation would inevitably subside. In other words, a certain bilateral antagonism is necessary in order to sustain the authorities' involvement. In the opposite extreme, Norwegian-Russian relations could deteriorate to a point where independent cultural contacts become unsustainable. This would involve a retreat into a regime based on cultural diplomacy in a narrow sense, characterized by initiatives from above and instrumentalized cultural cooperation.

Conclusion

For the past 70 years, no matter how strained Norwegian-Russian relations have been, a cultural boycott of the USSR or the Russian Federation has to the best of our knowledge never been mentioned.[7] Russia has a long tradition of responding in kind to initiatives of cultural exchange. At its simplest and most rudimentary, this means cultural diplomacy in a narrow sense, where culture is instrumentalized by foreign policy aims. In this form, culture focuses consciously on common ground, shared values and universal ideals, complementing and reinforcing rather than diversifying and developing existing notions of the other. Synergy effects are preplanned and controlled. Yet, even when culture is harnessed by political demand, it brings in a surplus of meaning. Often, this meaning becomes apparent only with the passing of time and with variations in the political and cultural climate. Therefore, once regular cultural exchange has been established, the resonances of shared works begin expanding the space of their experience, effect, interpretation and discussion. This expanded space resonates with voices and initiatives from below which, given the right circumstances, help expand cultural cooperation far beyond the limits of its formal, instrumental use.

In this chapter, we have presented two examples of Norwegian-Russian cultural cooperation and how they have facilitated the expansion of shared cultural discourse in various ways, challenging stereotypes and forcing both sides to confront and perhaps to adjust their notions of the other. As life itself, such experiences are never only pleasurable, but they are instrumental in producing a more profound and multifaceted understanding of the other, of ourselves, and of our cross-border relationship. The continued existence of border-transgressive cultural cooperation will be a decisive factor in Norwegian-Russian and, we believe, Russian-Western relations in times to come. Even in a severe Cold War scenario, a crucial role will be played by initiatives for cultural cooperation. This is one of three possible scenarios for Russo-Norwegian cultural relations toward 2035: a new Cold War scenario, where cultural cooperation again becomes restricted to cultural diplomacy in a narrow sense. A more optimistic and, we hope, probable scenario based on historical precedence is a continuation in some form of open cross-border cultural cooperation, supported to some extent politically and economically by the authorities on both sides. The least likely scenario in the short term and under the present circumstances would be a full normalization of Russo-Norwegian relations, where culture becomes more or less fully independent from government policies and initiatives.

Most importantly, our brief study suggests, where there is a will for cultural cooperation, border-transgressive initiatives from below will to some extent be not only a part of the picture, but also an accessible trajectory for overcoming differences.

December 2021

Notes

1 The Hangzhou Declaration "Placing Culture at the Heart of Sustainable Development Policies" (UNESCO, 2013) represents a milestone in this respect. It recommends "[a] specific Goal focused on culture be included as part of the post-2015 UN development agenda, to be based on heritage, diversity, creativity and the transmission of knowledge and including clear targets and indicators that relate culture to all dimensions of sustainable development" (UNESCO, 2013).
2 Interviews quoted are listed in a separate section of the bibliography with key information on each respondent. Interview quotes are referenced using interviewee initials.
3 The documentary film "From Vardø with love" (2006) describes this voyage.
4 See Steinholt (2005, p. 38).
5 The name also expresses the theater's self-image: a samovar (Russian tea kettle) is "warm, beautiful to look at, something that you can fill up, drain off and it is constantly simmering" (Samovarteateret, n.d.-a).
6 Coauthor Steinholt moderated this part of the event. The account is based on his observations, cross-checked with the interviewees present.
7 Thanks to interviewee MT for pointing this out.

References

Ang, I., Yudhishthir, R. I. and Mar, P. (2015). Cultural Diplomacy: Beyond the National Interest? *International Journal of Cultural Policy*, 21(4), pp. 365–381.

Arctic Voices (2009). Directed by Andersen, B.S.

Avtale mellom Norge og Russland om samarbeid innen kultur, utdannelse og vitenskapelig forskning. [Agreement Between Norway and Russia on Cooperation in Culture, Education and Scientific Research] (1994).

Avtale om vennskapsforhold mellom Sør-Varanger kommune og Pechenga Rayon. [Agreement on Friendship Between Sør-Varanger municipality and Pechenga Rayon] (1973).

Breit, E. L., Rowe, L., and Skedsmo, P. W. (2014). *Evaluering av støtteordningen BarentsKult. [Evaluation of the Support Scheme BarentsKult]*. FNI-rapport 11/2014. Oslo: Fridtjof Nansen Institutt.

British Council. (2020). *The Missing Pillar: Culture's Contribution to the UN Sustainable Development Goals*. Report. British Council.

Copland. D. B. (2012). "Border Show, Business and Performing States" in Wilson, T. M. and Donnan, H. (eds.), *A Companion to Border Studies*. Chichester: Wiley-Blackwell, pp. 507–521.

Finnmarken (1959). Tre minnerike dager i Murmansk. [Three Memorable Days in Murmansk]. *Finnmarken*. 5 September, p. 4.

Fors, B. S. (2014). "SAMOVAЯTEATERET – Drama på grensen" [Samovar Theater – Drama on the Border] in Viken, A. and Fors, B. S. (eds.), *Grenseliv*. Stamsund: Orkana Akademisk, pp. 161–174.

Fors, B. S. (2019). *Border Performances: Politics, Art and Tourism Where Norway Meets Russia*. Tromsø: UiT The Arctic University of Norway.

From Vardø with Love (2006). Directed by Ischenko D., and Korsaeth, H. [Documentary] Norway: Tundra Film.

Gorbachev, M. (1987). *Speech in Murmansk at the Ceremonial Meeting on the Occasion of the Presentation of the Order of Lenin and the Gold Star to the City of Murmansk*. Moscow: Novosti Press Agency, October 1, pp. 23–31.

Hojem, R. (2014). "Skolemusikken og turen til Nikel i 1973 – fritt etter hukommelsen" [School Music and Tour to Nikel in 1973] in Gamnes, O. B. (ed.), *70 år i fred ved en grense 1944–2014: Antologi fortellinger fra tidsvitner. [70 Years in Peace at the Border 1944–2014: Anthology Stories from Contemporary Witnesses]* Kirkenes: Dagfinn Hansens Trykkeri AS, pp. 284–286.

Hønneland, G. (2005). *Barentsbrytninger: Norsk nordområdepolitikk etter den kalde krigen. [Barents Breaches: Norwegian High North Policy after the Cold War]* Kristiansand: Høgskoleforlaget.

Moon, V. B. (1964). Soviet-Norwegian Relations since 1945. *The Western Political Quarterly*, 17 (4), pp. 659–670.

Norsk-russisk handlingsplan for kultursamarbeid i Nordområdene. [Norwegian-Russian Action Plan for Cultural Cooperation in the High North] (2009).

Prieto, A. (1999). *Border Art as a Political Strategy*. ISLA – Information Services Latin America.

Re-Aligned Art website. Available at: https://www.re-aligned.net/re-aligned/?lang=en (Accessed: 28 February 2021)

Samovarteateret. (n.d.-a). *What* [Online]. Available at: https://www.samovarteateret.com/ (Accessed: 10 January 2021).

Samovarteateret. (n.d.-b). *Arctic Voices* [Online]. Available at: https://www.samovarteateret.com/ (Accessed: 10 January 2021).

Steinholt, Y. B. (2005). *Rock in the Reservation: Songs from the Leningrad Rock Club 1981–86*. Bergen and NY: MMMSP

Steinholt, Y. B. (2018). "Cats, Punk, Arson and New Media: Art Activism in Russia 2007–2015" in Meikle, G. (ed.), *The Routledge Companion to Media and Activism*. Milton: Routledge, pp. 153–161.

Stokke, O. S. and Tunander, O. (eds.) (1994). *The Barents Region: Cooperation in Arctic Europe*. Lysaker: The Fridtjof Nansen Institute PRIO Sage.

Tjelmeland, H. (2012). "Border as Barrier and Bridge: The Norwegian-Soviet/Russian Border as a Political and Cultural Construction" in Katajala, K. and Lähteenmäki, M. (eds.), *Imagined, Negotiated, Remembered: Constructing European Borders and Borderlands*. Zürich: LIT Verlag, pp. 167–187.

UNESCO (2013). *The Hangzhou Declaration: Placing Culture at the Heart of Sustainable Policies*. Paris: UNESCO.

United Nations (n.d.). *The 17 Goals* [Online]. Available at: https://sdgs.un.org/goals (Accessed: 30 September 2021).

Viken, A. (2014). "Barents Spektakel; Estetisering og sublimering i grensepolitikken" in Viken, A. and Fors, B. S. (eds.), *Grenseliv*. Stamsund: Orkana Akademisk, pp. 145–160.

Vodka, Water and Glasnost – A Foxtrot for Three (2013). Arvola, I. Directed: Andersen, B.S.

Interviewees quoted

BSA: Andersen, B. S. (director, the Samovar Theater), Kirkenes, Norway, 2.24.21.

MA: Alnes, M. (grant program coordinator, the Norwegian Barents Secretariat), email 8.4.21.

SH: Hlebowicz, S. (visitor, Re-Aligned Art 2013), online, 2.19.21.

SP: Pedersen, S. (former intendant, Tromsø Kunstforening), online, 2.23.21.

AR: Rogatchevski, A. (seminar contributor, Re-Aligned Art 2013), email, January 2021.

MT: Traavik, M. (artist, curator and board member of Barents Spektakel), online, 2.23.21.

Part IV

Environment, Shipping and Emergency Response

12 Issues of Environmental Monitoring and Management in the Arctic

Alexei Bambulyak, Lars-Henrik Larsen, Rolf Rødven, Denis Moiseev and Salve Dahle

Introduction

Environmental management is a complex exercise in exploiting natural resources and safeguarding the environment. This chapter scrutinizes and compares the systems applied in environmental monitoring and management in two Arctic countries, Russia and Norway, exemplified by the joint management of the Barents Sea. The historical development and current content of environmental management routines and practices, and the adequacy of the systems are discussed. Norway and Russia have been sharing and exploiting the marine resources of the Barents Sea for centuries, and joint environmental monitoring and management systems related primarily to rich fisheries have gradually evolved (Wienerroither et al., 2011; Eriksen, 2014). This corresponds to the Sustainable Development Goal (SDG) 14 Life Below Water framework of the UN 2030 Agenda and is a good example of how bilateral and international cooperation can be built and contribute to achieving SDGs.

In the early 1990s, Norwegian-Russian and circumpolar collaboration in general took a major step forward and a new paradigm of Arctic cooperation rapidly emerged. The novel openness in the Arctic, with military threats being downplayed and cooperation on resource management, environmental protection, and nature conservation led to the establishment of several international initiatives and platforms. As early as in June 1991, the first ministerial conference including all eight Arctic countries, along with three indigenous people's organizations, was held in Rovaniemi in Finland. This conference adopted the Arctic Environmental Protection Strategy (AEPS) (Arctic Environment, 1991). The five basic objectives of AEPS were formulated as:

1 to protect the Arctic ecosystem, including humans;
2 to provide for the protection, enhancement, and restoration of environmental quality and the sustainable utilization of natural resources, including their use by local populations and indigenous peoples in the Arctic;
3 to recognize and, to the extent possible, to seek to accommodate the traditional and cultural needs, values, and practices of the indigenous peoples as determined by themselves, related to the protection of the Arctic environment;

DOI: 10.4324/9781003246015-16

4 to regularly review the state of the Arctic environment;
5 to identify, reduce, and, as a final goal, eliminate pollution.

The Norwegian-Russian cooperation had, prior to 1991, primarily dealt with managing fishery resources and establishing agreements on catch quotas. In the early 1990s, this cooperation was broadening and extended to include environmental monitoring and ecosystem management.

The 1991 AEPS initiative gradually evolved into today's Arctic Council (AC). The content and mandate for the AC are provided for in the Declaration on the Establishment of the AC signed in Ottawa, Canada in 1996. The AC is today the most prominent cooperation arena for the eight Arctic nations and representatives from the indigenous people's associations. The work is carried out in permanent working groups, currently six, addressing key issues in the management of the Arctic. The working groups are the Arctic monitoring and assessment program (AMAP), the Protection of the Arctic Marine Environment (PAME), the Conservation of Arctic Flora and Fauna (CAFF), the Emergency Prevention, Preparedness, and Response, the Sustainable Development Working Group, and the Arctic Contaminants Action Program.

The inclusion on the agenda of global climate change issues in the late 1990s and early 2000s, the increased pressure for hydrocarbon exploration in Russian and Norwegian Barents Sea, and the 2011 agreement on the delimitation of the formerly disputed area, all highlight the need for strengthened bilateral cooperation. However, Norway and Russia have so far not developed a jointly approved management plan for the Barents Sea, and today the principles of sustainable development and the ecosystem approach in management are implemented differently in the two countries while referring to the same international frameworks and goals.

Environmental monitoring in the Arctic

Environmental monitoring in the Arctic is a complex and diverse issue. Arctic and non-Arctic states, international organizations, industrial companies, and non-governmental organizations (NGOs) develop strategies and plans reflecting the issues of climate change, industrial developments, environmental and social risks, and impacts in the High North. The AC is now the leading intergovernmental forum promoting cooperation, coordination, and interaction among the Arctic states and other stakeholders on common Arctic issues, focusing on sustainable development and environmental protection.

The coordination of environmental monitoring activities under the AC is organized through its working groups. AMAP in its monitoring activities focuses on abiotic ecosystem components to assess the status of the Arctic region with respect to pollution and climate change issues. CAFF elaborates frameworks and guidelines for biodiversity and habitats monitoring in the Arctic.

AMAP designs its coordinated monitoring program and guidelines, bases its work on national and international monitoring network and research programs, and aims to harmonize this work. AMAP's assessments are based on information and results from monitoring and research work that generates a large amount of data. To optimize the work for gaining access to, processing, reviewing, analyzing, and storing the data, they have established a number of thematic data centers located at the expert institutes and organizations, such as the International Council for the Exploration of the Sea, the Norwegian Institute for Air Research, and the Norwegian Radiation and Nuclear Safety Authority. These centers also conduct data handling work for other international monitoring programs and facilitate harmonized data processing and reporting.

The Circumpolar Biodiversity Monitoring Program (CBMP) of CAFF is intended to harmonize and integrate efforts to monitor living resources. The CBMP is carried out by an international network of scientists, governments, indigenous peoples' organizations, and nature conservation groups. The program organizes its activities around the major ecosystems of the Arctic: marine, freshwater, terrestrial, and coastal. For each of these ecosystems, international steering and expert groups have been established to lead monitoring efforts and develop the biodiversity monitoring plans to guide these efforts. The CBMP emphasizes data management through the Arctic Biodiversity Data Service, capacity building, reporting, and integration of Arctic monitoring as well as communication, education, and outreach. The Arctic Marine Biodiversity Monitoring Plan of 2011 is the first of four pan-Arctic biodiversity monitoring plans developed by the CBMP to improve the ability to detect and understand the causes of long-term change in the composition, structure, and function of Arctic ecosystems. The Arctic Biodiversity Monitoring Plan for Freshwater Ecosystems came out in 2012, for terrestrial in 2013, and the fourth and final one for coastal ecosystems in 2019. All four Arctic monitoring plans of the CBMP apply an integrated ecosystem-based approach to monitoring. The ecosystem-based approach integrates information on land, water, and living resources, and lends itself to monitoring many aspects of an ecosystem within a geographic region (Gill et al., 2011).

AMAP is coordinated with and complements the CBMP managed by the CAFF, and both programs contribute to the Sustaining Arctic Observing Networks (SAON).

SAON is a joint initiative of the AC and the International Arctic Science Committee established in 2011 that aims to strengthen multinational engagement in pan-Arctic observing. SAON's ten-year strategy (2018–2028) addresses present and future Arctic observation needs. It sets priorities to fulfill the network's mission to facilitate, coordinate, and advocate for coordinated international pan-Arctic observations and mobilizes the support needed to sustain them. SAON itself has no part in the research, observations, or funding of these efforts; however, it encourages and promotes collaboration among ongoing observation networks and systems (SAON, 2018).

Ecosystem-based approach in Arctic environmental monitoring and management

The AC has developed a strategy for stimulating the ecosystem-based management of the Arctic Seas specified in the 2004 Arctic Marine Strategy Plan. The PAME working group of the AC applies a definition of Large Marine Ecosystems (LMEs) by ecological criteria, including bathymetry, hydrography, productivity, and tropically linked populations. PAME has identified 18 LMEs in the Arctic, each with unique features needing attention in a management plan. The Council has also developed guidelines for applying ecosystem-based management in a pan-Arctic perspective (Hoel, 2009; Skjoldal and Mundy, 2013).

Several of the Arctic LMEs fall entirely within the Exclusive Economic Zone (EEZ) of one single country, for example, LME 6, 7, and 8, are located entirely in the Russian EEZ. Some LMEs are shared among two or more Arctic coastal states. The Barents Sea LME is shared between Norway and Russia (Figure 12.1) (Skjoldal and Mundy, 2013; Matishov et al., 2003).

Managing Arctic ecosystems requires clear targets defining what ecosystem services are to be obtained and what impacts are to be tolerated. It requires input data of good spatial and temporal coverage and quality. Based on this, threshold values for selected indicators are specified. Threshold values are to be accompanied by definitions of trends or specific values which will trigger appropriate actions. Understanding the sensitivity and vulnerability of the ecosystem to any

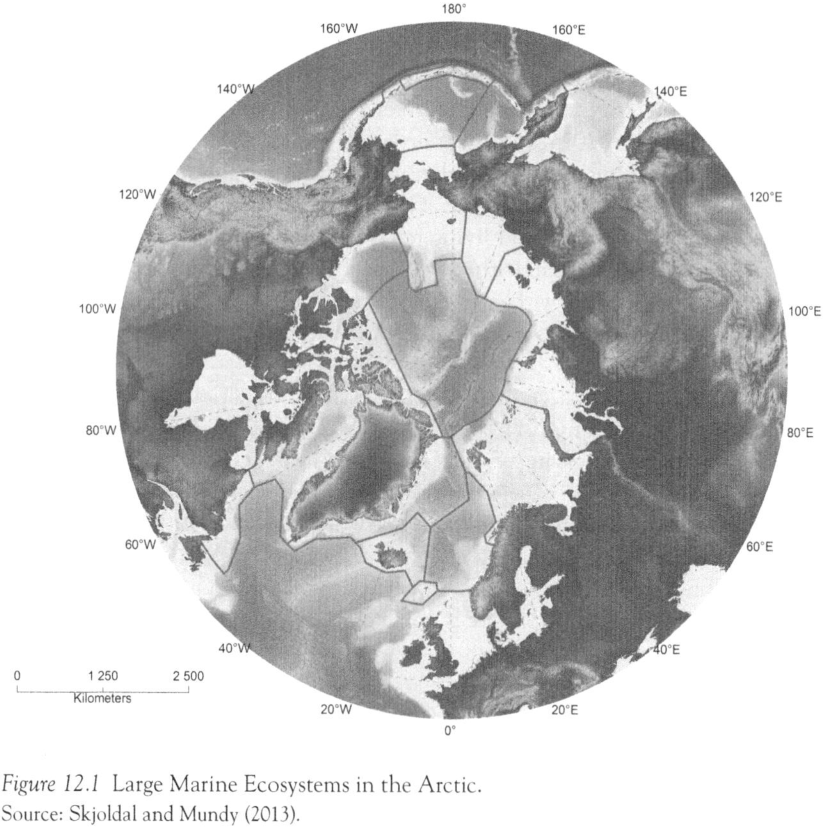

Figure 12.1 Large Marine Ecosystems in the Arctic.
Source: Skjoldal and Mundy (2013).

kind of pressure is essential. Thus, the key objectives are the collection of data to obtain knowledge and from knowledge to generate understanding, which again is turned into management.

Collecting monitoring data

Environmental monitoring is the process of collecting data on environmental variables, with the intrinsic intent to actually enforce measures. In the Barents Sea, collection of monitoring data has historically been directed at monitoring harvestable fish stocks. Data are collected through surveys or through information supplied by the commercial fisheries. Fishery research institutes – IMR in Norway and PINRO in Russia – have since 2003 undertaken coordinated ecosystem surveys, providing an annual picture of the Barents Sea. Combined summer and winter surveys generate insight into annual variations in the ecosystems and commercial fish populations (Mehl et al., 2013; Eriksen, 2014).

The extensive data gathering related to fisheries management in the Barents Sea has since 1998 been supplemented by the seabed and sediment monitoring carried out on behalf of the petroleum industry. Thus, a new commercial interest entering the Barents Sea triggered the collection of targeted data for monitoring this activity.

Environmental studies in the Barents Sea are carried out by national (Norwegian, Russian) and international public and private institutes, generating data for scientific and management purposes.

Provision of data and information for ecosystem-based management in the Arctic

Research is a curiosity-driven process, intrinsically including the risk of not delivering the evidence or conclusions requested by the party doing the management and commissioning the research. However, science is a key to achieving a basic understanding and a solid, unbiased, quality assured and reliable knowledge base.

Research addresses the nonroutine questions and issues relevant to ecosystem-based monitoring and management. This is done through sampling in the field, supplemented by experiments and analyses. Based on research, models are developed and verified through time series of data. Technological methods for data gathering and surveillance have improved rapidly in recent decades, enabling researchers to accumulate vast amounts of data on a number of previously nonmeasurable parameters. These include the atmospheric composition over the North Pole, counts of bacteria in samples from deep Arctic ice cores, and photographs from the bottom of the Polar Ocean – topics unmeasurable and beyond mapping capabilities just a few decades ago.

Not all areas of the Arctic are as accessible and amenable to monitoring and research as the Barents Sea. Despite satellites passing over the Polar regions and measuring, recording, and accumulating data day and night year-round, Arctic indigenous, local, and traditional key knowledge is disappearing with a generally more remote relationship to nature. Knowledge accumulated through generations is losing relevance as climate changes and traditional lifestyles in remote areas and settlements decline. Indigenous and traditional knowledge is not peer reviewed, it is not necessarily available online, but in many instances, it can be of more local significance than any other sources of information.

Indicators for environmental monitoring in the Barents Sea

Indicators for environmental monitoring in Norway

To monitor natural development and from the effects of actions by various actors, Norway defined a set of 40 indicators in the first ecosystem-based management plan for the Barents Sea issued in 2006 (Table 12.1). Data on each indicator are collected through several institutions regularly surveying the Barents Sea, supplemented by data from research and mapping projects.

The list of indicators is revised and adjusted regularly, but in order to maintain time series – monitoring is, after all, related to gaining knowledge on changes over time – one should be careful of excluding any chosen indicator from the list. The latest revision of the Management Plan (April 2020) presents an updated indicators list (Meld. St. 20, 2019–2020).

Indicators for environmental monitoring in the Russian Arctic Seas

In 2015, the Ministry of Nature Resources and Environment of Russia approved the list of species – indicators of the sustainable state of the Arctic marine ecosystems (Ministry of Nature Resources and Environment, 2015). The list includes 61 taxa – 22 of flora and 39 of fauna. Groups and examples of indicator species are presented in Table 12.2.

In 2020, Rosneft, the Russian oil major, published the first scientific report of its kind, Species – Biological Indicators of the Status of the Marine Arctic Ecosystems (Mokievskiy et al., 2020) – in the companies' series of environmental atlases of the Russian seas. The report summarizes existing knowledge on 61

Table 12.1 Original indicators for environmental monitoring in the Norwegian part of the Barents Sea (2006) and indicators included in the 2020 update

Theme	*Indicator*
Ocean climate	Distribution of ice in the Barents Sea
	Temperature, salinity, and nutrients at fixed transects
	Inflow of Atlantic water into the Barents Sea
Phytoplankton	Species composition
	Biomass and production expressed as chlorophyll
	Timing of the spring bloom
Zooplankton	Species composition
	Biomass

Non-harvested fish stocks	Blue whiting biomass and distribution
	Young herring biomass and distribution
Commercial fish stocks	Population of mature capelin
	Spawning stock biomass of North-East Atlantic cod
	Spawning stock biomass of Greenland halibut
	Spawning stock biomass of redfish
	Spawning stock biomass of beaked redfish
Benthic organisms	Species composition and number of benthic animals in research trawls
	Distribution of coral reefs, Gorgonaceans, and marine sponges
	Distribution of red king crab
Seabirds and mammals	Spatial distribution of seabird communities
	Population development in puffin
	Population development in common guillemot
	Population development in Brünnich's guillemot
	Population development in black-legged kittiwake
	Spatial distribution of whales
	Bycatch of harbor porpoise (excluded from 2020 list)
Introduced species	Introduced species (only king crab monitored regularly)
Vulnerable and threatened species	Red-listed species and habitat types
Contamination	Intertidal waste accumulation (in Svalbard, specified in 2020 list)
	Atmospheric contribution of contaminants
	Riverine contribution of contaminants
	Contaminants in sediments
	Contaminants in macroalgae (radioactivity, specified in 2020 list)
	Contaminants in blue mussels
	Contaminants in shrimp
	Contaminants in cod
	Contaminants in capelin
	Contaminants in polar cod
	Contaminants in seabirds (Brünnich's guillemot)
	Contaminants in ringed seal
	Contaminants in polar bear
Human activity	Fish mortality (included in 2020 list)

Table 12.2 Species of flora and fauna – indicators of the sustainable state of marine ecosystems of the Russian Arctic

	Category	*Species*
I	Seaweed, algae, plants	Twenty-two species
	• Brown algae	• Six species
	• Red algae	• Two species
	• Green algae	• Four species
	• Diatoms	• Six species
	• Dinophytic algae	• Three species
	• Vascular plants	• One species
II	Invertebrates	Eighteen species
	• Ascidians	• One species
	• Chaetognatha	• One species

	Category	*Species*
	• Polychaete	• Two species
	• Echinoderm	• Three species
	• Molluscs	• Five species
	• Crustacean	• Six species
III	Vertebrates	Twenty-one species
	• Fish	• Six species (e.g., polar cod, navaga, three-spined stickleback)
	• Birds	• Ten species (e.g., common eider, ivory gull, Brünich's guillemot, black-legged kittiwake)
	• Mammals	• Five species (Polar bear, walrus, ringed seal, beluga whale, bowhead whale)

Source: Created by authors based on Ministry of Nature Resources and Environment (2015).

indicator species in the Russian Arctic Seas, including the Barents, Kara, Laptev, East-Siberian, and Chukchi seas.

Joint Norwegian-Russian indicators for environmental monitoring

Norwegian-Russian cooperation in fisheries management has since the 1960s included environmental monitoring and several annual surveys. The results of this work have been published in joint reports and papers, including the regularly updated joint environmental status report for the Barents Sea (e.g., Aanes et al., 2009, 2016). Research institutes and environmental authorities from both countries have also been working under the Joint Norwegian-Russian Environmental Commission toward the harmonization of environmental monitoring tools and the development of joint indicators. From 2002 to 2011, several bilateral projects on the harmonization of monitoring tools were carried out. The basis was the Convention for the Protection of the Marine Environment of the North-East Atlantic Joint Assessment Monitoring Program/Coordinated Environmental Monitoring Program guidelines.

In 2005, the Marine working group of the Joint Norwegian-Russian Environmental Commission launched the project on Ecosystem Based Joint Monitoring of the Barents Sea. In 2015, the bilateral project group proposed the set of 22 joint indicators for environmental monitoring (see Table 12.3), including 11 indicators originating from the Norwegian-Russian fishery cooperation and ecosystem surveys by IMR and PINRO and 11 new indicators in a bilateral context (Korneev et al., 2015a). The work on joint indicators development is ongoing within the Action Plan of the Norwegian-Russian bilateral Environmental Commission.

Integrated management plans of Norway

In the last 15 years, Norway has managed the marine offshore areas following the ecosystem-based management principle. In Report No. 12 to the Norwegian Parliament (*Storting*) (2001–2002) – Protecting the Richness of the Seas – the

Table 12.3 The 22 proposed joint indicators and information regarding monitoring in Russia and Norway and status of ongoing monitoring in Russia and Norway (Yes* means monitoring of not all parameters/sub-parameters included in the existing monitoring programs)

Indicator	*Monitoring*	
	Russia	*Norway*
Sea ice cover in the Barents Sea	Yes*	Yes*
Meteorological conditions	Yes	Yes
Oceanographic conditions	Yes*	Yes*
Water masses properties and volume transport in the Barents Sea	Yes*	Yes*
Ocean acidification and ocean CO2 uptake	No	Yes*
Phytoplankton diversity, abundance, and biomass	Yes*	Yes
Zooplankton diversity, abundance, and biomass	Yes	Yes
Benthos diversity, abundance, and biomass	Yes*	Yes*
Microbe biomass and diversity	No	No
Sea ice biota, diversity, and abundance	No	No
Fish and shrimp biomass	Yes*	Yes*
Fishing pressure	No	Yes*
Introduced species	Yes*	Yes*
Seabird communities/assemblages at sea	Yes*	Yes*
Population development and demography of seabirds	Yes*	Yes
Dynamics of ice-associated marine mammals	Yes*	Yes
Vulnerable and endangered species	Yes	Yes*
Pollution levels in the physical environment	Yes*	Yes*
Contaminants levels in the biota	Yes*	Yes*
Bottom substrate	Yes	No
Demersal fauna biodiversity	Yes	Yes

Source: Korneev et al. (2015a), adopted by the authors.

concept of the ecosystem-based management of Norwegian Sea areas was introduced following the international 'Malawi principles' of the UN Convention on Biological Diversity (CBD, 1998). The aim was to establish three management plans, covering all Norwegian Sea areas, one for the Barents Sea, including Lofoten, one for the Norwegian Sea, and one for the North Sea.

The Integrated Management Plan for the Barents Sea and Lofoten was established in 2006 (Report No. 8 to the *Storting* 2005–2006). The first plan for the Norwegian Sea was approved in 2009 and the management plan for the North Sea was composed in 2013. The first update of the Barents Sea-Lofoten management plan was released in 2011 and the second update was issued in 2015.

Originally, all three management plans were to be fully updated every four years, but new data and information were not generated at a pace justifying the use of resources to implement such frequent updates. The current approach is that the management plans are revised every eighth year, and minor adjustments are made every fourth year (updates), and that all three plans are reviewed simultaneously. In April 2020, the first complete management plan covering all Norwegian Sea areas was published (Meld. St. 20, 2019–2020).

Two advisory groups have been established to develop the scientific basis for the marine management plans: the Forum for Integrated Marine Management and the Advisory Group on Monitoring.

Particularly valuable and vulnerable areas

Particularly valuable and vulnerable areas (SVO – in Norwegian *Særlig Verdifulle og Sårbare Områder*) are geographically defined areas which, on the basis of scientific assessments, have been identified as areas of significant importance for biological diversity and production within and often also outside the area (Eriksen et al., 2021).

For the assessment of valuable and vulnerable areas, that is, areas that are important as living areas for species at different times of the year in Norwegian waters, the Norwegian Environment Agency applies seven CBD scientific criteria for ecologically or biologically significant areas (EBSAs). Areas are defined as valuable according to the spatial and temporal distributions of certain species and life stages of fish, seabirds, marine mammals, and habitat types. In addition, several larger areas are defined as particularly valuable and vulnerable, based on their importance for biological diversity and production, and where disturbance could potentially cause long-lasting or irreversible damage. Detailed descriptions of the criteria, methodology, and documentation used to define valuable areas are provided for fish, seabirds, marine mammals, and habitat types. In the maps presented in the Norwegian Environment Agency's web portal, fish, seabirds, and marine mammals are determined on a monthly basis, whereas habitat types (particularly valuable areas) stay the same throughout the year.

A number of particularly valuable and vulnerable areas have been identified in the Barents Sea – Lofoten area. In the border area between Norway and Russia, these include 50 kilometers of coastal zones along Troms and Finnmark counties and a sea area north of the ice edge (Figure 12.2).

According to the integrated management plan, the designation of areas as particularly valuable and vulnerable does not have any direct effect on the form of restrictions on commercial activities, but indicates that these are areas requiring special caution. They have been used as a basis for setting an overall framework for activities, to make activities in such areas subject to special requirements using the current legislation. Such requirements may apply to the whole or part of a particularly valuable and vulnerable area and must be considered on a case-by-case basis for specific activities. For example, petroleum activities are not to be implemented or initiated in these areas, and additional seasonal limitations on exploratory drilling in the Barents Sea have been established for extended coastal areas along Troms and Finnmark counties and around Bjørnøya (Figure 12.3).

Ice edge – marginal ice zone: an example of an SVO

A marginal ice zone (MIZ) has been defined as an SVO in all versions of the Norwegian integrated management plans for the Barents Sea-Lofoten area. This SVO includes the sea areas that are most important for biological production and diversity related to the ice edge zone as a natural phenomenon. For the purposes of the management

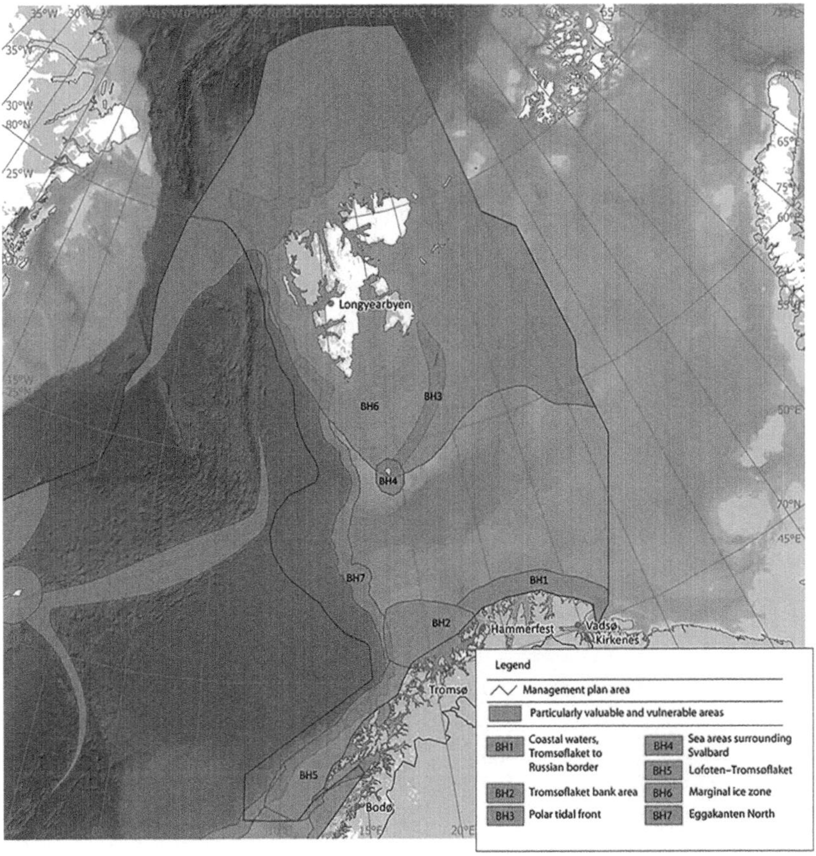

Figure 12.2 Particularly valuable and vulnerable areas (SVO) in the Norwegian part of the Barents Sea according to the integrated management plan.
Source: Meld. St. 20 (2019–2020).

plan, MIZ as a particularly valuable and vulnerable area has been delimited using statistical methods of expressing satellite observations of variations in sea ice extent throughout the year for a series of years. MIZ, as an SVO in the management plans issued in 2006, 2011, and 2015, was based on ice data for a recent observations period spanning a series of years (over 20 years) and the delimitation of the ice edge was set where sea ice occurred on 30% of the days in April (30% ice frequency).

The Forum for Integrated Marine Management has recommended that mapping MIZ should continue based on the presence of sea ice in April using the latest available 30 years' time series with satellite observations of ice cover extent. However, delimitation of the MIZ as an SVO was adjusted in terms of ice frequency from 30% to 15% of the days in April, thereby extending this particularly valuable and vulnerable area. This limit can be updated in the later versions of the management plans (Meld. St. 20, 2019–2020).

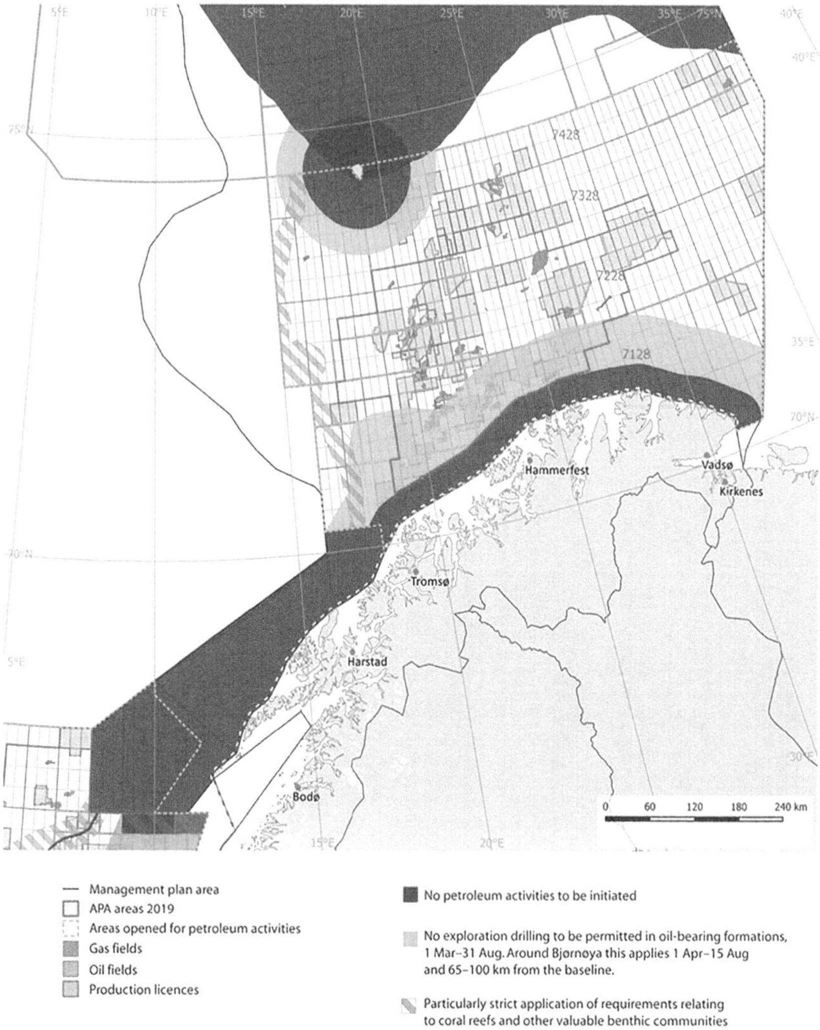

Figure 12.3 Petroleum license areas in the Barents Sea by 2020 and particularly valuable and vulnerable areas (SVO) with associated limitations in petroleum activities.

Source: Meld. St. 20 (2019–2020).

Integrated management plans of Russia

There is so far no approved integrated management plan for the Russian part of the Barents Sea. The Russian authorities have initiated the process of developing an integrated management plan for marine nature resources for the Russian part of the Barents Sea, launching research projects and processes of elaboration of relevant laws and regulations. Research institutes and consortia, tasked with the elaboration of concepts and drafts of the integrated management plan, have used international experiences and criteria, for example, LME concept, the International Union for Conservation of Nature (IUCN) and CBD EBSA criteria, for mapping environmental values and ecologically valuable areas in the Barents Sea (Bambulyak et al., 2021).

In 2013, the Zubov State Oceanographic Institute (SOI) in Moscow, at the behest of the Ministry of Economic Development of Russia, published a report on results of their research work on the elaboration of the methodology of marine spatial planning (MSP) and a comprehensive (integrated) management plan of marine nature use management in the Barents Sea, taking into account international experience and best practices of transborder resource utilization (Zemlyanov, 2013). The research team led by SOI used the LME concept to define borders for MSP in the Barents Sea, and referred to three basic publications, among them to the Norwegian Integrated Management Plan for the Barents Sea-Lofoten area 2005–2006.

In 2015, Sevmorgeo from St Petersburg, assigned by the Ministry of Natural Resources and Environment of Russia, issued a report called project of a plan for the comprehensive (integrated) management of marine nature use for the Russian part of the Barents Sea based on an ecosystem approach (Korneev, 2015b). The authors concluded that the UN Educational, Scientific and Cultural Organization (UNESCO) Intergovernmental Oceanographic Commission (IOC) Manual and Guide on MSP was a basic international document to be followed for developing the integrated management plan for Russian seas and pointed out that the Norwegian Integrated Management Plan for the Barents Sea-Lofoten area provided a good platform for MSP in the Russian part of the Barents Sea.

In 2016, the Ministry of Natural Resources and Environment of Russia presented the pilot project of the integrated management of nature resources use in the Russian part of the Barents Sea prepared on the basis of the report issued by Sevmorgeo. The UNESCO IOC MSP Manual and Guide and the Norwegian Integrated Management Plan for the Barents Sea-Lofoten area were used for setting aims and compiling this pilot project (Figure 12.4).

The plan for the pilot project implementation included elaboration and approval of the Federal Law on Marine Spatial Planning. The concept of the Law was elaborated by the Ministry of Regional Development of Russia. The Ministry of Regional Development was abolished in 2014 and responsibility for elaboration of the Law transferred to the Ministry of Economic Development of Russia. Activities related to the elaboration and implementation of the pilot project of the integrated management plan for the Russian part of the Barents Sea are currently coordinated by the State Commission on the Development of the Arctic.

Particularly valuable and vulnerable areas in Russia

SOI in their 2013 report proposed to define and map existing and planned special nature protected areas, areas of spawning, feeding, and fishing of aquatic resources as valued nature areas for MSP within the LME of the Barents Sea. Giving protection priority to key identified areas with high seasonal concentrations of birds and marine mammals has also been proposed (Zemlyanov, 2013).

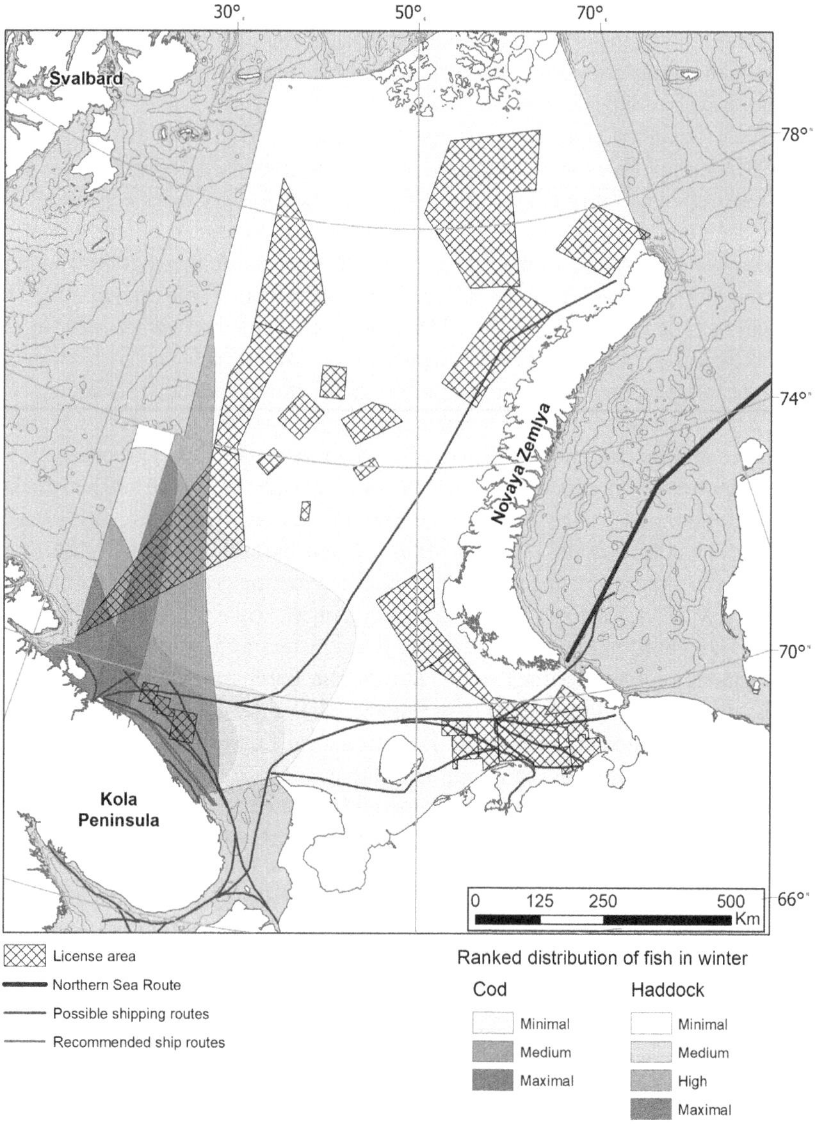

Figure 12.4 Areas of possible conflict of interests of fishery, shipping, and petroleum industries in the Barents Sea for the first quarter according to the 2015 project of an integrated management plan for the Russian part of the Barents Sea.

Source: Korneev (2015b), adopted by MMBI.

The research team led by Sevmorgeo defined and mapped biological or ecological values and valuable and vulnerable areas using CBD EBSA (Korneev, 2015b). Maps with visual presentations of the spatial and seasonal distribution of biotic components were created from monitoring data. The biotic components of the Barents Sea and their values were assessed and mapped for the four seasons

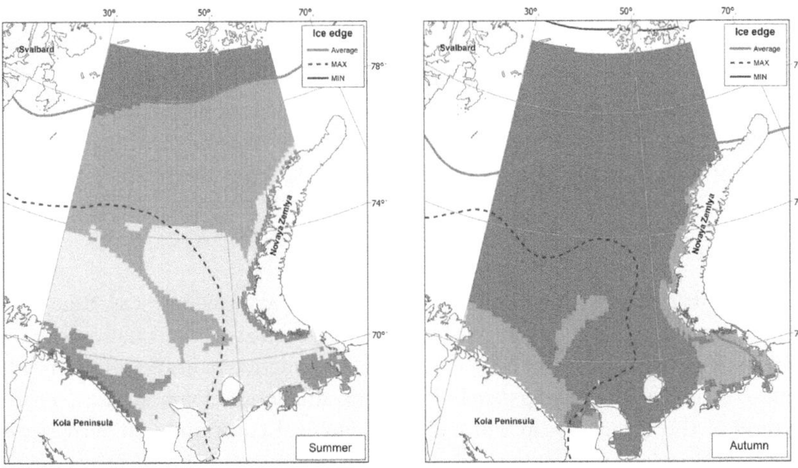

Figure 12.5 Example of the maps of integral vulnerability of the Barents Sea ecosystems to oil spills in summer (left) and autumn (right). Values of vulnerability are ranged from maximum to minimum.

Source: Shavykin and Ilyin (2009).

(winter, spring, summer, autumn) with ranks (from 0 to 3) in the respective units. The authors of the report suggested using mapping of (integral) vulnerability to specific impact (e.g. oil pollution) or integrated impact of human activity in addition to mapping biological values in the MSP process (see examples in Figure 12.5). This approach resembles those used in the Norwegian integrated management plans for defining particularly valuable and vulnerable areas (SVO).

Ice edges or MIZ were not proposed as particularly valuable and vulnerable areas in the abovementioned projects of the integrated management plans for the Russian part of the Barents Sea (Zemlyanov, 2013; Korneev, 2015b).

Integrated management plan for the Barents Sea – the Norwegian-Russian perspective

The integrated management plan for the Barents Sea-Lofoten area was introduced in Norway in 2006 and has been updated three times, including the recent update released in 2020. Russia is still in the process of developing and introducing an integrated management plan, including the elaboration and approval of norms and regulations. Despite the differences in the management plans' status on the two sides of the border, it is important to note that both the Norwegian and Russian integrated management plans and/or plan proposals are built on the same international approach for MSP with ecosystem-based management principles and the application of CBD EBSA criteria. Moreover, the Norwegian plan was used as one of the key reference documents for drafting the Russian 2015 plan (Korneev, 2015b; Aune et al., 2017). We also perceive certain differences in the possible practical implementation of international principles and criteria regarding the management of petroleum activities and in the definition of particularly valuable and vulnerable areas in the Barents Sea in Norway and Russia, initially those related to MIZ and ice-covered waters.

While it is unclear when the integrated management plan for the Barents Sea will be introduced in Russia, we find a good basis for developing practical cooperation on environmental monitoring, impact, and risk assessment, and also on the implementation of up-to-date and harmonized tools across the border.

Conclusions

The Arctic basin, including most of the bordering epicontinental seas, except the Barents Sea, has so far been difficult to access. It holds few commercially attractive fish resources and has not yet been the target of extractive industries. This may be about to change with growing demand in energy and food resources. The expected receding of the Polar ice cap will make the physical constraints on activities in the Arctic Seas less prominent, while with the commercial industries entering the scene in great number, conflicts of interest will inevitably arise and so also the need for management, pursuing common goals, and following mutually agreed rules.

If human resource exploitation is considered the main driver necessitating the development of management, the role of science in providing the needed understanding of Arctic ecosystems should also be emphasized. Even in the relatively well-explored Barents Sea, there are ecosystems of poorly known sensitivity and resilience to man-made stimuli, one example being the MIZ ecosystem with its unique species composition and production properties.

Human impact on the Barents Sea ecosystem through removal of biomass has taken place for centuries. Examples range from the harvesting of whales from the 15th to 17th centuries to the extreme harvesting of capelin in the 1970s and the continuous harvesting of the world's largest cod stock. Managing this type of influence requires large amounts of data and computation capabilities. With new impacts becoming eminent, such contaminations and petroleum activities, the scientifically justified ecosystem-based approach in management becomes vital. A leading role for the AC as an intergovernmental forum and coordination platform is to be encouraged.

The Barents Sea LME can be a driving force for the introduction of ecosystem-based management in a pan-Arctic perspective. Norway and Russia share and manage biological fisheries and other resources in the Barents Sea and carry out industrial activities, including oil and gas exploration and production, which entail environmental impacts and risks in the cross-border context. Two countries manage resources and control industries according to national laws and regulations, following international conventions (e.g., UNCLOS or CBD) and bilateral agreements. Norway and Russia have some similar procedures in managing the offshore petroleum industry, such as environmental impact assessment and monitoring and protecting the marine environment. However, no harmonized system has been established between Norway and Russia in industrial control and environmental protection in the Barents Sea (Bambulyak et al., 2015), and the principles of sustainable development and the ecosystem approach to management are implemented differently in the two countries.

For example, Norway that introduced an integrated management plan for the Barents Sea in 2006 has updated this plan regularly and addressed the issue of establishing a joint coordinated management plan with Russia. In the last ten years, there have been several initiatives in Russia aimed at the elaboration of an integrated plan for natural resources management in the Barents Sea, but no such plan has so far been established.

In Norway and Russia, the oil and gas industry has been moving northwards. In Norway, with its long experience in offshore exploration and production, the northernmost licensed areas are limited to 74°30' N, and certain limitations on petroleum activities have been established for operations in ice-covered waters, defining these as particularly valuable and vulnerable areas. Russia has already gained experience in year-round petroleum operations in areas covered by seasonal ice (e.g., the Varandey terminal and Prirazlomnaya platform in the south-eastern Barents Sea) and has granted licenses for operations above 74°30' N (Bambulyak et al., 2021).

Either an integrated management plan for the Barents Sea needs to be introduced in Russia in the near future or its development and approval will take a long time, so we can expect that oil and gas activities will not be managed in Russia in the same way as in Norway, also when it comes to operations in ice-covered waters. Nevertheless, certain steps can be implemented to contribute to the establishment of more harmonized and coordinated procedures in environmental management, monitoring, impact, and risk control between the two countries. These steps are to be coordinated through the programs and projects of the working groups of the AC and bilateral Norwegian-Russian Fishery and Environmental Commissions. This will also serve to build a better understanding of the Barents Sea ecosystem and support the development of a joint coordinated management plan for the entire Barents Sea that can be promoted and extended to other Arctic areas – LMEs defined by the AC.

December 2021

Ex-post reflections

This chapter was written in 2021 by a team of authors representing Norwegian and Russian environmental research institutes and the secretariat of the AMAP. This chapter describes the status of environmental monitoring and management in the Arctic, focusing on international and interregional cooperation, looking at Norwegian and Russian experiences as an example, and a driving force for circumpolar collaboration. This chapter reflects much of the authors' personal experiences in building up cooperation across borders during the last 30 years. In our conclusions, we looked positively to the future and emphasized the importance of science in providing knowledge about the changing Arctic, developing circumpolar environment monitoring systems, and moving step-by-step toward sustainable and ecosystem-based management.

The year 2022 brought tragic events and dramatic changes. Today, we cannot foresee how environmental monitoring and management in the Arctic can be implemented or what the role of the AC or international and bilateral agreements will be. However, what we can say for sure is that environmental monitoring of the

Arctic, ecosystem-based management of resources, and sustainable development of the circumpolar regions is of ever-growing importance and cannot be implemented without cooperation and trust between countries, institutes, and people.

April 25, 2022

Acknowledgments

The authors acknowledge RN Nordic Oil, Norway, for providing financial support for the study.

References

Aanes, S., Aars, J., Alekseev, G., Anders, J., Anisimova, N.A., Arneberg, P., Baik, B. et al. (2016). *Joint Norwegian-Russian Environmental Status 2013*. Report on the Barents Sea Ecosystem. Complete Report, Part II. IMR/PINRO Joint Report Series: Bergen.

Aanes, S., Alekseev, G., Anisimova, N., Arneberg, P., Baik, B., Bambulyak, A., Belikov, S. et al. (2009). *Joint Norwegian-Russian Environmental Status 2008*. Report on the Barents Sea Ecosystem. Complete Report. IMR/PINRO Joint Report Series: Bergen.

Arctic Environment (1991). Arctic Environmental Protection Strategy. Declaration on the Protection of Arctic Environment. Rovaniemi, Finland, 14th of June 1991, p. 45.

Aune, M., Bambulyak, A., Sagerup, K., Aniceto, A.S., Moiseev, D., Vaschenko, P., Kalinka, O., and Dukhno, G. (2017). Report OCEAN-1: Valuable Areas in the Barents Sea Phase 1. Ak-vaplan-niva AS 2017. Rapport 8328. p. 68.

Bambulyak, A., Golubeva, S., Sydnes, M., Sydnes, A.K., Larsen, L.H., and Streletskaya, V. (2015). Resources Management Regimes in the Barents Sea. In: Bourmistrov, A., Mellemvik, F., Bambulyak, A., Gudmestad, O.T., Overland, I. and Zolotukhin, A. (eds.) *International Cooperation and Arctic Petroleum: Barents Sea Scenarios*. Routledge. pp. 53–76

Bambulyak, A., Skeie, G.M., Moiseev, D., and Vaschenko, P. (2021). Norwegian and Russian Border Areas in the Barents Sea – Environmental Data for Petroleum Industry. Akvaplan-Niva Report 62178.03. p. 36.

CBD (1998). Report of the Workshop on the Ecosystem Approach. Lilongwe, Malawi, 26–28 January 1998. Convention on Biological Diversity. UNEP/CBD/COP4/Inf.9

Eriksen, E. (eds.) (2014). *Survey Report from the Joint Norwegian/Russian Ecosystem Survey in the Barents Sea and Adjacent Waters*, August–October 2014. IMR/PINRO Joint Report Series, No. 1/2015, p. 153.

Eriksen, E., van der Meeren, G.I., Nilsen, B.M., von Quillfeldt, C.H., and Johnsen, H. (eds.) (2021). Særlig verdifulle og sårbare områder (SVO) I norske havområder – Miljøverdi. En gjennomgang av miljøverdier og grenser i eksisterende SVO og forslag til nye områder. (Particularly Valuable and Vulnerable Areas (SVO). In *Norwegian Sea Areas - Environmental Value. A Review of Environmental Values and Boundaries in Existing SVO and Proposals for New Areas)*. Havforskningsinstituttet rapportserie: Rapport fra havforskningen 2021–26.

Gill, M.J., Crane, K., Hindrum, R., Arneberg, P., Bysveen, I., Denisenko, N.V., Gofman, V., Grant-Friedman, A., Gudmundsson, G., Hopcroft, R.R., Iken, K., Labansen, A., Liubina, O.S., Melnikov, I.A., Moore, S.E., Reist, J.D., Sirenko, B.I., Stow, J., Ugarte, F., Vongraven, D., and Watkins, J. (2011). *Arctic Marine Biodiversity Monitoring Plan* (CBMP-MARINE PLAN), CAFF Monitoring Series Report No. 3, April 2011, CAFF International Secretariat, Akureyri, Iceland.

Hoel, A.H. (eds.) (2009). Best Practices in Ecosystems-Based Ocean Management in the Arctic. *Norwegian Polar Institute Report 129*, p. 114.

Korneev, O. (eds.) (2015b). *Design of a Plan of Comprehensive (Integrated) Management of Marine Nature Use for the Russian Part of the Barents Sea based on an Ecosystem Approach.* Scientific-Research Report. Sevmorgeo, St Petersburg. p.1105 (in Russian).

Korneev, O., Titov, O., van der Meeren, G.I., Arneberg, P., Yu, Tchernova, and Jørgensen, N.M. (2015a). *Joint Russian-Norwegian Monitoring Project – Ocean 3.* Final Report 2012–2015. Brief Report Series, Norwegian Polar Institute Report 030.

Matishov, G.G., Denisov, V.V. and Dzhenyuk, S.L. (2003). Contemporary State and Factors of Stability of the Barents Sea Large Marine Ecosystem. In: *Large Marine Ecosystem of World: Trends in Exploration, Protection, and Research.* Elsevier, pp. 41–74

Mehl, S., Aglen, A., Alexandrov, D.I., Bogstad, B., Dingsør, G.E., Gjøsæter, H., Johannesen, E., Korsbrekke, K., Murashko, P.A., Prozorkevich, D.V., Smirnov, O.V., Staby, A., and de Lange Wenneck, T. (2013). *Fish Investigations in the Barents Sea Winter 2007–2012.* IMR-PINRO Joint Report Series 1–2013, p. 97.

Meld. St. 20. 2019–2020. Report to the Storting (White Paper). Norway's Integrated Ocean Management Plans. Barents Sea-Lofoten area; the Norwegian Sea; and the North Sea and Skagerrak. Norwegian Ministry of Climate and Environment.

Ministry of Nature Resources and Environment of Russia (2015). On Approval of the List of Species of Flora and Fauna as Indicators of the Sustainable State of the Marine Ecosystems of the Arctic Zone of the Russian Federation. Resolution Dated 22.09.2015 # 25-r.

Mokievskiy, V.O, Tsetlin, A.B., Azovskiy, A.I., Naumov, A.D., Kosobokova, K.N., Kuzischin, K.V., Sapozhnikov, F.V., Vedenin, A.A., Gavrilo, M.V., Isachenko, A.I., Ilyushin, D.G., Simakova, U.V., Zalota, A.K., Shabalin, N.V., Maksimova, O.V., Glazov, D.M., Mardashova, M.V., Yu, O., Kalinina, A.S., Valieva, D.R., Zagretdinova, R.E., Lazareva, and Udovik, D.A. (2020). *Species – Biological Indicators of the State of Arctic Marine Ecosystems.* Rosneft and Arctic Research Centre Report. Moscow, NIR Foundation, p. 386.

Report No. 8 to the Storting. 2005–2006. *Integrated Management of the Marine Environment of the Barents Sea and the Sea Areas off the Lofoten Islands.* The Royal Norwegian Ministry of the Environment.

Report No. 12 to the Storting. 2001–2002. *Protecting the Riches of the Seas.* The Royal Norwegian Ministry of the Environment.

SAON (2018). *Sustaining Arctic Observing Networks Strategy: 2018–2028.*

Shavykin, A.A. and Ilyin, G.V. (eds.) (2009). *Assessment of the Integral Vulnerability of the Barents Sea Area to Oil Pollution.* Murmansk: MMBI, p. 99 (in Russian)

Skjoldal, H.R. and Mundy, P. (2013). Large Marine Ecosystems (LMEs) of the Arctic Area. Revision of the Arctic LME Map 15 May 2013. PAME.

Wienerroither, R., Johannesen, E., Dolgov, A., Byrkjedal, I., Bjelland, O., Drevetnyak, K., Eriksen, K.B., Høines, Å., Langhelle, G., Langøy, H., Prokhorova, T., Prozorkevich, D., and Wenneck, T. (2011). *Atlas of the Barents Sea Fishes.* IMR/PINRO Joint Report Series.

Zemlyanov, I.V. (ed.) (2013). *Elaboration of Methodology of Marine Spatial Planning and Comprehensive (Integrated) Management Plan of Marine Nature Use Management in the Barents Sea taking into Account International Experience and Best Practices of Trans-Border Resource Utilization.* Scientific-Research Report. Moscow: Zubov State Oceanographic Institute, p. 388 (in Russian)

13 International Shipping and the Northern Sea Route

Björn Gunnarsson and Arild Moe

Introduction

The Northern Sea Route (NSR) is the Russian term for the waterways north of Siberia. They form a part of the Northeast Passage (NEP), which is the historical term for the Arctic Sea passage between the Atlantic and the Pacific. Whereas NEP is a loose term, without strict geographical boundaries, the NSR is now precisely defined, as starting with the entry to the Kara Sea and stretching all the way to the Bering Strait. Northwards, it extends 200 nautical miles from the coast. In the Russian legislation, it is referred to as a 'water area' – *akvatoriya*. Within this area, there are several alternative shipping lanes which can be chosen depending on the ice situation or a vessel's water depth requirements (Figure 13.1).

The Russian regulations for shipping within this area are contested by some states, notably the United States. They hold that the regulations go further than permitted under the law of the sea. Nonetheless, commercial users respect the Russian administration (Solski, 2020).

During Soviet times, the NSR was open to foreign shipping only exceptionally. This changed with the speech by Mikhail Gorbachev in Murmansk in 1987, where he called for international cooperation in the Arctic generally and in shipping specifically (Åtland, 2008). In early 1991, the NSR was officially opened to international shipping. The decision was spurred by a reassessment of the security situation and the expectation of economic benefits. However, political declarations alone do not spur commercial interest. Use of the sea route, which peaked in 1987, plummeted after the collapse of the Soviet Union. The comprehensive International Northern Sea Route Programme was carried out in the 1990s analyzing the conditions and potential for international use of the sea route, but the international shipping industry generally felt that the ice situation made regular commercial navigation unpredictable and unsafe (Ragner, 2000).

This mood changed with the publication of reports documenting a receding ice cover (ACIA, 2005) at the same time as the US Geological Survey assessment for 2008 of the hydrocarbon potential in the Arctic attracted worldwide attention (Gautier et al., 2009). The Arctic Council initiated the Arctic Marine Shipping Assessment produced by its Protection of the Arctic Marine Environment (PAME) working group (AMSA, 2009). The comprehensive report, including

DOI: 10.4324/9781003246015-17

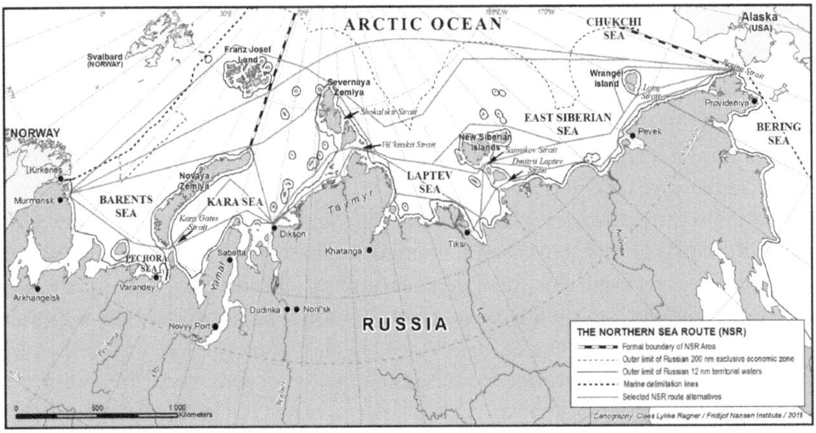

Figure 13.1 Map of the Northern Sea Route.
Source: Fridtjof Nansen Institute.

both opportunities and challenges from Arctic shipping and with strong emphasis on marine safety and environmental protection, was published in 2009. The NSR started to attract considerable commercial attention and the Russian government made development of the NSR a high priority (Moe, 2014).

New regulations and procedures were adopted, including swift processing of applications to navigate the sea route and a reformed fee system. Changes were explicitly intended to increase interest among international users. Expectations were high since rapidly melting sea ice made use of the NSR for international transits between the Pacific and the Atlantic look increasingly viable.

In parallel, reflecting the global interest in the Arctic environment, the International Maritime Organization negotiated an International Code for Ships Operating in Polar Waters, which came into force in 2017. Russia is a party to the Polar Code, which covers both marine safety issues and environmental protection and was active in the negotiations. Chircop and Czarski (2020) conclude that there is a "substantial degree of harmonisation" between the Code and the Russian regulations. The main criticism of the Code is that parts of it exempt large portions of Arctic shipping, namely ships, used only for domestic voyages.

In this chapter, we will look at the role international cooperation and participation came to play in shipping on the NSR and the outlook for the continued involvement of the international shipping industry. First, we review the international shipping activity on the NSR over the last decade. Then we discuss driving forces and forces of deterrence, followed by necessary preconditions for wider usage of the route seen from the perspective of international shipping, and evaluate Russian plans and policies affecting international shipping. In the concluding section, we briefly summarize the key points and outline the most likely development over the next several years.

International shipping on the NSR 2010–2019[1,2]

Regular transit shipping on the NSR began in 2010. From 2011, several international transit voyages between European and Asian Pacific ports as well as destination voyages from NW Russia (Murmansk) to the Asian Pacific were organized annually (Table 13.1). In 2010–2013, most of the cargo was liquid hydrocarbons (gas condensate, naphtha, *liquefied natural gas* [LNG], jet fuel, gas oil, and heavy oil) in addition to coal and iron ore.

The main cargo owners were Russian (Novatek and Eurochem), followed by South Koreans, Norwegians, and Canadians. These early voyages enjoyed considerable support from the Russian government (through the state icebreaker company Atomflot) and many of them were mainly meant to test the technical feasibility of shipping on the NSR by Arc4 ice-class cargo vessels during the summer-fall season with assistance from Russia's nuclear icebreakers.[3] In addition to tankers, bulk carriers, and a few LNG carriers, other types of vessels transiting the NSR were reefers, research vessels, icebreakers, and passenger vessels. Nordic shipping companies operating several ice-class A1 (Arc4) tankers and bulkers in the Baltic Sea during the winter had an advantage over other shipping companies, including many Russian companies. Several of these vessels were used in international shipping on the NSR during the summer-fall season. Thus, 49% of all transit voyages via the NSR 2010–2013 were made by Nordic shipping companies (Danish, Swedish, and Finnish) or 64% if we exclude voyages involving Russian companies. During this time, companies from 13 countries were participating in international shipping via the NSR.

Table 13.1 Annual number of voyages and cargo volume for international transits on the NSR between the Atlantic and Pacific during the period of 2010–2019.

Year	*International Transit*	*International Transit Cargo (t)*	*Destination Voyage*	*Destination Voyage Cargo (t)*
2010	1	41,000	1	70,165
2011	4	185,243	14	590,102
2012	9	337,371	17	793,315
2013	14	633,791	14	484,097
2014	4	72,472	2	0
2015	6	34,938	1	0
2016	8	201,946	5	0
2017	12	154,415	4	20,253
2018	17	339,070	2	144,499
2019	14	285,245	8	361,094
TOTAL	89	2 285,491	68	2 463,525

Note: The same information is also shown for those destination voyages that took place between NW Russia (outside the western border of the NSR, mainly Murmansk) and ports in the Asian Pacific region, sailing through both the western and eastern boundaries of the NSR. The annual number of all destination voyages during 2016–2019 is shown in Figure 13.3.

Source: Gunnarsson and Moe (2021).

The global economic recession of 2014 had a major impact on the NSR. The transportation of liquid hydrocarbons and iron ore along the NSR to the Asian Pacific market that had dominated transport on the NSR during 2010–2013 came to a standstill. The only hydrocarbon transport was one shipment of coal from Vancouver in Canada to Finland. The remaining cargo was some general cargo and frozen fish and meat. Freight rates were depressed as shipping companies struggled with overcapacity of tonnage. This meant that time saved using the Arctic route became less important for the economy of transporting cargo. Commodity prices of raw materials fell sharply due to declining demand, especially in Asia, and the previous price differences between European and Asian markets were evened out. This dampened the interest in more costly transport of Arctic commodities to Asian markets. Instead, the decreased value-to-weight ratio of transported goods put emphasis on "economy of scale", making it more profitable to transport commodities on very large vessels going through Suez or around the Cape of Good Hope. Reduced bunker fuel prices also meant that lower fuel consumption on shorter voyages via NSR compared to southern routes was less significant for the economic calculations of shipping operators and cargo owners.

Also contributing to the lack of interest in international shipping on the NSR during the period of 2014–2015 were the US/EU economic sanctions against Russia starting in 2014 in the aftermath of the Ukraine crisis and the subsequent countersanctions from Russia. The ensuing geopolitical tensions did not encourage international shipping companies to become involved in NSR ventures that would require long-term investments in new ice-strengthened vessels.

The Russian authorities realized that their effort to rapidly increase international transit traffic on the NSR was not bearing fruit. Much larger numbers of voyages would be required to make international transits a source of income for Russia and justify the costly operations of its nuclear icebreakers.

Whereas the general interest of the international shipping industry in transit on the NSR seemed to be fading, there were exceptions. Having actively declared their interest in Arctic shipping for some years, China's COSCO Shipping established its own dedicated Arctic shipping business (Moe and Stokke, 2019), becoming the most active player in NSR transit shipping in recent years. This started with project cargo using general cargo and heavy-lift carriers. But the company also constructed a series of ice-strengthened combined bulk and container ships of a size suitable for the Arctic. With these ships, they opened a multi-purpose-vessel cargo route through the Arctic. In 2016–2019, 45% of international transits were done by COSCO, followed by several German companies with 25% (Table 13.1). This change underscores the fact that the cargo base is not a given and that new players may see new opportunities. Much has been made of the Chinese political interest in Arctic shipping and the potential for cooperation with Russia. However, as shown by Kobzeva (2020), there are considerable discrepancies in the interests of the two countries, and Chinese shipping infrastructure investments have not materialized. Despite the Chinese engagement, international transits have been totally dominated by spot-market deliveries of

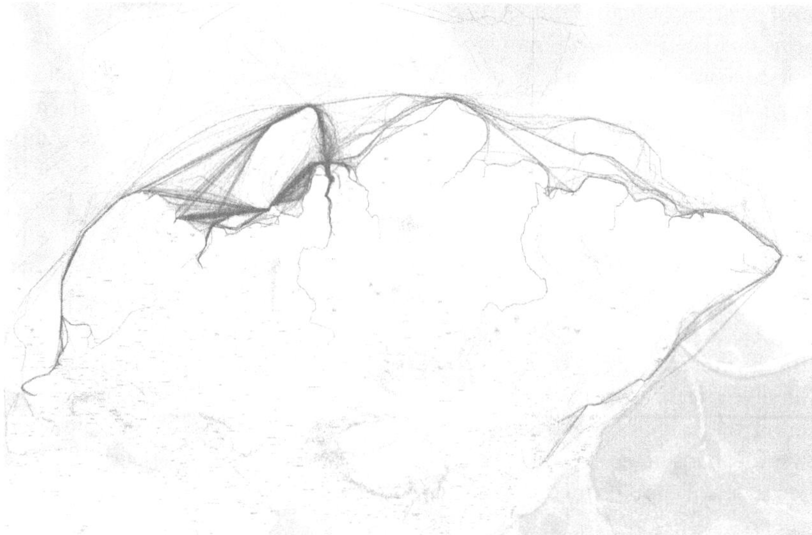

Figure 13.2 Sailing tracks of vessels operating on the NSR in 2018, based on Automatic Identification System (AIS) data. Most of the shipping activity on the NSR occurs in the SW Kara Sea.

Source: Gunnarsson and Moe (2021).

commodities and transport of project cargo, and vessel repositioning between the Atlantic and Pacific markets. All these reflect short-term decisions by shipping companies and cargo owners and not long-term strategies.

At the same time as the outlook for growth in international transits was looking more uncertain, Russia focused on domestic and destination shipping on the NSR servicing resource extraction projects in the Ob Bay/Yenisey area (Figures 13.2 and 13.3). By 2016, Atomflot had signed contracts for icebreaker support with all current project developers in the area. For international shipping companies involved in transit shipping, it therefore became clear that Russian natural resource projects would increasingly occupy the capacity of Russia's nuclear icebreakers.

The development of energy projects, however, presented other opportunities for international shipping. In 2016–2017, Norwegian companies provided support and supply vessels for offshore operations in the Ob Bay and Kara Sea, and companies from the Netherlands, Belgium, and Luxembourg became engaged in extensive dredging operations in the Ob Bay. Other non-Russian companies provided general cargo vessels, bulkers, heavy-lift carriers, and drilling rigs in support of Russian natural resource project development. The largest number of non-Russian companies in domestic shipping (cabotage) was in 2016 and 2017, with up to 23 companies operating each year. Most of these voyages were between Murmansk and Sabetta in the Ob Bay.

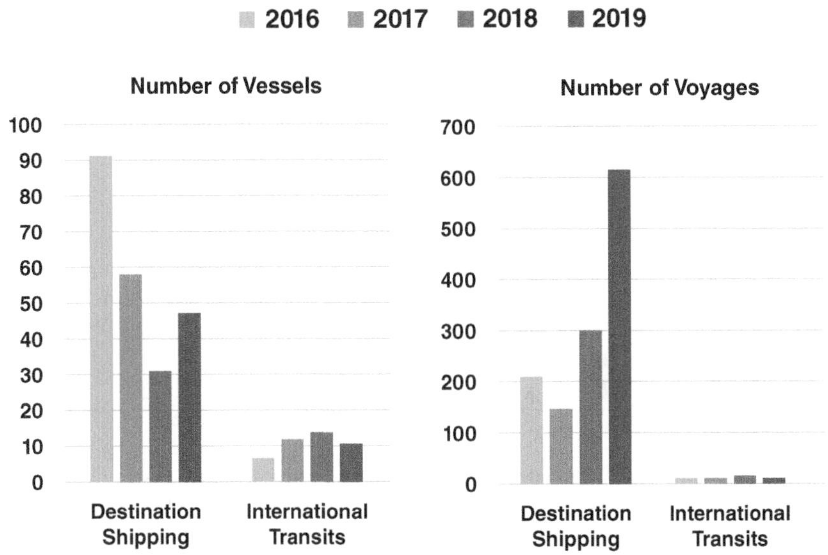

Figure 13.3 Number of vessels and voyages involved in destination shipping and international transit shipping on the NSR 2016–2019. The port of departure or arrival in destination shipping was almost exclusively Sabetta on the Yamal peninsula.

Source: Gunnarsson and Moe (2021).

European shipping companies were also involved in destination shipping in 2016 and 2017, transporting prefabricated LNG modules and other project cargo on heavy-lift carriers and general cargo vessels to the Yamal LNG plant at the port of Sabetta (Figure 13.3). The LNG modules originated from several construction yards in China and Indonesia and were shipped via the Suez route, first to Zeebrugge in Belgium and other European ports before onward shipment to Sabetta. Several shipments also came through the Bering Strait during the summer-fall season. Most of the European shipping companies transporting heavy project cargos to Sabetta were from the Netherlands, followed by Germany.

Shipments of LNG from Sabetta started in December 2017. This is carried out by three non-Russian companies in addition to Russia's Sovcomflot, on long-term charter contracts. The foreign-owned LNG carriers are operated by Dynagas (Greece), Teekay Shipping LNG (operated from the UK) and Mitsui O.S.K. Lines (Japan). These companies established joint ventures with subsidiaries of China's COSCO Shipping to finance the construction of their new fleet of 15 Arc7 LNG carriers for Yamal LNG at a price of some 300 million USD each. The first shipments involved direct transports from Sabetta to western European ports for unloading, or ship-to-ship transshipment to conventional vessels near Honningsvåg off the northern coast of Norway (Figures 13.2 and 13.3). The first

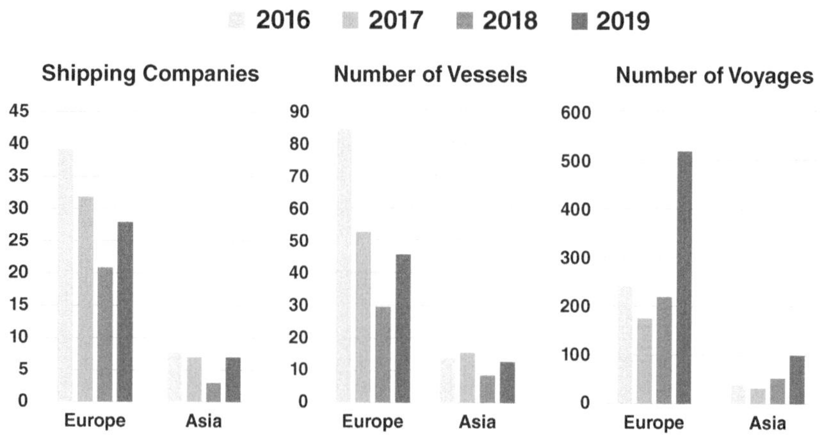

Figure 13.4 Number of European and Asian shipping companies/operators on the NSR in 2016–2019 and the number of their vessels and voyages on the NSR.

Source: Gunnarsson and Moe (2021).

direct shipment eastwards reached China in July 2018, followed by three additional voyages later that year. In 2020, 35 voyages went in the eastern direction, but the vast majority of shipments, some 219, went west to Europe according to the CHNL Information Office. Additionally, shipping companies from Greece and Germany were chartered to transport gas condensate from Sabetta to European ports.

As shown in Figure 13.4, there have been far fewer Asian shipping companies working on the NSR than European companies. Much more frequent voyages have also taken place between the NSR and European ports than ports in the Asian Pacific region.

In conclusion: International use of the NSR increased over the past decade but not in the way or to the extent many had expected. International transit shipping between the Atlantic and the Pacific saw only modest growth and did not become a significant component in international shipping. Real growth was in destination shipping between the Russian Arctic and ports outside the region, primarily between the Ob Bay and European ports, conducted by Russian and non-Russian companies, to an extent hardly foreseen ten years ago. Here we see a concerted effort of Russian companies and the Russian government to develop huge resource projects with maritime logistics where there are no alternative modes of transport.

Driving forces and forces of deterrence

A key condition for further development in the Russian Arctic is efficient and innovative Arctic logistics, largely based on maritime transportation.

The build-up of a new maritime infrastructure and transport and logistics system on the NSR, and along the whole Eurasian Arctic coast, will take many years and will be costly. Without cost sharing, the up-front capital costs are prohibitive and too high for Russia to take on alone. Russia is therefore hoping that international shipping companies and other foreign investors will take an active part in establishing the needed maritime infrastructure along the Arctic route. It is common for big shipping companies to invest heavily in port facilities in support of their own logistics operations along established transport routes (Falck, 2018). However, international shipping companies already using traditional routes will not easily alter their own long-established logistics operations.

Based on expectations of a future increase in trade volumes between the European and Asian markets, several shipping companies are closely monitoring the traffic and infrastructure development along the Arctic route (e.g., Beveridge et al., 2016; Milaković et al., 2018). This interest is further fueled by ongoing ship traffic congestion in the Strait of Malacca; instability in the Middle East and along the Strait of Hormuz that could impact shipping through the Suez Canal, and the persistent threat of piracy off the coast of Somalia.

For international transit shipping the obvious advantage of using the NSR instead of the traditional route through the Suez Canal is the reduction in the transport distance (30–50%) and sailing time (10–15 days) between ports in Scandinavia/NW Europe and NE Asia, assuming the same sailing speed on both routes. This can lead to substantial cost savings during the summer-fall season (July-November), when sea ice conditions are most favorable on the NSR and transport can proceed without icebreaker assistance. The precise distance advantage depends on the location of the departure and receiving ports – the further north ports are located, the greater the distance advantage of the NSR.

Yet, as discussed above, shipping costs are also impacted by the price of fuel, freight rates, and global market developments. When freight rates and bunker prices are low, the economic advantage of using the NSR compared to southern routes can be quickly lost, as occurred in 2014. Such factors will be considered by so-called opportunistic users of the NSR, that is, users who evaluate conditions on a short-term basis and compare them with other transport options. Potential investors using the sea route over the longer term need to consider several other factors.

Geopolitics and environmental-climate politics play a role. Tensions, sanctions, and countersanctions, higher trade tariffs, and regionalization are not conducive to the development of international shipping via the NSR. Perceptions of increased militarization – real or imagined – along Russia's northern coast are also likely to hold back foreign investments (Melino and Conley, 2019). The global push for reduced greenhouse gas emissions and more environmentally friendly operations, as reflected in the UN Sustainable Development Goals (SDGs), in particular SDG 13, on the urgent need to combat climate change and its impacts, is making large companies reluctant to engage in Arctic operations requiring long-term commitments. Several major shipping companies and owners of international brands, concerned about projected environmental risks having reputational consequences, have already declared that they will not use the NSR or other Arctic routes (Schreiber, 2019). On the other hand, Russia can claim that shorter sailing distances using the NSR translate into reduced emissions. Moreover, if enough LNG-powered vessels are introduced, climate-related arguments in favor of the NSR will be strengthened.

Preconditions for increased international shipping on the NSR

An important prerequisite for the NSR's integration into the global transportation system is regular year-round shipping along the entire length of the route. Now, year-round shipping is only taking place in the western part of the NSR to European ports led by Arc7 LNG carriers, oil tankers, and container/multipurpose vessels. Year-round transport would need to be extended eastwards along the NSR to the Asian Pacific ports and include high ice-class bulk carriers, general cargo vessels/heavy-lift carriers, and larger container vessels with a high level of winterization.

To facilitate such a transport scheme, powerful icebreakers are essential in assisting transiting vessels and in keeping the Arctic route open year-round at acceptable commercial speed regardless of the sea ice conditions. This will include strategic deployment of several icebreakers along the whole length of the NSR, minimizing the consequences of accidents and transport delays due to sea ice.

To be of interest to commercial shipping, the NSR also needs to provide an acceptable level of predictability and punctuality of cargo transportation on a year-round basis. Regularity of supply of goods is no less important than the cost of transportation. This is particularly true for containerized cargo. The amount of bulk cargo shipped between NE Asia and Northern Europe is limited; the big trade volume is containers. Large-scale container shipping is problematic for the NSR but a prerequisite for the route's full integration into the global transportation system. Obstacles include unpredictability due to delays caused by unexpected sea ice conditions, draft limitations along the Arctic coast, and lack of markets along the route (e.g., Cariou et al., 2019).

To justify investments in expensive ice-class vessels, round-trip shipments with cargo in both directions between NE Asia and NW Europe would need to be the norm. An additional prerequisite is therefore the identification of a sizeable and sustainable cargo base, including containerized cargo, for trade between markets at opposite ends of the NSR. No such permanent cargo base has so far been identified.

High ice-class Arctic cargo vessels designed to operate under severe Arctic conditions during the winter-spring season and which can break sea ice up to two meters thick should not sail long distances in ice-free waters. A solution is to establish transshipment terminals located in ice-free waters at each end of the Arctic route and have conventional feeder vessels bring cargo to the terminals and deliver cargos from these to their final destinations. The establishment of transshipment terminals would mean that specialized Arctic shuttles could be fully utilized in the most efficient way. However, transshipment obviously involves extra costs and time. The economics of such a scheme for the Arctic route is not yet clear.

An effective and predictable administrative and management system serving international shipping is also required. This would include connecting shipping companies with the best available navigational, sea ice, and communications services, providing traffic coordination and route optimization, marketing, and future traffic analysis and strategies. The icebreaker tariff system also needs to be user-friendly and fees must be able to compete with Suez and Panama Canal fees.

International commercial shipping companies will only use the route if it is considered safe and if all available means are in place to minimize impacts on the environment following strict risk mitigation measures. At the same time, the NSR management also needs to find ways to reduce risks of shipping delays due to sea ice by improving ice forecasting and ice reconnaissance. Before a voyage, detailed assessment and forecasting of ice conditions and other operational conditions *en route* needs to be accomplished and presented to NSR users.

Given the long distances, well-equipped land-based and offshore emergency stations must be strategically placed along the whole length of the NSR, enabling timely response to all kinds of maritime emergencies. Emergency services include refuge assistance and support for ships in distress, search and rescue operations, oil spill response, and salvage. Icebreakers and multipurpose emergency and rescue vessels will play an important role as floating support units in case of accidents.

For safety reasons and due to the remoteness and harsh climatic conditions that ships face on the NSR, year-round transit navigation needs to be supported by stable high-bandwidth satellite communication coverage throughout the NSR Water Area. This needs to include high resolution and near real-time satellite imagery of developing sea ice conditions along the vessels' sailing paths. The vessels should also receive analyzed (processed) satellite sea ice images and ice charts from public and private ice service providers. Such ice monitoring technology will assist vessels in choosing the optimal route through the sea ice in real time and limit operational risks and voyage delays.

At the end of the day, it is the global maritime industry that will decide when the shorter Arctic route is safe, efficient, reliable, environmentally sound, and economically viable in comparison with other routes across the world's oceans.

Russian policies impacting international shipping on the NSR

It is readily admitted by Russian authorities that the NSR needs comprehensive infrastructure improvements. The most recent official document is the "Plan for Infrastructure Development on the NSR until 2035", adopted by the Russian government in December 2019 (Plan, 2019). It stipulates measures to improve emergency communication, navigation infrastructure, to build new powerful icebreakers, to enhance the rescue capacity of both vessels and bases, and to develop port infrastructure. No official total budget exists, but estimates put total investments at approximately USD 20 billion, half of which will cover the const ruction of a series of new nuclear icebreakers (Burmistrova, 2019; Moe, 2020b).

The government counts on a substantial share being covered by maritime users and investors. Achieving the goals will clearly improve the conditions for shipping, including attractiveness to foreign users. However, the high costs involved and the ensuing financial requirements make postponements probable.

Rapid development of cargo transportation on the NSR has become a key goal in Russia's Arctic policy. A target of 80 million tons by 2024 (and continued increase thereafter) was first announced by President Putin in 2018, later repeated in other official documents, including the new Arctic Strategy of 2020 (Strategy, 2020). The commercial enterprises expected to implement this herculean task, both as cargo owners and transport companies – namely Novatek, Rosneft, Gazprom, Nornickel, Rosatom, and Sovcomflot – by necessity have become key players in the development of the NSR. These industrial stakeholders have a shared interest in an operative NSR which can bring inputs to large Arctic project developments as well as transporting the output to market. The Russian government is providing direct export access and long-term favorable tax conditions (Henderson and Moe, 2019; Moe, 2020a). An important question is how these companies will impact international cooperation in the shipping sector. One might expect that companies involved in natural resources extraction would prefer a liberal shipping regime which allows for competition between shipping companies offering transport services. However, the resource extraction industries are all closely intertwined with Russian state development policies. The logistical solutions for all these extraction projects are likely to be developed as large package deals, where long-term transport arrangements, state support, and ice-breaking services are included.

To ensure state control of all infrastructure developments and maritime operations on the NSR, Russia enacted a law in late December 2018 giving the state nuclear power corporation Rosatom control over current operations of the NSR and the management of state property and assets in ports. This came in addition to the nuclear icebreakers operated by its subsidiary Atomflot. Rosatom would coordinate and distribute state investments and collect state income. Navigational and hydrographical support would be the joint responsibility of Rosatom and the Ministry of Transport. The Ministry would be responsible for developing legislation and regulations and for ensuring their implementation and process applications for the use of the NSR through its NSR Administration, but important decisions would be made in consultation with Rosatom (Moe, 2020a). Thus, Rosatom and Atomflot have become responsible for the execution of state policy on the NSR at the same time as they conduct the running operations. Monopolization of services, in particular icebreaker services, may discourage international users planning long-term investments, as they may fear becoming totally dependent on Atomflot as regards specification of services as well as prices.

When Novatek's Yamal LNG project was developed, it seemed that international shipping companies would come to play a key role in the transportation of LNG, with 14 of the 15 custom-built carriers owned and operated by foreign companies, as noted above. However, as early as in 2018, Russia enacted legislation demanding that hydrocarbons from within the NSR be transported exclusively on Russian-flagged vessels, and shortly thereafter a requirement to build new vessels for this purpose in Russia was introduced (Moe, 2020a). Neither regulation could be implemented immediately and completely. Exemptions from the flag requirement were granted to the carriers already delivered for Yamal LNG and the Russian government has also accepted that some of the 40–50 new Arc7 LNG carriers required for other planned LNG projects will be built abroad (Vedeneeva, 2020).

The room for international shipping companies in destination shipping is now less than expected some years ago. The key cargo owners in the Russian Arctic will to a large extent rely on their own shipping fleets to transport their cargo, as do Nornickel, Gazprom Neft, and Rosneft (through its subsidiary Rosnefteflot), and as Novatek is planning to do. Novatek set up a joint venture with Sovcomflot to own and operate 15 ice-breaking Arc7 LNG carriers for the upcoming Arctic LNG-2 in the Ob Bay. The carriers, to be built at Russia's new Zvezda Yard (near Vladivostok), will transport LNG from within the NSR to transhipment hubs in Murmansk or Kamchatka (Dyatel, 2019). Novatek also signed an agreement with Sovcomflot and China's COSCO Shipping and the Silk Road Fund to transport the LNG with regular carriers from the transhipment hubs to global markets (Novatek, 2019). Other extraction companies are also likely to enter into long-term contracts with designated foreign shipping companies. The rapid build-up of capacities at Russia's Zvezda Yard, where Rosneft is the key founder, is contingent on extensive cooperation with South Korean yards while non-Russian companies play an important role in the financing and operation of the new fleets of LNG carriers. The Zvezda Yard is seen as the key to the revival of Russia's shipbuilding industry and Rosneft is a staunch supporter of measures that can guarantee customers for the Yard.

A more troublesome development for international shipping is Rosatom's decision to establish its own container shipping company (Humpert, 2019), which would operate in competition with other users of the NSR, shipping cargos that could benefit from the advantages of the NSR, either directly or via transhipment hubs. Large investments in such a company will be a strong argument for shielding it from competition. In such a situation, Rosatom may become less interested in encouraging other users. As infrastructure operator, it would be able to effectively set the terms for international transits.

Conclusion

The developments discussed in this chapter indicate that Russia's policies for the NSR are becoming more inward-looking: current support for international use is not high on the agenda. Maritime infrastructure development along the NSR is being developed as a necessity to bring large quantities of energy and mineral resources from the Russian Arctic to the global market and support Russia's own domestic shipping and shipbuilding industry. Resource extraction companies in the Russian Arctic are building their own shipping fleets to bring commodities out of the remote areas.

Over the last ten years, international transits have been totally dominated by spot-market deliveries of commodities and transport of project cargo and for vessel repositioning between the Atlantic and Pacific markets, all reflecting short-term decisions by shipping companies and cargo owners rather than long-term strategies.

The Russian government has concluded that only when regular year-round navigation – serving resource extraction projects in the Russian Arctic – is established will the international shipping industry start to show real interest in the NSR. Russia is predicting that regular year-round transportation will be the norm on the NSR already during the second half of this decade. But even if the infrastructure materializes, a booming international transit business will not start automatically. Russia cannot decide the potential for international transit shipping; on the other hand, positive international market developments and reduced tensions will not help if supportive Russian policies and framework conditions are not in place.

Conditions for large-scale investments by the international shipping industry and cargo producers for use of the NSR are still not in place. Besides, there are uncertainties about Russia's longer-term policies. It remains an open question whether Russian preferences will support a de facto monopoly on Arctic transits rather than encourage competition from international shipping companies.

International shipping on the NSR has been dominated by European shipping companies with only a handful of Asian companies being involved each year. Transport of commodities from the NSR to European ports takes place much more frequently than to ports in the Asian Pacific region. Companies based in NW Europe have extensive experience of operating ice-strengthened cargo vessels in sea ice during the winter in the Baltic Sea, and some of these have been used on Arctic voyages during the summer-fall season. European companies also have extensive experience in the transport of heavy-lift cargos and in offshore support as well as in dredging operations in European coastal ports and rivers. This expertise and equipment have been in demand in the Russian Arctic and will likely continue to be so for several planned extraction projects.

A major limitation on the use of the NSR as an international shipping route has been the availability of ice-strengthened vessels of different segments and sizes for use on Arctic voyages. However, international shipping companies will not invest in expensive ice-class vessels only for spot-market deliveries of goods during parts of the year. They need confidence in stable framework conditions if they are going to undertake large investments in ice-strengthened vessels. International tensions and protectionism do not inspire confidence in stable conditions on the NSR. Some companies may be reluctant to engage in Arctic transport due to public concerns about perceived environmental risks. Yet Russia can claim that shorter sailing distances using the NSR translate into reduced emissions, and if enough LNG-powered vessels are introduced, climate-related arguments in favor of the NSR will be strengthened.

The last ten years have also highlighted the sensitivity of international transit shipping to conditions on the global commodity market and the freight market. The cost of chartering a vessel and the price of bunker fuel also influence the choice of route. When freight rates and bunker prices are low, the economic advantage of using the NSR compared to southern routes can be quickly lost. Russia is relying on the Asian Pacific market to provide the highest future demand and prices for its commodities, justifying expensive infrastructure development along the more icebound eastern part of the NSR. If the price difference in commodities between the European and Asian markets is evened out, the rationale for sending such commodities eastwards through the Arctic to Asia disappears.

In the immediate future, however, the development path seems quite certain. Continued growth in destination shipping is connected to the development of resource extraction projects and transport of the output from these projects to foreign markets. The speed of development is likely to be affected by price developments in the international markets. Over the slightly longer term, many factors will play a role, as discussed in this chapter. The market outlook will be much affected by climate policies and the energy transition.

December 2021

Ex-post reflections

Even if a major conclusion in our study was that Russia's NSR policies had become more inward-looking, the economic underpinning of further expansion of navigation rests on integration with the outside world in two ways: for the provision of technology and investment in the huge extraction projects and the associated specialized transportation fleet, and as market for the products, mainly LNG and oil. A new international situation following the war in Ukraine as well as the economic sanctions imposed on Russia will have negative consequences for both, but exactly how much it is too early to tell. Already LNG development plans are being scaled back because of technology sanctions and the withdrawal of western companies. Possibly, western technologies can be replaced by domestic Russian or Chinese technology after some time, but investors will make new risk assessments. Big Asian importers of LNG and oil, notably China and India, will remain interested in Arctic oil and gas. Nevertheless, questions remain; for instance, will there be sanctions against companies transporting hydrocarbons from Russia? How comprehensive and effective will western efforts to deny Russia export revenue be? This again will depend on the political situation in Russia emerging in the aftermath of the war. The outlook for international cooperation in the development of the NSR looks very different today than when the study was concluded. All the same, the commercial experience from the use of the sea route, which is a major theme in this chapter, will also be relevant for future discussions about NSR in a new political environment.

April 13, 2022

Notes

1 The material presented in this section is partly based on Moe (2020a), Gunnarsson (2021), and Gunnarsson and Moe (2021). The traffic data was provided by Atomflot

(data for 2010–2012), the Northern Sea Route Administration (for 2013–2019), and the CHNL Information Office in Murmansk (for 2016–2019). Detailed descriptions of data sources, methodology, and definitions are to be found in these three journal articles.

2 The following definitions are important for the analysis: a *voyage on the NSR* is a voyage that originates from within the NSR, arrives in the NSR area, or transits the NSR. *International shipping on the NSR* is a voyage that departs from or arrives at a foreign (non-Russian) port and/or is conducted by a foreign shipping company. This includes *international transit voyages* between the Pacific and the Atlantic (between two foreign ports) and *destination voyages* between a Russian port and a foreign port. In both these cases, voyages can be made by either foreign or Russian companies. The third category comprises foreign companies involved in Russian *domestic shipping*.

3 The Russian Marine Register of Shipping (RMRS) ice classes are divided into non-Arctic, Arctic, and icebreaker classes. The ice-class notation is followed by a number denoting the level of ice strengthening: Ice1–3 for non-Arctic ships; Arc4–9 for Arctic ships, and Icebreaker 6–9 for icebreakers.

References

Arctic Climate Impact Assessment (ACIA) (2005). *Overview Report*. Cambridge University Press.

Arctic Marine Shipping Assessment (AMSA) (2009). *Arctic Council*. Protection of the Arctic Marine Environment (PAME).

Åtland, K. (2008). Mikhail Gorbachev, the Murmansk Initiative, and the Desecuritization of Interstate Relations in the Arctic. *Cooperation and Conflict* 43(3), pp. 289–311.

Beveridge, L., Fournier, M., Lasserre, F., Huang, L., and Tetu, P.L. (2016). Interest of Asian Shipping Companies in Navigating the Arctic. *Polar Science* 10(3), pp. 404–414.

Burmistrova, S. (2019). Проект Северного морского пути подорожал почти на 150 млрд руб (The Plan for the Northern Sea Route Has Gone Up in Price by 150 Billion Roubles). *RBK*, 1 February.

Cariou, P., Cheaitou, A., Faury, O., and Hamdan, S. (2019). The Feasibility of Arctic Container Shipping: The Economic and Environmental Impacts of Ice Thickness. *Maritime Economics & Logistics* 23(4), pp. 615–631.

Chircop, A. and Czarski, M. (2020). Polar Code Implementation in the Arctic Five: Has Harmonisation of National Legislation Recommended by AMSA Been Achieved? *The Polar Journal* 10(2), pp. 303–321.

Dyatel, T. (2019). Танкеры для «Арктик СПГ-2» построят без иностранцев (Tankers for Arctic LNG-2 Will Be Built Without Foreigners). *Kommersant*. 4 September.

Falck, H. (2018). Arctic Shipping Perspective in the Context of Global Shipping. In: Corell, R.W., Kim, J.D., Kim, Y.H., Moe, A., Van der Zwaag, D.L. and Young, O.R. (eds.) *The Arctic in World Affairs: A North Pacific Dialogue on Arctic 2030 and beyond: Pathways to the Future*. North Pacific Arctic Conference Proceedings. Jointly published by the Korea Maritime Institute and East-West Center.

Gautier, D. L., Bird, K. J., Charpentier, R. R., Grantz, A., Houseknecht, D. W., Klett, T. R., Moore, T.E., Pitman, J.K., Schenk, C.J., Schuenemeyer, J.H., Sørensen, K., Tennyson, M.E., Valin, Z.C., and Wandrey, C. J. (2009). Assessment of Undiscovered Oil and Gas in the Arctic. *Science* 324(5931), pp. 1175–1179.

Gunnarsson, B. (2021). Recent Ship Traffic and Developing Shipping Trends on the Northern Sea Route—Policy Implications for Future Arctic Shipping. *Marine Policy*, 124(2021), 104369.

Gunnarsson, B. and Moe, A. (2021). Ten Years of International Shipping on the Northern Sea Route: Trends and Challenges. *Arctic Review* 12, pp. 4–30.

Henderson, J. and Moe, A. (2019). *The Globalization of Russian Gas - Political and Commercial Catalysts*. Cheltenham: Edward Elgar.

Humpert, M. (2019). Rosatom Looks to Invest $7 Billion to Create an Arctic Shipping Service. *High North News*, 26 November.

Kobzeva, M. (2020). Strategic Partnership Setting for Sino-Russian Cooperation in Arctic Shipping. *The Polar Journal* 10(2), pp. 224–352.

Melino, M. and Conley, H.A. (2019). *The Ice Curtain: Russia's Arctic Military Presence*. Washington, DC: Center for Strategic and International Studies.

Milaković, A.S., Gunnarsson, B., Balmasov, S., Hong, S., Kim, K., Schütz, P., and Ehlers, S. (2018). Current Status and Future Operational Models for Transit Shipping along the Northern Sea Route. *Marine Policy* 94, pp. 53–60.

Moe, A. (2014). The Northern Sea Route: Smooth Sailing Ahead? *Strategic Analysis* 38(6), pp. 784–802.

Moe, A. (2020a). A New Russian Policy for the Northern Sea Route? State Interests, Key Stakeholders and Economic Opportunities in Changing Times. *The Polar Journal* 10(2), pp. 209–227.

Moe, A. (2020b). Russian Policies for Development of the Northern Sea Route: An Assessment of Recent Developments and Implications for International Users. In: Brigham, L., Corell, R.W., Kim, J.D., Kim, Y.H., Moe, A., Morrison, C.E., VanderZwaag, D.L. and Young, O.R. (eds.), *The Arctic in World Affairs: Will Great Power Politics Threaten Arctic Sustainability?* North Pacific Arctic Conference Proceedings. Jointly published by the Korea Maritime Institute and East-West Center.

Moe, A. and Stokke, O.S. (2019). Asian Countries and Arctic Shipping: Policies, Interests and Footprints on Governance. *Arctic Review* 10, pp. 24–52.

Novatek (2019). Novatek, Cosco Shipping, Sovcomflot and Silk Road Fund Sign an Agreement in Respect of Maritime Arctic Transport LLC. Press Release from Novatek. 7 June.

Plan (2019). План развития инфраструктуры Северного морского пути на период до 2035 года (Plan for development of the Northern Sea Route in the period until 2035). Confirmed by Order of the Government of the Russian Federation, 21 December 2019, No. 3120.

Ragner, C.L. (ed.) (2000). *The 21st Century – Turning Point for the Northern Sea Route?* Proceedings from the Northern Sea Route User Conference, Oslo, 18–20 November 1999. Dordrecht, The Netherlands: Kluwer Academic Publishers.

Schreiber, M. (2019). Nike, Other Retailers Vow Not to Use Arctic Shipping Routes. *Arctic Today*, 31 October.

Solski, J.J. (2020). New Russian Legislative Approaches and Navigational Rights within the Northern Sea Route. *The Yearbook of Polar Law Online* 12(1), pp. 228–250.

Strategy (2020). Стратегии развития Арктической зоны Российской Федерации и обеспечения национальной безопасности на период до 2035 года (Strategy for Development of the Arctic Zone of the Russian Federation Until 2035). Presidential Decree No. 645, 26 October 2020.

Vedeneeva, A. (2020). Новатэк сможет заказать газовозы за рубежом (Novatek May Order Gas Carriers Abroad). *Kommersant*. 23 January.

14 Polar Ship Design and Operations

Past, Present, and Future

Ove Tobias Gudmestad

Introduction

From an early emphasis on geographic exploration and exploitation of the resources in the polar offshore area (by hunting for walrus ivory teeth, seals, and whales), the focus is currently shifting toward the sustainable use of the Arctic's resources. Developments in the Antarctic are mainly limited to fisheries, cruise traffic, and scientific expeditions. The focus in the Arctic is currently on using the Arctic offshore for fisheries, transport of oil and gas products, cargo traffic, and leisure (cruise traffic) in a safe and environmentally sustainable way. During this process, maritime operations have become relatively safe due to the introduction of international codes for the design and strengthening of polar vessels (ice class), the rules of the International Maritime Organization (IMO), and in particular the requirements for training of polar crew members. The continuous work to align the classification societies' rules for ships in polar regions is a step toward improved safety for sailing in these regions. Safety for crew members and passengers has also improved through the use of modern communications systems (particularly satellite navigation) and the availability of ships and helicopters to support search and rescue (SAR) operations.

The international society has implemented the IMO Polar Code (IMO, 2017) for the design and operation of ships in the polar regions (Arctic and Antarctic waters). The Polar Code is a functional code stipulating functional requirements beyond the requirements of the IMO Safety of Life at Sea (SOLAS) Convention (IMO, 2001) and the IMO Marine Pollution (MARPOL) Convention (IMO, 1983). The countries of the Arctic have, furthermore, arranged for close cooperation regarding operations, SAR in Arctic waters, and splitting the Arctic into zones where each of the countries has specific responsibilities.

Polar shipping has developed from a situation of high risk to ships and crew members, where once ships might easily be crushed in the ice and where the survival of crew was a matter of good luck. The first rules for navigating in ice were the Finnish ice class rules issued in 1890 (Kärmäräinen and Riska, 2018), and later (in 1948), international collaboration was enhanced with the establishment of IMO, issuing regulations and standards providing for safer vessels and increased probability of rescue should an operation fail. It is, however, conceded that polar

DOI: 10.4324/9781003246015-18

shipping needs further development to reduce the probability of ship damage and to further mitigate the consequences of adverse events.

The objective of this chapter is therefore to document in the next Section how shipping operations in the Arctic developed from small wooden boats designed according to experience and best practice, to strong double hull ice-resistant vessels through international cooperation, with emphasis on ice class notations, as introduced by the classification societies, and how international cooperation through IMO has further improved the safety of polar shipping, in particular by introducing the International Polar Code (IMO, 2017). In Section 14.3, the safety of the seafarers is briefly discussed with reference to another chapter in this volume (Borch and Andreassen, 2021). The necessity to point out that shipping always requires SAR capabilities must never be forgotten and polar shipping is particularly vulnerable due to the lack of infrastructure in many regions. Section 14.4 highlights the needs to limit emissions from ships in polar waters, and Section 14.5 recounts the challenges encountered through international work toward improved safety standards for ships, crew, and passengers with a view to the needs of the future, where autonomous ships may be seen in the polar regions. Section 14.6 summarizes the content of the paper by pointing out the needs for further international cooperation in the design of polar vessels, for improved operational assistance, for better training of crew members, and for improved means to ensure rescue of all involved in case of vessels in distress. It must be conceded, however, that there will still be emergency situations regardless of safer vessels, increased number of international regulations, and improvements in operational support.

Maritime operations in the Arctic – developments from past to the present

The history of Arctic shipping

Arctic shipping has long traditions; the sailing of Ottar from Hålogaland (possibly from Kvaløya in the Troms area) of Norway to Bjarmeland in the White Sea area was followed by his report to the King of England around the year 870. The next significant development to note here is the expeditions of the Vikings to Greenland and Vinland (modern Newfoundland) and possibly further south, around the year 1000. Their low ships with sails (the "Viking ships", Figure 14.1) represented the state of the art of ship design at that time as they were designed for speed and stability (wide ships), although they had no specific design feature to resist ice. The Vikings reached America, and the "Vinland Map", possibly the first map showing America, is currently housed in Yale University's Beinecke Rare Book and Manuscript Library. The paper on which the map is drawn dates to approximately 1434, which is nearly 60 years before Christopher Columbus arrived in the West Indies.

Thereafter, expeditions by the Dutch, led by Wilhelm Barents (from 1594 and onwards), reached high latitudes using small wooden sailing ships, which were typical of that time with a much higher freeboard than the Viking ships.

Figure 14.1 Viking ship. A digital reconstruction shows that the *Tune* Viking Ship that must have been a fast-sailing vessel that could also be rowed. Illustration by 7Reasons for NIKU (Nikel, 2020).

The Barents Sea is named after the expeditions he led. Following Barents's expeditions, whale hunting and the extraction of oil from whales were developed into an industry on Spitzbergen with many seasonal settlements. Ivory from walruses was also in high demand at that time. Smeerenburg on Amsterdam Island (79°40′N 11°00′E), for example, was founded by Dutch hunters and traders as early as in 1620, north of present-day Ny Aalesund. Many historical documents provide accurate information about these explorations.

F. Nansen's polar expedition of 1893–1896, with the vessel *Fram* (Figure 14.2), when Nansen aimed to freeze the vessel into the ice and drift across the North Pole (Nansen, 1897), is well known to all those interested in the history and technological development of the Arctic, in particular the design of the expedition research vessel *Fram* was specially adapted to the polar regions. She had a rounded form that caused her to lift up onto the ice when subjected to ice pressure. The force caused by the horizontal ice pressure has a component normal to the hull and an uplifting component. Since then, this design has been the standard for all polar vessels. Note that when a ship moves into the ice, it is lifted up, and a vessel with *Fram* geometry exhibiting will thus break the ice due to its weight and can thereafter move forward. Evacuation was at that time onto the ice or to open lifeboats.

The fishermen making their living from fishing along the Norwegian coast used smaller vessels, and it was not until seal hunting expeditions moved into the ice that vessels were strengthened for Arctic conditions (Alme, 2009; Gudmestad and Alme, 2015). The vessels generally hunted for the seals together, so a "buddy effect" was present in case any of the vessels got into distress. The ice pressure exerted on the vessels during the closing of open leads, however, caused many vessels to be squeezed, so that they eventually sank. Even ships made of steel were damaged by drifting ice floes, to the extent that they had to be abandoned. In these situations, the crew gathered on the ice and were, in most cases, rescued by other vessels in the area. On the way to the hunting grounds, however, all vessels were exposed to the same meteorological conditions, and the loss of lives in some situations was exceptionally high. Weather conditions in 1917, when seven vessels (and 79 persons) were lost in Vestisen (northwest of Iceland), were characterized by a heavy storm from the northeast, together with cold temperatures and a polar low pressure; similar conditions were encountered in 1952 when five vessels were lost. The ships were probably hit by severe icing conditions and thus lost stability. Following these events, nothing was heard from those on board and nothing was found. It can be commented that during heavy sea spray icing, the only way to regain stability is to remove the ice, in particular ice accumulated high above the deck.

The crew of the sealer *Kapp Flora* met a better fate. In April 1924, 13 vessels were lost in the mouth of the White Sea due to forces from moving ice. The *Kapp Flora* (Figure 14.3) was on her way out of the White Sea with a full catch when she

Figure 14.2 Fram. A model of the ship used by F. Nansen in the Arctic 1893–1897, by O. Sverdrup in the Arctic 1898–1902 and by R. Amundsen 1910–1912 in Antarctica. With permission from the Bergen Maritime Museum.

was cracked open between ice floes. At the evacuation, another sealer, the *Godøy*, was only 500 meters away. The crew of the *Kapp Flora* failed to locate the *Godøy* in the snow and walked and dragged the lifeboats across the ice and came ashore at the rescue station at Cape Orlov on the Murmansk Coast (Vestlandsnytt, 1962). Later, they returned home aboard another vessel.

Drifting ice is a particular concern when sailing in polar waters. Outside the ice edge, the ice floes drift on the waves and can cause great harm to equipment on board vessels. Glacier ice floes or multiyear ice floes may furthermore have very damaging effects, as the "ice foot" is hidden below the waterline and may not be seen by the person on watch on the bridge. The seal hunters were particularly concerned about the risk of colliding with such ice. Further to the historical records by Alme (2009), Marchenko (2009) has prepared an exhaustive review of the experience of Russian Arctic navigation, listing the loss of vessels navigating the route from the Kara Gate to the Bering Strait during the last hundred years. She also discusses the reasons for the accidental losses of vessels and documents concerns for present shipping due to uncertainties in ice loading during ice navigation.

Figure 14.3 The *Kapp Flora* in the ice off Cape Orlov on the White Sea. Note the heavy ice and the open lifeboat. (With permission from the owner of the picture, T. Nærland, Nærbø, Norway, the daughter of P. O. Paulsen, one of the survivors.) For pictures, see also Ishavsmuseet, Brandal's Home Page.

The accidents called for design standards for vessels navigating polar waters. Ice class notations for ships were developed by several classification societies (see Figure 14.4 based on The Baltic Sea Ice Class Service). The first Finnish ice class rules were developed in 1890 for vessels sailing in the Baltic Sea. Note that the ice is harder in this area than elsewhere in the polar regions due to the lower salt content of the Baltic Sea. The Finnish ice class rules issued in 1932 introduced ice classes 1A, 1B, and 1C for ships strengthened for navigation in Baltic ice, Ice Class 2 for ships classified for unrestricted service but not strengthened for navigation in ice, and Ice Class 3 for other vessels. All class societies later developed their own ice class rules, which are compatible with and accepted by the IMO. Note that the IMO Guidelines for Ships Operating in Arctic Ice-Covered Waters (IMO, 2002) include no direct technical requirements; however, the classification societies have issued additional requirements depending on the actual thickness of the ice. Note also that ship classification is an international competitive business; however, all classification societies work according to the agreed IMO regulations. The ice class requirement has considerably reduced the risk in navigating the Arctic Seas. For modern ice class rules, see, for example, DNV (2013) and Mejlænder-Larsen (2015).

Classification Society	Ice Class				
Finnish-Swedish Ice Class Rules	**IA Super**	**IA**	**IB**	**IC**	**Category II**
Russian Maritime Register of Shipping (Rules 1995)	UL	L1	L2	L3	L4
Russian Maritime Register of Shipping (Rules 1999)	LU5	LU4	LU3	LU2	LU1
Russian Maritime Register of Shipping (Rules 2008)	Arc 5	Arc 4	Ice 3	Ice 2	Ice 1
American Bureau of Shipping	Ice Class I AA	Ice Class I A	Ice Class I B	Ice Class I C	D0
Bureau Veritas	ICE CLASS IA SUPER	ICE CLASS IA	ICE CLASS IB	ICE CLASS IC	ID
CASPPR, 1972	A	B	C	D	E
China Classification Society	Ice Class B1*	Ice Class B1	Ice Class B2	Ice Class B3	Ice Class B
Det Norske Veritas	ICE-1A*	ICE-1A	ICE-1B	ICE-1C	ICE-C
Germanischer Lloyd	E4	E3	E2	E1	E
IACS Polar Rules	PC6	PC7	-	-	-
Korean Register of Shipping	IA Super	IA	IB	IC	ID
Lloyd's Register of Shipping	Ice Class 1AS FS (+) Ice Class 1AS FS	Ice Class 1A FS (+) Ice Class 1A FS	Ice Class 1B FS (+) Ice Class 1B FS	Ice Class 1C FS (+) Ice Class 1C FS	Ice Class 1D Ice Class 1E
Nippon Kaiji Kyokai	NS* (Class IA Super Ice Strengthening) NS (Class IA Super Ice Strengthening)	NS* (Class IA Ice Strengthening) NS (Class IA Ice Strengthening)	NS* (Class IB Ice Strengthening) NS (Class IB Ice Strengthening)	NS* (Class IC Ice Strengthening) NS (Class IC Ice Strengthening)	NS* (Class ID Ice Strengthening) NS (Class ID Ice Strengthening)
Polski Rejestr Statków	L1A	L1	L2	L3	L4
Registro Italiano Navale	ICE CLASS IA SUPER	ICE CLASS IA	ICE CLASS IB	ICE CLASS IC	ID

Figure 14.4 Correspondence between the Baltic ice class (Finnish-Swedish Ice Class Rules) and the ice classes of other classification societies.

Source: HELCOM Recommendation 25/7. © HELCOM.

The present status of ship operations in the Arctic

In the Russian part of the Arctic, the authorities are becoming concerned about the strength of vessels using the Northern Sea Route (NSR), and the Russians are strengthening the requirements regarding vessels' ice-breaking capabilities, or icebreaker assistance is requested by the NSR administration (NSR's Home Page, no date). This relates in particular to the large Liquefied Natural Gas (LNG) tankers sailing from LNG plants in Ob Bay. These are of "Double Acting Tanker" design (Figure 14.5), whereby the ship goes stern first into the ice. The stern and the Azipod propellers break the ice (Aker Arctic's Home Page). The bow is a normal bulb bow designed for speed in ice-free waters. It should be noted that the NSR from the Kara Gate to the Bering Strait is open to large ice-breaking vessels for several months of the year. The Russian NSR administration imposes strict regulations for the use of the route.

While sailing in the Matisen Strait (one of the two major channels separating the groups of the Nordenskiöld Archipelago north of Russia) on September 4, 2013, an ice floe hit the "sea-river" type tanker, Nordvik. The tanker sustained a hole in one of the ballast tanks on her port side. The vessel, with an ice class of Ice 1, was loaded with 4,944 tons of Arctic diesel fuel and was following the route from Ob Bay to Khatanga (Bellona, 2013). The captain of the vessel was in

Figure 14.5 Double Acting Tanker. The Christophe de Margerie-class ice-breaking LNG carriers are built by DSME (Daewo Shipbuilding Marine Engineering) for the Yamal LNG project. Image courtesy of Dmitrii Lobusov.

violation of the permit's requirements when the ship entered the water area with medium ice conditions.

In Antarctica, the M/V *NordNorge* was involved in rescuing the crew and passengers of the M/V *Explorer* in 2007 (Gard News, 2008). The *Explorer* had a hole in her starboard side, and the *NordNorge* was 40 nautical miles away and so was able to come to the rescue. It should be noted that the *Explorer* only had open lifeboats, even though the voyage was an Antarctic cruise. According to Bignell (2011), "an 'inexperienced and overconfident' captain drove his ship too fast towards a 'wall of ice'" causing the ship to eventually sink. This event was a wake-up call, reflecting the danger of polar cruises; luckily, a "buddy-ship" was close by for rescue. The event also serves to show the need for the Polar Code for polar shipping.

Recent years have witnessed an increase in use of the Arctic navigation routes. The cruise ship *Crystal Serenity* sailed through the Northwest Passage in 2016 with more than 1,000 people on board. The Maersk container vessel *Venta Maersk* sailed from South Korea to Europe in August/September 2018 (*High North News*, 2019). In July 2018, the LNG carrier *Christophe de Margerie* set a record of 18.5 days for a non-escorted ship to transit from Sabetta in Ob Bay to China (gCaptain, 2018). However, in 2018, an Arctic cruise ship ran aground in the Canadian Northwest Passage (Humpert, 2018). The local community close to the location of the grounding site had to empty all reserves to accommodate those who were evacuated. The event made it clear that Arctic shipping is very vulnerable in case of accidents.

The IMO Polar Code (IMO, 2017), developed to ensure safer sailing in polar waters, came into force on January 1, 2017. The Polar Code sets common standards for vessels and services to navigate the polar regions. It was developed from the SOLAS Convention (IMO, 2001) regarding safety for navigating in the polar regions and the MARPOL Convention (IMO, 1983) for the prevention of pollution from ships navigating the polar regions.

It should be noted however that the Polar Code does not apply to fishing vessels. This gives rise to discussion on the safety of fishing vessels in polar waters. These vessels are susceptible to sea spray icing, as they often have a low freeboard and a large amount of deck equipment (Johansen et al., 2020). On December 28, 2020, the Russian fishing vessel *Onega* sank near Novaya Zemlya due to icing. Seventeen of the crew lost their lives (Nilsen, 2020). The vessel was fishing beyond the range of any official rescue capability. Note that ships sailing in ice conditions with ice coverage less than 1:10 are also exempted from ice class rules.

The future of polar shipping and need for improved international norms

In the 21st century, Arctic shipping has increased considerably and the Arctic waterways have become global waterways. A large increase in marine activity in the Arctic region is expected in the coming years as the area covered by ice during the summer months has been diminishing considerably (National Snow and Ice Data Center's Home Page) in recent years:

- Increase in marine traffic due to oil and LNG transportation
- A general increase in transshipment of cargo due to reduction in ice cover
- Russian official policy regarding increased use of the NSR
- Increase in cruise traffic as Arctic cruises are advertised as "adventure tours"
- Increase in numbers of passengers on board cruise ships making rescue an increasingly challenging task
- Increased fishing activity in the region

After the recovery from the COVID-19 pandemic, an increase in Antarctic cruise traffic is also likely.

Due to this anticipated increase in sailings in the polar regions, IMO considered it necessary to strengthen the Polar Code by providing an Informative Annex on how the requirements of the Polar Code can be satisfied. (IMO, 2019). By issuing such a document, the IMO provides a recommendation that should be followed by the classification societies, reducing competition between the classification societies regarding the least expensive ways to satisfy the requirements of the code. Similar concerns are valid regarding the required training of ships' officers. In the reports from the SARex exercises conducted off northwest Spitzbergen (Solberg et al., 2016, 2017) and by Solberg and Gudmestad (2018; see also Gudmestad and Solberg, 2019), it is demonstrated that the competence of the leaders of the means of rescue play a major role in the safety of those being rescued.

Discussion is moreover needed on whether fishing vessels should be covered by the Polar Code requirements as the number of people involved in fishing, particularly in the Arctic region, is in the order of thousands, and their lives are as precious as those of personnel on commercial vessels and of the passengers and crews on board cruise ships. The responsibility for rescue operations should be further clarified.

The role of insurance companies is also important. How will the insurance companies ensure that they will not be liable to indemnify in the order of billions of dollars in the case of the loss of many passengers on a cruise ship? Only where the risk is small will the insurance industry agree to insure passengers and vessels as well as the environmental clean-up after a potential accident. This will necessitate improvements in the attitude of the cruise industry to rescue in case of accidents. The industry may decide to limit the number of passengers on board polar cruises.

Finally, the responsibility of the shipowner should be stressed. Note also that the captain of the ship, the commander, has the most important role in deciding whether the sailing entails an acceptable risk. It will take a brave captain to go against the decision of the vessel's owner; however, in case of an accident, the captain will eventually be taken to court to document that the risk of the sailing was as low as was reasonably practicable (ALARP).

The increased activities in the polar regions, in particular in the Arctic, call for further development of international cooperation. It is expected that the work in IMO related to polar shipping will continue with increased intensity, whereby rules and regulations will be even better aligned to ensure safe polar shipping. Also, the classification societies are encouraged to continue to cooperate to ensure that all ships sailing in polar regions maintain the same safety standards (International Association for Classification Societies, IACS, 2019).

Challenges for international cooperation related to search and rescue

The history of Arctic search and rescue from vessels in distress

Prior to the telecommunications era, SAR depended on the availability of vessels spotting the ship in distress or on the means for evacuation. The only option for rescue was evacuation into open lifeboats. The wait in the lifeboat was extremely challenging in cold and snowy weather, in darkness, and in stormy sea conditions. The loss of lives during World War II was huge, following torpedo attacks and subsequent evacuation into open lifeboats (see, example, e.g., Brekmoe, 2020). Survivors from shipwrecks often walked over the ice to reach inhabited areas; an example is the loss of the sealer *Kapp Flora* (see Figure 14.3 and subsequent discussion). The crew of the *Kapp Flora* dragged the lifeboats across the ice and came ashore at Cape Orlov on the Murmansk Coast.

The present and the future status of search and rescue in the Arctic Seas

On the present status regarding SAR in the Arctic Seas, see Borch and Andreassen (2021). It should be noted that there is a need to strengthen polar shipping SAR due to the increase in polar shipping in consequence of the diminished ice cover. There is no approved guidance on how the requirements of the code could potentially be satisfied; however, an IMO work group (IMO, 2019) has developed "*draft interim guidelines* on life-saving appliances and arrangements for ships operating in polar waters" (see Gudmestad et al., 2019). In June 2019, the Norwegian Maritime Administration incorporated these suggestions into new regulations specifically for the Svalbard waters (Norwegian Maritime Administration, 2019).

However, the capability of coastal nations to rescue large groups of people, particularly from cruise ships navigating "exotic polar locations", is inadequate. Captains deciding to sail in uncharted waters (Sollid et al., 2018), where there is a risk of running aground, far away from means of rescue and without contact to other vessels, cannot expect rescue vessels to arrive within days. The Polar Code requirement for five days' survival while the rescue means are on their way is generally realistic. However, in some cases, the time to rescue may even exceed five days (Solberg et al., 2020) should there be a huge number of survivors to attend to. It may be advisable for the coastal nations to issue a disclaimer note, clarifying the limitations in rescue capabilities and the estimated rescue time for ships sailing in polar regions. This will limit the possibility of legal actions against coastal nations because of limited rescue capacity and capability.

The need to limit pollution of the polar region by shipping

The depositing of soot and dust on ice and snow accelerates the melting of the polar regions' white cover of ice and snow. This situation accelerates global warming and measures are being taken to limit the "blackening" of the Arctic region in particular. The main concern is with the emission of black smoke and particles from shipping due to burning of heavy oil as fuel. Norway, for example, has enacted that all vessels within the economic zone of Svalbard must use low sulfur oil as fuel (HFO-free Arctic, 2020). "The eight nation Arctic Council has set targets to limit black carbon (or soot) emissions between 25 and 33 percent below 2013 levels by 2025 in a bid to slow Arctic warming" (Climate and Clean Air Coalition, 2017).

Further to pollution of the air, there is an urgent need to limit pollution of the sea. The disposal of bilge into sea is a problem, in particular in the Arctic, where degradation processes are slow. Disposal of plastics from fishing vessels is of grave concern as plastics degrade into nanoparticles which enter the food chain, eventually ending up in the human body. Fisheries must be regulated so as to end the dumping of obsolete/broken equipment. IMO (2018) has adopted an Action Plan to address marine plastic waste from ships.

Cleaning of ballast water has been an issue in the maritime industry. International regulations are in place (IMO, 2004). Foreign species brought from warmer waters may thrive in the fresh water of the Arctic, the more so with the warming trend in the Arctic Seas. There is thus a need to keep ballast water cleaning high on the international agenda.

International norms for Arctic ship design and operations and the future outlook

International norms have been developed by the IMO. The organization includes most of the UN's member countries. These countries have different interests; some countries register ships for shipowners so as to minimize tax and maintenance costs. In addition, shipowners' associations and certifying bodies are represented in IMO meetings. The broad membership makes it difficult to pass the necessary regulations for safer and cleaner shipping. The work to obtain approval for the International Polar Code (IMO, 2017) took decades. However, regulations with a broad mandate and agreed with general consensus are strong and will be adhered to. The key to success was to develop a functional code stating objectives that can be achieved in different ways (Engtrø et al., 2020). The classification societies will thereafter develop detailed rules or standards, whereby the IMO regulations can be fulfilled. This approach works well, provided the classification societies develop rules which are at the same safety level and which are adhered to in the approval of new designs and class reviews. On the other hand, it must be noted that the classification business is a commercial undertaking and classification societies might be tempted to approve vessels that should have been upgraded rather than being deemed seaworthy.

Individual countries may also impose specific restrictions on vessels sailing in their territorial waters. Norway imposed specific requirements on vessels within the 12-mile limit at Svalbard (NMA, 2019) and several countries (e.g., around the Mediterranean) have requirements on use of low sulfur fuel. There are also examples where the Arctic Council has imposed specific requirements on Arctic shipping (Arctic Council, 2011). It has been suggested that this body, representing all Arctic countries, should impose a low sulfur fuel requirement on all Arctic shipping within their areas of responsibility. The NSR may benefit from such a requirement, whereby the market and the companies shipping goods may consider the NSR a "greener" choice for their shipping services.

The future may see changes in shipping through the use of modern technology. The trend to minimize the number of crew members onboard ships continues, with a goal of developing autonomous ships. This may not mean that the ship is completely unmanned, but rather that the ship should be in a position to sail without supervision for certain distances. An autonomous ship is considered to be supported by a digital twin, whereby the operation of the ship is supported and monitored from an office possibly far away. The risk of cyber collapse or cyberattack, however, must be taken into account. One real concern is communication collapse due to increased solar activity and particles that would damage the satellite communication network. This threat is particularly great in the Arctic as the solar particles are attracted toward the magnetic North Pole. In this respect, it is necessary to carry out risk analysis to document that the risk of disaster is exceptionally low, as the collapse of communication would affect all relevant vessels.

When it comes to designing vessels for the Arctic, a trend toward risk-based design is being promoted (Kujala et al., 2019). This approach could save investment costs; however, it is necessary for all relevant risks to be included in the analysis, including possible future risks. The evidence of failure to consider risks is a matter of grave concern, that is, the "black swans" (Aven, 2014). Furthermore, the economic impact of wet and dry bulk shipping in Arctic waters is being discussed (Solakivi et al., 2018). It must, however, be realized that the pressure to limit costs must be compared to a potential increased risk for vessel and crew. It may be difficult to disregard the impacts of floating ice in the Arctic, even if the area covered by ice is diminishing. The future will see larger waves in the Arctic due to longer stretches of open water. The ice floes will have more impact power, including those thrown against vessels during "green sea" submergence of the bows of vessels.

The present-day international norms are in general ready to regulate activities in the future too, but the next section lists challenges where future international regulations are required.

Conclusions and recommendations regarding international cooperation related to shipping in the Arctic

This chapter has reviewed progress in shipping in the polar regions from the early explorers via the daring seal hunters and whalers and from the ice-strengthened ships to the modern design of large vessels for commercial and leisure traffic in the polar regions. The recent focus is on passenger safety and the requirement to minimize pollution of the sea and air. The important role of the IMO in this respect has been highlighted, and a scenario of shipping in polar waters moving in the direction of sustainability and further international cooperation of the traffic is suggested as the main learning point.

Of key concern is the climate impact, regarding which the NSR could be presented as a green intercontinental transportation route, with reference the concerns of the Intergovernmental Panel on Climate Change (IPPC, 2021), by prohibiting the use of heavy fuel oil and by highlighting the support of icebreakers to ensure that schedules for all traffic can be maintained while still reducing fuel consumption compared to southern routes and by ensuring the safety for ships sailing along the route. The total climate impact of using the NSR should be compared to that of alternative routes and be communicated as a key opportunity for sustainable intercontinental sea transportation.

The near future will see new challenges caused by the rapid introduction of new requirements related to clean fuel and the safety aspects associated with the potential explosion risk of utilizing these new fuel types. This will be reflected in the operations of the vessels. Furthermore, interest in autonomous ships will introduce needs for relevant international standards, taking into account the cyber threats to which ships are exposed, particularly when sailing in Arctic waters. It is also expected that larger fishing vessels will need to be covered by the Polar Code, given the yearly losses of fishing vessels and crews. The reduction of the polar ice cap will in particular necessitate such a move, as the fishing vessels will enter even more remote areas. Finally, grave concern is raised in relation to the safety of passengers and crew onboard large cruise liners and a limitation in the size of cruise liners could avoid scenarios with the greatest loss of life. Cooperation between all members of the international maritime community is called for, whereby insurance companies also become involved in ensuring safer design and operations.

December 2021

Ex-post reflections

This chapter was prepared in 2021, at a time of peaceful cooperation in Europe. An increase in international traffic along the NSR was predicted. The recent challenging relations between Russia and other nations operating in Arctic waters would probably limit the international trade along the NSR.

Regarding the design of vessels, cooperation related to ship transport and on issues related to SAR, few changes are envisaged as the Arctic countries all benefit from the present international agreements.

April 12, 2022

References

Note: All links were accessed on 16 September 2021.

Aker Arctic, Helsinki, Finland. Accessed from: https://akerarctic.fi/en/reference/yamalmax/.

Alme, J.B. (2009). *Ishavsfolk si erfaring. Boka om is, isens menn, storm og forlis.* (Arctic people tell about experience. The book about ice, the men of ice, storm and shipwreck) Trondheim: Tapir akademisk forlag. New edition with English summary by Alme and Gudmestad, 2013.

Arctic Council (2011). Agreement on Cooperation on Aeronautical and Maritime Search and Rescue in the Arctic. Accessed from: http://library.arcticportal.org/1474/.

Aven, T. (2014). *Risk, Surprises and Black Swans. Fundamental Ideas and Concepts in Risk Assessment and Risk Management.* Routledge.

Bellona (2013). Accessed from: https://bellona.org/news/arctic/2013-09-foundering-vessel-in-arctic-finally-rescued-by-icebreakers-russian-maritime-authorities-not-pleased-by-incident.

Bignell, P. (2011). The sinking of the explorer. *The Independent*, October 22, 2011. Accessed from: https://www.independent.co.uk/travel/news-and-advice/sinking-explorer-1667532.html.

Borch, J. and Andreassen, N. (2021). International cooperation within emergency/ disaster response in the Arctic Sea areas. *Chapter 14 of present book*, draft September 2021.

Brekmoe, L. (2020). Døden lurte under overflaten. (Death lurked beneath the surface). *Framtid*, June 13, 2020. Accessed from: https://www.framtidinord.no/nyheter/2020/06/13/D%C3%B8den-lurte-under-overflaten-22067968.ece.

Climate and Clean Air Coalition, CCAC (2017). Arctic countries commit to reduce black carbon emissions by as much as a third. Accessed from: https://www.ccacoalition.org/en/news/arctic-countries-commit-reduce-black-carbon-emissions-much-third.

DNV (2013). Ships for navigation in ice. *DNV Rules for Classification of Ships*, Part 5, Chapter 1. Norway: DNV.

Engtrø, E., Gudmestad, O. T. and Njå, O. (2020). Implementation of the polar code: Functional requirements regulating ship operations in polar waters. *Arctic Review*, 11, pp. 47–69. https://doi.org/10.23865/arctic.v11.2240.

Gard News (2008). In November 2007 the M/V NORDNORGE was involved in the dramatic salvage of the crew and passengers of the M/V EXPLORER in the Antarctic. *Gard News*, 190, May/June 2008. Accessed from: https://www.gard.no/web/updates/content/53579/mv-nordnorge-antarctic-rescue-of-mv-explorer.

gCaptain (2018). Accessed from: https://gcaptain.com/icebreaking-lng-carrier-christophe-de-margerie-sets-new-northern-sea-route-record/.

Gudmestad, O.T., and Alme, J. (2015). Implementation of experience from the Arctic seal hunter expeditions during the late 19th and the 20th century. *Ocean Engineering*, 111, 1–7. Accessed from: https://doi.org/10.1016/j.oceaneng.2015.10.029.

Gudmestad, O.T., and Solberg, K.E. (2019). Findings from two Arctic search and rescue exercises North of Spitzbergen. *Polar Geography*. DOI: 10.1080/1088937X.2019.1597394

Gudmestad, O.T., Solberg, K.E. and Barane, E. (2019). Search and rescue exercises in Polar Waters. Presented at Arctic Coast Guard Forum, Turku, April 2019.

HFO-free Arctic (2020). Reaction to Norway's proposal for arctic ban on polluting heavy fuel oil in Svalbard. Accessed from: https://www.hfofreearctic.org/en/2020/11/09/reaction-to-norways-proposal-for-arctic-ban-on-polluting-heavy-fuel-oil-in-svalbard/.

High North News (2019). Accessed from: https://www.highnorthnews.com/nb/maersk-tilbake-til-den-nordlige-sjoruten-i-sommer.

Humpert, M. (2018). Arctic cruise ship runs aground in Canada's Northwest Passage. *High North News*, August 28, 2018. Accessed from: http://www.highnorthnews.com/arctic-cruise-ship-runs-aground-in-canadas-northwest-passage/.

IMO (International Maritime Organization). (1983). *International Convention for the Prevention of Pollution from Ships (MARPOL)*. London: International Maritime Organization. Accessed from: https://www.imo.org/en/About/Conventions/Pages/International-Convention-for-the-Prevention-of-Pollution-from-Ships-(MARPOL).aspx.

IMO (International Maritime Organization). (2001). *SOLAS: International Convention for the Safety of Life at Sea, 1974, and 1988 Protocol Relating Thereto: 2000 Amendments, Effective January and July 2002*. London: International Maritime Organization.

IMO (International Maritime Organization). (2002). *Guidelines for Ships Operating in Arctic Ice-Covered Waters*. London: International Maritime Organization.

IMO (International Maritime Organization). (2004). *The International Convention for the Control and Management of Ships' Ballast Water and Sediments*. London: International Maritime Organization.

IMO (International Maritime Organization). (2017). *Shipping in Polar Waters. The Polar Code*. London: International Maritime Organization. Accessed from: https://www.imo.org/en/MediaCentre/HotTopics/Pages/Polar-default.aspx.

IMO (International Maritime Organization). (2018). *Marine Litter*. London: International Maritime Organization. Accessed from: https://www.imo.org/en/MediaCentre/HotTopics/Pages/marinelitter-default.aspx.

IMO (International Maritime Organization). (2019). *Interim Guidelines on Life-Saving Appliances and Arrangements for Ships Operating in Polar Waters*. Accessed from: https://www.register-iri.com/wp-content/uploads/MSC.1-Circ.1614.pdf.

International Association for Classification Societies, IACS (2019). Requirements concerning polar class. IACS, London. Accessed from: https://www.iacs.org.uk/download/1803

IPPC (2021). The Int. Climate Panel, Sixth Assessment Report Issued 9th August 2021. Accessed from: https://www.ipcc.ch/.

Ishavsmuseet, Brandal, Norway. Accessed from: https://www.ishavsmuseet.no/skutekatalog/kap-flora/.

Johansen, K., Sollid, M.P. and Gudmestad, O.T. (2020). Stability of vessels in an ice-free Arctic. *TransNav, the International Journal on Marine Navigation and Safety of Sea Transportation*, 14(3), pp. 663–671. doi:10.12716/1001.14.03.19

Kärmäräinen, J. and Riska, K. (2018). Ice class rules and international regulations. In Carlton, J., Jukes, P. and Choo, Y-S. (eds.) *Encyclopedia of Maritime and Offshore Engineering*, Part 2, Chapter 136, Wiley

Kujala, P., Goerlandt, F., Way, B., Smith, D., Yang, M., Khan, F. and Veitch, B. (2019). Review of risk-based design for ice-class ships. *Marine Structures*, 63, pp. 181–195.

Marchenko, N. (2009). Experiences of Russian Arctic navigation. *Proceedings of the 20th International Conference on Port and Ocean Engineering under Arctic Conditions*, June 9–12, 2009, Luleå, Sweden.

Mejlænder-Larsen, M. (2015). Vessels for possible use on the NSR. Presentation by DNVGL October 15, 2015. Accessed from: https://pdf4pro.com/amp/view/vessels-for-possible-use-on-the-nsr-chnl-2faedc.html

Nansen, F. (1897). *Fram over Polhavet* (Fram above the Arctic Ocean). Kristiana: Aschehoug Publishing House.

National Snow and Ice Data Center's Home Page. Accessed from: http://nsidc.org/arcticseaicenews/.

Nikel, D. (2020). The fastest viking ship ever found. *Forbes*, 1st April, 2020. Accessed from: https://www.forbes.com/sites/davidnikel/2020/04/01/the-fastest-viking-ship-ever-found/.

Nilsen, T. (2020). Icing believed to cause sinking of fishing boat in Barents Sea, 17 missing. *Barents Observer*, December 28, 2020. Accessed from: https://thebarentsobserver.com/en/2020/12/icing-believed-cause-sining-fishing-boat-barents-sea-17-missing.

Norwegian Maritime Administration (NMA) (2019). Ny forskrift om bygging, utrustning og drift av passasjerskip i territorialfarvannet ved Svalbard. (New regulations on construction, equipment and operation of passenger ships in the territorial waters of Svalbard). Haugesund. Accessed from: https://www.sdir.no/sjofart/regelverk/rundskriv/ny- forskrift- om-bygging-utrustning-og-drift-av-passasjerskip-i-territorialfarvannet-ved-svalbard/.

NSR, Northern Sea Route Administration's Home Page. Accessed from: http://www.nsra.ru/en/home.html.

Solberg, K.E., and Gudmestad, O.T. (2018). *SARex3: Evacuation to Shore, Survival and Rescue*. Report 75 from University of Stavanger, Norway. Accessed from: https://brage.bibsys.no/xmlui/handle/11250/2578301.

Solberg, K.E., Gudmestad, O.T. and Kvamme, B.O. (2016). *SARex Spitzbergen: Search and Rescue Exercise Conducted off North Spitzbergen: Exercise Report*. Report 58 from University of Stavanger, Norway. SARex1. Accessed from: https://uis.brage.unit.no/uis-xmlui/handle/11250/2414815.

Solberg, K.E., Gudmestad, O.T. and Skjærseth, E. (2017). *SARex2: Surviving a Maritime Incident in Cold Climate Conditions. Exercise Report*. Report 69 from University of Stavanger, Norway. Accessed from: https://brage.bibsys.no/xmlui/handle/11250/2468805.

Solberg, K.E., Jensen, J.E., Barane, E., Hagen, S., Kjøl, A., Johansen, G. and Gudmestad, O.T. (2020). Time to rescue for different paths to survival following a marine incident. *Journal of Marine Science and Engineering*, 8(12), p. 997. https://doi.org/10.3390/jmse8120997 (registering DOI)

Solakivi, T., Kiiski, T. and Ojala, L. (2018). The impact of ice class on the economics of wet and dry bulk shipping in the Arctic waters. *Maritime Policy & Management*, 45(4), pp. 530–542.

Sollid, M.P., Gudmestad, O.T. and Solberg, K.E. (2018). Hazards originating from increased voyages in new areas of the Arctic. I: *Proceedings of the 24th IAHR International Symposium on Ice*, Vladivostok, Russia June 4–9, 2018, pp. 272–280.

The Baltic Sea Ice Class Service. Accessed from: http://www.bsis-ice.de/material/table_iceclasses.pdf.

Vestlandsnytt (1962). Då 'Kapp Flora' gjekk ned i Kvite-havet i 1924, Paul Olai Paulsen fortel. (When 'Kapp Flora' went down into the White Sea in 1924, Paul Olai Paulsen told). *Vestlandsnytt*, November 12, 1962.

15 International Cooperation in Emergency Response in the Arctic Sea Areas

Natalia Andreassen and Odd Jarl Borch

Introduction

The commercial activities in the Arctic, including both air and sea transport, have raised concerns about the preparedness and response capacities related to various emergencies in the region. Emergency response operations in the Arctic can be challenging due to severe climate, vast distances, and the technical limitations of communication and rescue equipment (Borch et al., 2016; Sydnes et al., 2017). Search and rescue (SAR) operations at sea and in ice may prove extremely difficult. Rescue efforts may demand a wide range of physical and human resources which are scarce in the Arctic. A rapid response is crucial for survival in cold temperatures and freezing waters. However, it may take a long time before the resources arrive on scene (Marchenko et al., 2018). Moreover, challenges of information sharing and communication infrastructure in the Arctic environment may hamper the coordination of emergency responses (Andreassen et al., 2020; Chen et al., 2008).

Coordination of maritime SAR operations may involve several layers of contributors, including volunteers, government emergency response actors, military support, and SAR units on scene. The external resources will often belong to different agencies with different institutional frameworks, expertise, and practices (Moynihan, 2009).

Depending on the magnitude, nature, and complexity of an incident, the Arctic countries' resources may not suffice to deal with incidents in remote areas, and cross-border cooperation may be crucial (Borch et al., 2016).

This cooperation is facilitated by a range of formal institutional arrangements, including regulatory documents and an established system of organizations and command in maritime SAR (Kapucu, 2005). However, complexity and the need for costly resources call for additional mechanisms in disaster response governance. The experiences of actors, the nature of the emerging international cooperation, and informal relations between institutions are of special interest. Governance mechanisms may bring all resources together in a well-coordinated joint operation with highly efficient use of scarce resources. Formal, administrative, and trust-based governance mechanisms may facilitate connecting actors into one multiagency team.

DOI: 10.4324/9781003246015-19

In this chapter, we elaborate on the role and configuration of these governance mechanisms in combined Norwegian-Russian SAR operations. Our data sample includes primary and secondary sources – interviews with key players, including SAR mission coordinators, incident reports, and articles describing the cases.

This chapter starts with a background description of the governance and formal institutional arrangements of Arctic maritime SAR operations. We then proceed with descriptions of the two cases, reflecting on what happened, the key actors involved, and their roles in the operation. We describe the processes of international cooperation and aspects that facilitated success and challenges in the response operation. Finally, we reflect on the outcomes of international cooperation and on how international cooperation experiences may have implications for the future development of cross-border emergency response in the Arctic seas.

Governance of maritime SAR operations in the Arctic

Maritime SAR operations are governed by several international laws and agreements. The 1979 International Convention on Maritime Search and Rescue at Sea (the SAR Convention) aimed at developing an international SAR system so that rescue operations are coordinated by cooperation between neighboring SAR organizations. The 1974 International Convention for the Safety of Life at Sea (SOLAS) enshrined in law the obligation of ships to go to the assistance of vessels in distress. The 1982 UN Convention on the Law of the Sea (UNCLOS) requires coastal states to maintain an adequate and effective SAR service regarding safety at sea.

The International Aeronautical and Maritime Search and Rescue (IAMSAR) manual is jointly developed by the International Maritime Organization (IMO) and the International Civil Aviation Organization. It offers guidelines for a common aviation and maritime approach to management and standard operational patterns (IMO and ICAO, 2016).

In addition, SAR operations in the Arctic are governed by multilateral and bilateral SAR agreements. The 2011 Agreement on Cooperation on Aeronautical and Maritime Search and Rescue in the Arctic (Arctic SAR Agreement) is the first legally binding treaty signed under the auspices of the Arctic Council. The responsibility is shared between all the eight Arctic states, and each member state has a particular SAR area of responsibility.

The bilateral 1995 Agreement between the Russian Government and the Norwegian Government on cooperation in SAR in the Barents Sea stipulates that the parties shall provide assistance in SAR in the Barents Sea, outlines the competent national authorities responsible for the implementation and their tasks, and specifies how requests for help are forwarded and procedures for information exchange (Elgsaas and Offerdal, 2018).

The cooperation through the Arctic Council working groups and the Arctic Coast Guard Forum is valuable for exchanging best practices, discussing challenges and how to overcome them. The cooperation in the Barents Euro-Arctic Region launched in 1993 provides for relations on two levels within the intergovernmental Barents Euro-Arctic Council (BEAC) and interregional Barents Regional Council (BRC), the overall objective being sustainable development. Also, within this agreement, multiagency field exercises combining resources from Finland, Norway, Russia, and Sweden take place every third year. The bilateral and multilateral agreements constitute an important formal governance framework for emergency response and provide meeting places for aligning managerial structures as well as trust-enhancing arenas (Borch et al., 2018).

Cases of cross-border emergency response

The descriptions start with an overview of the SAR incident, describing the key actors involved. Then the processes of international cooperation and aspects that challenged or facilitated cross-border cooperation are summarized.

The case of the Victor Koryakin *Shipwreck*

The incident

This case describes a successful joint SAR operation saving a crew of 12 from the sinking vessel. MV *Viktor Koryakin* was a dry cargo vessel registered in Russia transporting timber along the Northwest coast of Russia. On December 18, 2007, the vessel ran aground and later broke in two and sank off the Rybachiy Peninsula in the Petsjenga region, Murmansk *oblast*. The location was 55 kilometers from the Norwegian border and the town of Vardo. The 88-meter long vessel had a crew of 12. The weather was severe with a strong northwesterly gale up to 35 meters/second (Aamo et al., 2017; Udin et al., 2016).

The operation was coordinated by Murmansk Maritime Rescue Coordination Center (MRCC). Support from the Norwegian Joint Rescue Coordination Center (JRCC) was requested, and a Norwegian SAR helicopter rescued all the crew members of the Russian dry cargo vessel only minutes before the vessel broke up and sank (*Barents Observer*, 2010).

Key actors, their roles, and interests in the operation

The rescue of the MV *Viktor Koryakin* crew was a combined operation led by Murmansk MRCC. At 08.58 Moscow time, UTC -3, MRCC Murmansk received the mayday call with notification of a vessel in distress from the radio operator of the Murmansk Shipping Company. The distress signal was sent out by *very high frequency* as the vessel did not have satellite communication facilities. Murmansk MRCC informed State MRCC and began the coordination of the rescue operation. Fishing trawler *Scherbakov* was asked to relay communications with the vessel in distress as the radio connection was poor. Murmansk MRCC sent the rescue tug *Purga* to assist with estimated time of arrival at 02.30 Moscow time (UTC -3). A Finnish and a Russian vessel were approximately two hours away. Using life rafts was dangerous because of bad weather and shallow waters with wave height up to 9 meters. The captain of MV *Viktor Koryakin* reported that the engine room was flooded and the situation was critical.

The mayday message was also received by the Norwegian maritime radio in Vardo at 07.41 Oslo time, UTC-1. Vardo radio informed the JRCC Northern Norway (JRCC NN). After receiving this message, JRCC NN called MRCC Murmansk and offered help. MRCC Murmansk acknowledged the need for help as the situation was critical and stated that they had no helicopter capacity available for immediate action.

JRCC NN Bodo scrambled the *Sea King* rescue helicopter from Banak Airport in the Eastern part of Finnmark. JRCC asked the Air Coordination Center in Bodo to alert all airports along the route to be ready for landing and fueling. Also, to have clearance for crossing Russian airspace from the Murmansk air control, JRCC called the Norwegian Military Joint Headquarters (NJHQ) and asked them to inform the Northern Fleet Headquarters in Russia. The university hospital in Tromso and the regional hospital in Kirkenes were alerted. Four Norwegian ambulance planes were on standby. The rescue subcenter at the regional police station in Kirkenes was also mobilized.

At 10.18, Vardo radio lost contact with MV *Viktor Koryakin*. The *Sea King* rescue helicopter reported arrival time at 10.25. At 10.46, the helicopter reported back that the entire crew had been saved, three hours after receipt of the mayday signal. Soon after that, the vessel broke up and sank.

The helicopter captain landed at Kirkenes with the rescued crew. Landing at Murmansk Airport was considered but rejected due to the risk of delays with customs. If the SAR helicopter were to be held up, this would severely reduce the SAR capacity in Northern Norway until the helicopter was released. The crew was taken instead to Kirkenes, where the police helped with transport and accommodation in cooperation with the Russian Consulate General in Kirkenes.

There were about 50 tons of fuel oil and 20 tons of diesel oil in the tanks of the sunken vessel in area where it ran aground. It was impossible to recover the fuel due to stormy weather. The wind washed the leaked oil ashore. The Norwegian Coastal Administration through the vessel traffic coordination center in Vardo was ready to help with oil recovery response. Two days later, the tanker *Kotlas* and the boom-laying boat *Markap* arrived from Murmansk and collected recovered oil from the wreck.

The processes of international cooperation

The SAR operation included close coordination between the relevant Norwegian and Russian SAR authorities and institutions at different levels. First, the close cooperation of the JRCC NN and MRCC Murmansk contributed to the successful mobilization and coordination of incident response.

In addition, the military operational headquarters in both regions were in a dialog to discuss potential limitations due to military restrictions, due, among others, to the fact that the Norwegian SAR helicopters are operated by the Norwegian Air Force. At the tactical level, the air traffic controller of Northern Norway was in touch with the Murmansk tower and air traffic control in the Murmansk region to accept the flight plan and the route for the rescue helicopter entering Russian territory at short notice.

As for resources, vessels from Finland and Russia, the Norwegian rescue helicopter squadron, the airports of Finnmark county, the university and regional hospitals, and the air ambulance capacities of Northern Norway were mobilized. Kirkenes police served as a rescue subcenter. As there was a possibility of severe injuries among the crew, the hospital in Tromso prepared the facilities. The Consulates General in Kirkenes and Murmansk were also in the loop. In oil spill response, the Norwegian Coastal Administration offered their support to avoid negative environmental consequences due to oil pollution.

Aspects facilitating successful cooperation practices and challenges

Successful cooperation practices. The facilitators of the successful SAR operation in the case of the MV *Viktor Koryakin* can be found both within the formal agreements serving as a legal platform for cooperation, administrative-organizational coordination mechanisms facilitating dialog and capability selection, and also mutual relationships of trust in both civilian and military agencies.

Formal institutions. The prompt and adequate cross-border, multiagency response was based on the cooperative infrastructure of formal institutions and agreements. Regarding SAR at sea, the international legislation serves as a general platform for cooperative action. This includes the 1982 UNCLOS requiring coastal states to maintain an adequate and effective SAR capacity regarding safety on and over the sea. The 1979 International Convention on Maritime Search and Rescue (the SAR Convention) requires that the coastal states establish rescue coordination centers and enter into SAR agreements with neighboring states.

Administrative coordination. Both Norway and Russia have over the years built up a professional capacity with each country responsible for the coordination of SAR operations. They share training and experience related to air, land, and sea operations. Many of the communication, coordination, and control mechanisms are common, related to maritime radio in the Global Maritime System for distress calls and the IAMSAR manuals. Through annual meetings and exercises, the parties meet to exchange experiences and plan and conduct an annual joint Norwegian-Russian exercise involving civilian and naval SAR capacities (Staalesen, 2017).

Trust relations. Professional and personal relations of trust are an important aspect of the successful SAR operation. Over all these years, a close relationship of trust has been developed between the Norwegian and Russian SAR coordinators of the maritime coordination centers exchanging best practice and professional experiences. As military resources play an important role, these same relations are built through annual meetings between the coastguard institutions of the two countries and the military supreme command of Norway and the Russian Northern Fleet. Finally, the meetings of the Arctic Council and their expert groups provide an additional arena for building trust, respect, and a common operational ground.

Challenges. The case of the *Victor Koryakin* was a successful joint SAR operation saving a crew of 12 from the sinking vessel. The main challenges in this case relate to the operating conditions and military restrictions. Severe weather and high waves hindered the use of samaritan vessels. Scarcity of resources and time escalated the complexity of managing the operation among several command levels, between authorities as well as civil-military cooperation. Such complexity may prove a challenge. There is a need for all parties potentially involved to be notified and to obtain clearance for crossing borders. Potential risk of delays at customs can be a challenge. In the case reported, the risk was avoided by choosing another airport closer to the rescue subcenter where the operation could proceed more smoothly.

The case of the MI-8 crash

The incident

On October 26, 2017, a Russian MIL MI-8 helicopter with eight people on board disappeared outside Barentsburg in Norway's Arctic Archipelago Svalbard. The helicopter was on a route from the abandoned Russian settlement known as Pyramiden to the helicopter landing site at Barentsburg, Svalbard's second largest settlement. The helicopter pilot communicated with the air controllers at Longyearbyen Airport and with the Heerodden helipad just minutes before it disappeared. The weather was bad, with visibility down to 1.5 kilometers, temperature one degree below zero, and an easterly wind. The temperature in the sea was approximately two degrees Centigrade (Klanderud et al., 2017).

At 15.33, the Governor of Svalbard office received a distress call from the Heerodden helicopter landing site outside Barentsburg. A Russian MI-8 helicopter was missing. The county governor immediately launched a disaster alarm. The Governor of Svalbard alerted the JRCC North Norway. Several units from the local governor were directed to the area where the helicopter had disappeared, including SAR helicopters and vessels. The local hospital and the university hospital at Tromso on the mainland were mobilized.

JRCC NN took the lead in the operation and proceeded by alerting all relevant units, including the nearest coastguard vessel with significant experience in SAR cooperation. The vessel *Polarsyssel*, a part of the Governor of Svalbard capabilities, was assigned to take the role as on-scene coordinator at the probable incident site until the coastguard arrived. At 15.53, the JRCC called the NJHQ requesting SAR resources. JRCC also informed the Ministry of Foreign Affairs, which helped to inform the Russian authorities. In a very short time, authorities at all levels were informed and started responding.

The first helicopter started to search the area at 16.14, 30 minutes after the callout. An intensive search took place during the night with shore, sea, and air resources. This search included several vessels and helicopters. Due to the difficult conditions, a surveillance plane from the Norwegian Air Force and the Danish Arctic Command in Greenland were mobilized with advanced search sensors. However, no survivors were found, and it was assumed that the helicopter had crashed into the sea and sunk with crew and passengers. The operation was defined as search after deceased persons (SEAO) the next day, as there was no hope of finding any survivors. In Norway, this is a task for the police, at Svalbard represented by the Governor of Svalbard. Several additional resources were mobilized for this search effort. The Russian authorities also offered their help. A transport plane from the Norwegian Air Force arrived with an autonomous underwater vehicle to conduct an underwater search for the helicopter and the victims. The Norwegian marine research vessel from the Institute of Marine Research G.O. *Sars* also arrived and started a search. On Day 3, a Russian transport plane with a rescue team, including rescue divers, from the Russian emergency response ministry EMERCOM arrived. They brought a remotely operated underwater vehicle, scuba diving equipment, and dinghies. These resources were used to search along the shoreline on the seabed.

On Day 4, the helicopter was located on the seabed at a depth of approximately 200 meters, 2 kilometers from the shore. The next day one of the missing persons was found close to the wreck. The search for missing persons continued for two weeks. Thereafter, the organized search for deceased persons ended. The SAR crew from Russia returned home. The Governor of Svalbard's rescue helicopters and vessel regularly searched the sea and coastline around Barentsburg for bodies in the following months, but without success.

The Norwegian Safety Investigation Authority chartered the heavy lift vessel *Maersk Forza* to raise the helicopter to the surface and bring it ashore. The hulk of the helicopter was transported to the mainland for more thorough investigations by representatives from the Instate Aviation Committee in Russia and advisers from the Russian helicopter design company. The report on the investigation by the Norwegian Safety Investigation Authority was finalized in February 2020 with safety advice to the Russian Aviation Authority, the operator company, and the Civil Aviation Authority Norway.

The incident was investigated by the Norwegian police and no evidence of negligence was found. The case was closed after a thorough investigation lasting until May 2020.

Key actors, their roles, and interests in the operation

The response operation at Svalbard included several phases involving different actors and coordinating bodies. The SAR and SEAO operations and the investigations included several hundred persons and a wide range of air and surface units with advanced equipment from four countries. Several Norwegian ministries and directorates were involved during the different phases. The Governor of Svalbard spent significant resources over a long period, contributing to the search and investigation. All actors mobilized the available resources for the SAR operation.

Major resources were also mobilized to find the missing persons and to conduct the accident investigation.

The nature of the processes of international cooperation

Cooperation across borders in the case reported included all the main phases of an emergency and incident response. This included the SAR operation, the search for deceased persons (the SEAO operation), the police incident investigation, and a safety investigation. At each stage, there was cross-border cooperation coordinated by the Norwegian authorities (Governor of Svalbard, 2017). The SAR operation demonstrated the mobilization of significant resources at short notice, both locally and from the Norwegian mainland. This included civilian and military resources. Requests to neighboring countries for available resources yielded a patrol plane from the Danish Air Force stationed in Greenland. A Russian delegation of 40 persons arrived to assist in the operation (*Aftenposten*, 2017). The Russian emergency response ministry EMERCOM made specialized diving capacities available to Svalbard. The Russian and Norwegian aviation safety authorities cooperated on the investigation. Challenges were related to having such an unexpected incident in a small community and to cooperating with personnel from different countries at short notice using different communication tools.

Aspects that facilitated successful cooperation practices and challenges

Successful cooperation practices. The case reported demonstrates fast mobilization of national and international resources and close international cooperation at all phases of the incident response – putting out the alert, the SAR operation, the SEAO operation, and the investigation into the incident. The crisis staff management locally and nationally launched an immediate information channel to the strategic-level authorities and to Russia, which suffered a heavy loss. Compared to the first case from 2007, more formal institutions and coordinative mechanisms were in place by 2017 at all decision levels. Providing the necessary resources for responding to SAR operations is a challenging task in a small community far from mainland resources. However, the Norwegian government managed to provide significant resources over a short period of time. The governor has two SAR helicopters and a SAR vessel on call. Ambulance planes and paramedics are on call on the mainland and were dispatched to Svalbard. The rescue helicopters on the mainland were airborne in 15 minutes.

Formal institutions. As stated earlier, air as well as sea accidents are well provided for in international agreements. SAR operations are regulated through the bilateral Norwegian-Russian SAR agreements, together with the 2011 Agreement on Cooperation on Aeronautical and Maritime Search and Rescue in the Arctic between the member states of the Arctic Council – Canada, Denmark, Finland, Iceland, Norway, Russia, Sweden, and the US. These agreements, together with frequent exercises, as well as the Arctic Council working groups provide institutional platforms for the smooth running of SAR operations. For Svalbard, the Svalbard Treaty recognizes the sovereignty of Norway over the Arctic archipelago of Svalbard. The signatories to the Treaty are given equal rights to engage in commercial activities, and there are military restrictions as to bases and fortifications.

Administrative coordination. An improved communication infrastructure and well-established procedures for alerting and decision-making significantly reduced the time to mobilization and response in this case. The procedures are described in the IAMSAR manual developed jointly by the IMO and the International Civil Aviation Organization. This contributes to understanding the roles and responsibilities of each coordinator involved and ensures common ground for education and training. Use of liaison is one of the factors that contributes to better situational awareness and successful collaboration between agencies and resources.

Trust relations. Svalbard is an archipelago with a close-knit community with in total approximately 2,600 inhabitants in three settlements. Longyearbyen is the largest community with the seat of the governor. Approximately 500 Russians and Ukrainians live in the mining community of Barentsburg. Over the decades, there has been close cooperation and cultural exchange between the Russian and Norwegian communities, mostly consisting of miners. In recent years, cooperation in education and research has been significant, not the least through the Norwegian University Center at Longyearbyen and the research institutions in the small researcher community at New-Aalesund. The governor's administration frequently visits the other communities and discusses practical matters as to the living conditions at Svalbard. The close contact and good cultural relations between the Russian and Norwegian populations at Longyearbyen and Barentsburg provided a good platform of trust for close cooperation in this incident.

Challenges. This operation has faced several challenges that could potentially influence international cooperation processes. First, there were communication line challenges because many unfamiliar actors were connected in one critical operation involving considerable uncertainty. Second, limitations in the communication equipment caused uncertainties in disseminating information on the location of the incident.

Another challenge concerns the capacity to coordinate a large operation over time. Numerous resources arrived on scene to provide assistance and had to work under the command of an on-scene coordinator and cooperate with other parties on scene and afterwards in the search for missing persons. This called for significant efforts in coordination and communication between the agencies involved.

International cooperation practices in emergency response: outcomes and projections

Outcomes of international cooperation processes for global development and sustainable development goals

International coordination in disaster response operations may be a driving force for the development of a more resilient Arctic. There is a need to take care of the safety of the population and visitors to the Arctic through a very competent emergency preparedness and response system. The communities need to be robust and able to cope with severe crises. As most communities and regions in the Arctic are sparsely populated and have limited resources, cooperation is vital. As the military forces represent significant resources in the Arctic, their capacities should be capable of providing support in civilian emergencies. This calls for clear political signals, frequent meetings, and joint training and exercises. Keeping the channels for cooperation open in times of political and military tension is of critical importance for the Arctic. Here, there are many good examples of openness in the SAR relations between the Arctic countries, in particular between Norway and Russia.

The international cooperation in cross-border emergency management contributes to achieving the sustainable development goals (SDGs) of the UN by addressing challenges such as economic inclusion, dwindling natural resources, geopolitical instability, environmental degradation, and the multifaceted impacts of climate change.

In particular, SDG 11 "Make cities inclusive, safe, resilient and sustainable" within its target 11.5 has a focus on significantly reducing the number of deaths and the number of people affected by disasters, including water-related disasters, with a focus on protecting the poor and people in vulnerable situations. Better cooperation contributes to its target 11.8 to support positive economic, social, and environmental links between urban, peri-urban, and rural areas by strengthening national and regional development planning as well as 11.9, which focuses on resilience to disasters and the development of the holistic disaster risk management at all levels, in line with the Sendai Framework for Disaster Risk Reduction 2015–2030.

The cooperation on maritime emergency management also contributes to SDG 14 "Conserve and sustainably use the oceans, seas and marine resources". One of the goals, 14.1, focuses on preventing and significantly reducing marine pollution of all kinds. More effective cross-border response to emergencies causing oil spills contributes especially to target 11.2, to sustainably manage and protect marine and coastal ecosystems, to avoid significant adverse impacts, including strengthening their resilience, and to take action for their restoration to achieve healthy and productive oceans.

The cooperation on emergency management that helps to save and restore nature contributes to SDG 15 "Life on land". Especially relevant is the contribution to target 15.12, focusing, among others, on increasing the capacity of local communities to pursue sustainable livelihood opportunities.

Historical and present experience and a projection for the future

The Arctic is a region of considerable geopolitical and military importance. Improving national preparedness and international cooperation on emergency prevention, preparedness, and response have been on the agenda of all Arctic countries (Elgsaas and Offerdal, 2018).

In general, there is a broad political consensus to maintain the high priority of an effective SAR service with a marked focus on international support. The Host Nation Support (HNS) Guidelines, developed through international organizations, including the Red Cross, the UN's Office for the Coordination of Humanitarian Affairs, and the EU, has been established to ensure that possible assistance from other countries is managed in a satisfactory manner (DSB, 2014).

The socioeconomic development in the Arctic, including the infrastructure and increase of activities in offshore industries, indicates that international cooperation in production safety and emergency preparedness will remain on the agenda of the Arctic states. Environmental security, strengthening of SAR, and preventing disasters are therefore associated as one of the main directions for the idea of sustainable development in the Arctic (Andreassen, 2016).

Challenges and their potential impacts

Cross-border cooperation in crises is challenging. Rapid reaction is needed, significant resources are at stake, and advanced resources are needed for crisis solutions. In SAR operations, the obligation to contribute is clear. However, limitations in equipment, differences in training, language, and management culture may hamper cross-border emergency response (Andreassen et al., 2018).

Another factor that may cause barriers to the cooperation may be political and military tensions between two countries. Political and military disputes may destroy the spirit of partnership. Since the sanctions imposed on Russia after the Ukraine conflict, there have been signs of difficulties in relations. There have also been tensions between Norway and Russia regarding governance and restrictions that may hamper relations (Pedersen, 2017). These tensions may potentially cause challenges to obtaining clearance for crossing borders and getting assistance from other countries as fast as possible. At Exercise Barents, for instance, challenges with border crossing hampered adequate training.

Policing and cooperation on anti-terror response are also among challenges in the Arctic area (Myhrer, 2020). Here, there is limited cooperation due to the involvement of extremely secret special forces. Military tensions are very much present here as a barrier to closer cooperation.

In academia and in research on emergency management, tensions between countries may inhibit the exchange of knowledge and competence in emergency management. Russia and Norway have never had a break in cooperation in research and development (R&D), regardless of political fluctuations in political relations (Bye, 2020). However, strict security rules in several countries may limit student and knowledge exchange and impair research production. One example is the stricter control measures introduced in Russia on international educational and research cooperation (Danilov, 2021). In practice, this causes delays in research production and may in the future inhibit cooperation.

Success factors and their potential outcomes

Despite the challenges, the long history of cooperation between the two countries has been contributing to a more robust emergency preparedness and response system in the Arctic Sea areas, as illustrated by the two cases reported here.

Over the decades, there has been a strong feeling of commitment in the Arctic among both civilian and military responders as well as top management to adhere to the contractual agreements, keep the door open for communication, and maintain a strong cooperative relation based on trust.

In terms of cooperation, the mechanisms that facilitate interorganizational relations and mutual understanding merit acknowledgment. As part of the contractual theory, relational governance has been introduced as a framework for reducing hazards and barriers to cooperation (Dyer and Singh, 1998; Uzzi, 1997). The focus has been on formal contracts to mitigate the risk of increased dependency, reduce skepticism, and enhance coordination. However, a broader range of governance mechanisms is needed. Relational governance can complement the use of formal contracts (Poppo and Zenger, 2002). Both administrative coordination mechanisms as well as trust-enhancing mechanisms may provide platforms for increased interplay and smoother interaction. Borch (1994) emphasizes the interplay between these three mechanisms and how they may serve together as a platform for tackling higher system complexity, as in large-scale crises.

Formal institutions. Norway and Russia initiated systematic collaboration on SAR as early as in the 1950s. The first agreement was concluded in 1956 stipulating cooperation on SAR of anyone in distress in the SAR regions regardless of nationality. An extended and more detailed version of the first agreement was achieved in 1988, focusing, among others, on the cooperating institutions and their procedures for communication and entering each other's territories. The formal agreement serving as a platform for today's cooperation was established in 1995 with the agreement between the Russian Government and the Norwegian Government on cooperation in SAR of people in distress in the Barents Sea (SAR Agreement of 1995). The agreement includes procedures for assistance and calls for an annual exercise between the two countries named Exercise Barents.

Also, the cooperation in the Barents Euro-Arctic Region launched in 1993 specifies relations on two levels within the intergovernmental BEAC and interregional BRC with the overall objective of enhancing preparedness and sustainable development.

In 2011, the Arctic SAR Agreement was concluded under the auspices of the Arctic Council between the Arctic countries, specifying the main areas of responsibility of each country and including analyses of experiences and exercises. Formal agreements will be enhancing the cooperation in future, overcoming possible political tensions and contributing to faster cross-border assistance in emergencies.

Administrative coordination. There are several administrative coordination mechanisms that contribute to a stronger international emergency system.

The regularly updated IAMSAR manual describes the roles of the coordinators involved.

Senior personnel and management also meet through the Arctic Council cooperation platform.

The Emergency Prevention, Preparedness & Response (EPPR), and Protection of the Arctic Marine Environment (PAME) working groups are of special importance. Best practices are disseminated within expert groups on SAR, marine environmental response, and radiological and nuclear emergencies response. The Arctic Coast Guard Forum provides a useful arena for finding practical solutions involving both civilian and military capacities.

Improved resource data bases may provide administrative capabilities between the countries helping to find the best resources and solutions available as well as constructing fast response, cross-border teams.

Trust relations. The cooperation fora for the Arctic Council are flourishing. For the coastguards, a formal agreement creating joint meeting and training platforms was established in 2015 with the founding of the Arctic Coast Guard Forum, including the Norwegian Coast Guard and the Border Guard Service of the Russian Federation Federal Security Service. The cooperation within the Arctic Coast Guard Forum provides a meeting place for key decision-makers at all levels.

Especially valuable for building trust is the cooperation in training and exercises where "the grass roots" of international cooperation participate. The SAR Agreement stipulates an annual exercise between the two countries named Exercise Barents, including SAR and oil spill response. Barents Rescue exercises are held every two to three years and usually include several scenarios for a complex cross-border response. Such exercises help to test and improve standard operating procedures, learning from each other's experiences, and cultivating trust between agencies and countries. Arranging different types of exercises involving several countries will contribute to stronger international cooperation in emergency management (Nilsen, 2018).

In academia, close research cooperation has been established between the universities of Northern Norway and Northwest Russia. Several R&D projects have been conducted with a focus on maritime risk, preparedness, and emergency response. Exchange of knowledge and best practices as well as identifying gaps and limitations in the emergency response system is an important part of research cooperation. Several research projects, such as the project on maritime preparedness and international partnership (MARPART), have contributed to a better understanding of emergency preparedness and response. Universities in five Arctic countries and the managements of the maritime SAR coordination centers, both in Norway and Russia, have participated in this R&D activity.

This cooperation has been taken further into a permanent scientific and educational network on Arctic safety and security, where a broad range of universities is included. Through this network, students as well as researchers and professionals can meet and develop and also disseminate new knowledge, for example, through joint book projects (Andreassen and Borch, 2020). Cooperation between academia and professional SAR agencies will serve to improve the development of emergency management. Joint exercises, publications, and educational activities contribute to mutual competence development and trust.

Conclusions

This chapter focused on cooperation across borders in complex and resource-demanding fields. An important challenge in international relations is the question of reconciling the conflicts of interests, political tensions, uncertainty, and complexity related to the partnership in question. The cases demonstrated that the number of actors involved and the long response process call for a range of agencies to be involved and also adequate governance mechanisms. The cases described the significant impacts of international cooperation within emergency response. In the High North regions, distances are long and communities are small. There are and will continue to be limited local resources to meet crises and disasters. In major disasters, cooperation and joint operations across borders based on mutual understanding and trust are crucial.

December 2021

Ex-post reflections

As a result of Russia's invasion of Ukraine, Norway and other European countries and allies are reducing their contact with the Russian authorities. The Arctic Council has paused its work, the Nordic Council of Ministers has suspended all cooperation with Russia, and the BEAC meetings have been postponed until further notice. This will hamper the exchange of knowledge and joint policy development within emergency preparedness and response.

However, the Norwegian Ministry of Foreign Affairs claimed in news release on **05.03.2022** that cooperation essential to ensure Norway's safety and security would be maintained. Prime Minister Jonas Gahr Støre in his address to the *Storting* on Ukraine on **26.04.2022** stated: "We must ensure that we can maintain cooperation with Russia on search and rescue, emergency preparedness, and maritime safety in the vast sea areas in the north".

According to Tore Wangsfjord, acting Head of Division at the JRCC NN, "the operative cooperation about search and rescue has been exempt from the sanctions against Russia, and the interaction between the JRCC NN and the Maritime Rescue Coordination Centre in Murmansk runs as normal" (Edvardsen, 2022).

The chapter "International Cooperation within Emergency Response in the Arctic Sea Areas" has discussed the impacts of historical developments on the maritime emergency response operations that are relevant to consider in light of the current cooperation crisis.

June 07, 2022

References

Aamo, M.A., Helland, A.Ø., Medby, A. and Solvang, Ø. (2017). Helikopter med åtte passasjerer har styrtet i havet ved Svalbard, [Helicopter with eight passengers has crashed into the sea near Svalbard], Hovedredningssentralen, *Nordlys* (fremover.no).

Aftenposten (2017). Styrtet russisk helikopter funnet på 209 meters dyp utenfor Barentsburg, [A crashed Russian helicopter found at a depth of 209 meters outside Barentsburg] 29. okt. 2017, https://www.aftenposten.no/norge/i/8bvGx/styrtet-russisk-helikopter-funnet-paa-209-meters-dyp-utenfor-barentsbur.

Andreassen, N. (2016). Arctic energy development in Russia — How "sustainability" can fit?, *Energy Research & Social Science* 16, pp. 78–88.

Andreassen, N. and Borch, O.J. (eds.) (2020). *Crisis and Emergency Management in the Arctic. Crisis and Emergency Management in the Arctic. Navigating Complex Environments*, Routledge.

Andreassen, N., Borch, O.J. and Schmied, J. (eds.) (2018). *Maritime Emergency Preparedness Resources in the Arctic: Capacity Challenges and the Benefits of Cross-Border Cooperation between Norway, Russia, Iceland and Greenland*, MARPART Project Report 4, R&D Report 33, Nord University, http://hdl.handle.net/11250/2569868

Andreassen, N., Borch, O.J. and Sydnes, A.K. (2020). Information sharing and emergency response coordination, *Safety Science*, https://doi.org/10.1016/j.ssci.2020.104895.

Barents Observer (2010). Russian award to Norwegian rescue workers, *Barents Observer*, September 22, 2010, http://barentsobserver.com/en/sections/topics/russian-award-norwegian-rescue-workers.

Borch, O.J. (1994). The process of relational contracting, in Shrivastava, P., Dutton, J. and Huff, A. (eds.) *Advances in Strategic Management*, Greenwich, CT: JAI Press Inc.

Borch, O.J., Andreassen, N., Marchenko, N., Ingimundarson, V., Gunnarsdóttir, H., Jakobsen, U., Kern, B., Iudin, I., Petrov, S., Markov, S. and Kuznetsova, S. (2016). *Maritime Activity and Risk Patterns in the High North*. MARPART Project Report 2. Bodø: Nord Universitet 2016 (ISBN 978-82-7456-757-3), 124 p. Nord universitet/R&D report (4), http://hdl.handle.net/11250/2432922.

Borch, O.J., Andreassen, N., Sydnes, M. and Sydnes, A.K. (2018). Maritime preparedness: Institutional framework in Norway, in Elgsaas, I. and Offerdal, K. (eds.) *Maritime Preparedness Systems in the Arctic – Institutional Arrangements and Potential for Collaboration*, MARPART Project Report 3, FoU-rapport 27, Nord University, http://hdl.handle.net/11250/2501164.

Bye, H-G. (2020). Forskningssamarbeidet mellom Norge og Russland styrkes i nord [Research cooperation between Norway and Russia is strengthened in the north], *High North News*, 23.03.2020, https://www.highnorthnews.com/nb/forskningssamarbeidet-mellom-norge-og-russland-styrkes-i-nord.

Chen, R., Sharman, R., Rao, H.R. and Upadhyaya, S.J. (2008). Coordination in emergency response management, *Commun* ACM 51(5), pp. 66–73.

Danilov, P.B. (2021). New Russian restrictions on international educational cooperation, *High North News*, 26.03.21, https://www.highnorthnews.com/en/new-russian-restrictions-international-educational-cooperation.

DSB – Norwegian Directorate for Civil Protection (2014). *Guideline to Host Nation Support in Norway. A Generic Guideline for Norwegian Sectors*. Skien, Norway: Norwegian Directorate for Civil Protection (DSB).

Dyer, J.H. and Singh, H. (1998). The relational view: Cooperative strategy and sources of interorganizational competitive advantage, *Academy of Management Review* 23(4), pp. 660–679.

Edvardsen, A. (2022). Norwegian-Russian preparedness exercise in the Barents Sea cancelled, *High North News*, 31.05.22.

Elgsaas, I. and Offerdal, K. (2018). *Maritime Preparedness in the Arctic: Institutional Arrangements and Potential for Collaboration*, R&D Report No. 27, Bodo, Norway: Nord university.

Governor of Svalbard (2017). *Helikopterulykken på Svalbard 26.10.17*, [The helicopter accident on Svalbard] PPT, Hovedredningssentral webpage uploads, retrieved at https://www.hovedredningssentralen.no/wp-content/uploads/2018/11/Helikopterulykken-p%C3%A5-Svalbard.pdf.

International Maritime Organization and International Civil Aviation Organization (2016). *IAMSAR Manual, Volume I Organization and Management*. Tenth edition 2016. IMO: London and ICAO: Quebec.

Kapucu, N. (2005). Interorganizational coordination in dynamic context: Networks in emergency response management, *Connections* 26(2), pp. 33–48.

Klanderud, A., Eide, C., Salater, R-A. and Eliassen, H.H. (2017). Russisk helikopter styrtet på Svalbard [Russian helicopter crashes in Svalbard], *TV2 nyheter*, https://www.tv2.no/a/9451398/.

Marchenko, N., Andreassen, N., Borch, O.J., Kuznetsova, S., Ingimundarson, V. and Jakobsen, U. (2018). Arctic Shipping and risks: Emergency categories and response capacities, *International Journal on Marine Navigation and Safety of Sea Transportation* 12(1), pp. 107–114.

Moynihan, D.P. (2009). The network governance of crisis response: Case studies of incident command systems, *Journal of Public Administration Research and Theory* 19(4), pp. 895–915.

Myhrer, T.-G. (2020). The legal possibilities for policing the Arctic Ocean, in Andreassen, N. and Borch, O.J. (eds.) *Crisis and Emergency Management in the Arctic. Navigating Complex Environments*. Routledge.

Nilsen, T. (2018). Icy waters: Norwegian - Russian cooperation key to save lives as wave of new cruise ships sail north, *Barents Observer*, https://thebarentsobserver.com/ru/node/3670.

Pedersen, T. (2017). The politics and presence: The Longyearbyen Dilemma, *Arctic Review of Law and Politics* 8, pp. 95–108.

Poppo, L. and Zenger, T. (2002). Do formal contracts and relational governance function as substitutes or complements? *Strategic Management Journal* 23(8), pp. 707–725.

Staalesen, A. (2017). Northern Fleet takes on key role in Exercise Barents, *The Barents Observer*, https://thebarentsobserver.com/en/security/2017/06/northern-fleet-takes-key-role-exercise-barents.

Sydnes, A.K., Sydnes, M. and Antonsen, Y. (2017). International cooperation on search and rescue in the arctic, *Arctic Review Law Politics* 8, https://doi.org/10.23865/arctic.v8.705. ISSN 1891–6252. pp. 109–136.

Udin, S., Petrov, S., Markov, S. and Kuznetsova, S. (2016). Maritime activity risk patterns and types of unwanted incidents. The Russian Sea areas in the North-West Arctic, in: Borch, O.J. et al. (eds.) *Maritime Activity and Risk Patterns in the High North*. MARPART Project Report 2. Bodø: Nord Universitet 2016 (ISBN 978–82–7456–757-3), 124 p. Nord universitet/R&D Report (4).

Uzzi, B. (1997). Social structure and competition in interfirm networks: The paradox of embeddedness, *Administrative Science Quarterly* 42(1), pp. 35–67.

Part V

Integrating Chapter

16 International Cooperation for Global Development

What Can We Learn from The Arctic?

Anatoli Bourmistrov, Frode Mellemvik and Andrey Mineev

Introduction

The Arctic has always been a very special geographical area in the world. It is very often associated with inhospitable nature and difficult operating conditions; fragile, small, and dispersed local communities; and abundant natural resources much needed for the economic development of the world. The Arctic has attracted much attention in the last couple of decades, especially due to all its natural resources, concerns about climate change, and as a potential theater for various types of conflicts. Some see the Arctic as a treasure chest – others as an area that needs to be preserved.

In this chapter, we would like to shift the focus and examine the Arctic as a special place where we can learn about an interesting cooperation model that can be a template for cooperation aiming at sustainable global development. This can be of two important reasons. First, this is because of the unique history of Arctic cooperation. The Arctic actors, despite different views and limited resources, were able to develop and, with time, refine what appears to be a robust model for international cooperation. Thus, if such a model exists, it deserves separate attention – an analysis that can conceptually formulate this model – and then use it to explain successful or less successful areas of cooperation. Such a model can be a potential template for the rest of the world.

Second, achieving the sustainable development of the Arctic is not merely an ultimate goal that is a natural part of the global sustainable development agenda (e.g., guided by UN Sustainable Development Goals [SDGs]). Essentially, sustainable development as a process is quite indispensable to the very existence and future development of Arctic communities. Otherwise, the Arctic might be abandoned by humans. The strategic importance of the Arctic resources as well as the vulnerability of this region in the face of climate change and pressure from the extractive industries make it an eminently suitable spot to study international cooperation for sustainable development. Here, international cooperation can not only be realized, but also facilitate sustainable development in the Arctic and also elsewhere.

This chapter provides an analytical summary of all the chapters in this book. The chapters, each dealing with some aspect of international cooperation, were

DOI: 10.4324/9781003246015-21

reviewed in terms of identifying the common features of a potential Arctic cooperation model. The findings were then contrasted to establish theoretical approaches to cooperation and a potentially new way of understanding cooperation was identified.

International cooperation as a theoretical phenomenon – different perspectives

In the scientific literature, cooperation as a phenomenon is examined in light of different theoretical perspectives belonging to different disciplines. Let us consider some examples of how cooperation is understood as a phenomenon.

Anthropology looks at cooperation as a form of some distinctive human behavior in certain similar situations. For instance, it seems that in many cultures, cooperation is seen as *a moral obligation* based on certain universal moral rules, for example, family values, group loyalty, reciprocity, bravery, respect, fairness, and possession (e.g., Curry et al., 2019). By contrast, in economics (e.g., game theory), cooperation is considered as a process to gain (economic) advantages via the so-called "cooperative games". Based on assumptions of human rationality, the game theory of cooperation addresses what incentives actors have to act jointly in a "game" to gain something valuable, such as complementarities, reduction of risks, and economies of scale or/and scope (e.g., Peña and de Arroyabe, 2002). To some extent, something similar is also discernible in international relations research. The focus there is on international actors, their objectives, and their preferences to study their strategic interactions in a continuously changing environment in order to understand how those interactions play out for reciprocity, reputation, and trustworthiness (Dai et al., 2010). In this sense, cooperation can be also understood as a game – a continuous state of interaction among a limited number of state agents on the global arena who adjust their behavior because others do or intend to do something else. It seems that this "game-like" cooperation often appears as a zero-sum game – the benefit of one party results in losses for the other. Even in cases of "successful" international cooperation, parties manage to cooperate only to maintaining stability or the status quo.

Organizational studies take a slightly different approach. Cooperation between companies and organizations, for example, strategic alliances and joint ventures, is usually studied from the perspective of the Resource-Based View (RBV). Cooperation is a search for complementarities (e.g., in resources), compatibility (e.g., in routines/traditions), and different copying mechanisms related to insufficient levels of those complementarities and compatibilities achieved (e.g., Beerkens and Derwende, 2007). Specifically, a network model of organizational cooperation moves our attention away from focal companies and their resourced-based positions to the complex web of relationships in which the companies are embedded (Håkansson and Snehota, 1989). In that way, the relational context of an organization becomes more important than the organization itself, as the organization's relationships (continuous interactions with other parties) determine its identity. Cooperation in this sense is about a search for meaning and role by relating one's own activities and resources to those of the other parties in the cooperative context.

The theoretical approaches described above do not consider international cooperation as a process of contributing to global development, such as achieving the UN SDGs. Studying problems like that would require a set of assumptions other than those on which the conventional theories of cooperation are built. In particular, we have to move away from the actor-centric approach to cooperation and understanding it as a zero-sum game. However, some scholars attempt to create alternative models to capture the links between international cooperation and global development. For instance, Paulo (2014) describes a conceptual framework based on economic theory distinctions between global and domestic public goods and illustrates the role of international cooperation and its interdependence with domestic action. Others consider international cooperation as contributing to global development by looking at different aspects of the quality of cooperation processes, that is, procedural quality and outcome quality of cooperation processes (Bernauer et al., 2020).

To analyze how Arctic international cooperation can contribute to global development, we can extract some common elements found in all theoretical approaches. Who are the key actors and what are their interests in the Arctic and their reasons for international cooperation in the Artic? What is the nature of processes of international cooperation, for example, how does cooperation take place and develop over time? What is the nature of interactions between actors? What factors hamper international cooperation? Why is cooperation difficult? And finally, what are the outcomes of international cooperation processes for global development, exemplified by quantifying some of the UN SDGs?

Examples of Arctic cooperation in this book: actors, mechanisms, and outcomes

Besides this integrating overview, this book consists of 15 thematic chapters that describe developments in many areas of Arctic cooperation. Some of those developments seem to be quite successful, judged by the outcomes achieved (Table 16.1). However, some areas have not been so successful and have achieved less (Table 16.2). How can those successful and less successful areas of Arctic cooperation be characterized? What are the differences and how can such differences be explained?

Table 16.1 shows several features apparently common to quite many areas of cooperation. First, cooperation in many areas is not driven by any obvious economic goals, but rather by aspirations to achieve some kind of a common good which is an important value in itself (e.g., better knowledge, saving lives, challenging the existence of borders). Second, economic reasons do play a role but very differently, for instance, from typical economy-driven strategic alliances – cooperative projects are needed because there is always lack of resources to carry

Table 16.1 Characteristics of Arctic cooperation with a clear positive contribution to SDGs

Area	*Who?* *Major Actors*	*What?* *Interests Related to the Arctic*	*Why?* *Reasons for Cooperation*	*How?* *Cooperation Mechanism*	*Why not?* *Barriers*	*So What? Most Important SDGs*
Polar ship design	IMO/ classification societies/ shipowners	Standard/rules for efficient polar ship design	Safer/more robust vessels for polar conditions	IMO sets general code translated by classification societies into detailed rules	Shipowners sometimes reluctant to use more expensive polar vessels	Use of oceans (SDG 14)
Maritime search and rescue (SAR)	Rescue agencies/ military/police/ hospitals	Rapid response under crises	Coordination of limited resources to achieve best possible outcomes in maritime SAR	Combination of formal agreements and coordination functions with maintaining trust-enhancing arenas	Different institutional frameworks and border-crossing permissions	Safe, resilient, and sustainable cities (SDG 11); Use of oceans (SDG 14); Life on land (SDG15)
Cross-border culture cooperation	Cultural workers/local communities	Challenge the existence of physical/mental/ cultural borders of Arctic people	Culture belongs to no one and should be shared; positive curiosity	Cultural democracy supported by governmental agreements and formal institutions	Language, physical distance, knowledge	Lifelong learning (SDG 4); Gender equality (SDG 5); Improved partnerships (SDG 17)
Climate change cooperation	Researchers/ politicians	Producing relevant knowledge on Arctic climate change processes and for use in adaptation strategies	Joint scientific assessment of climate change and its consequences; cooperation as a necessity due to lack of resources	Partnerships between individual researchers and institutes leading to agreements between Arctic states	Unclear role of businesses in promoting the highest standards of sustainability in the Arctic	Potentially, all SDGs, especially improved partnerships (SDG 17)

Internationalization of higher education	Professors/administrative staff at universities	Cooperation as a social mission for academia in the Arctic	Few formal limitations; collegiality as a principle	Bottom-up processes: good projects lead to more projects	Location and attractiveness of the Arctic HEIs	Better quality of education (SDG 4) and improved partnerships (SDG 17)
Arctic Council cooperation	Arctic states/permanent participants/observers	Knowledge-sharing and specific problem-solving and not high politics in the Arctic	To address huge common challenges in the Arctic together	Joint work of experts leads to consensus to create legal norms	Different visions by actors for the economic future of the Arctic	Potentially all SDGs, especially improved partnerships (SDG 17)
Arctic security	Arctic/non-Arctic states	Security policies	Mutual dependence on creating favorable investments and economic development	Explicit intention to solve potential disputes through diplomacy	Importance of security and defense differs between regions	Potentially all SDGs, especially improved partnerships (SDG 17)
Environmental monitoring and assessment	Arctic/non-Arctic states; NGOs, international organizations, companies	Managing Arctic ecosystems using internationally accepted approaches	Research-driven management process of common Arctic resources	Working groups in the Arctic Council	No jointly approved Russian-Norwegian management plan for the Barents Sea	SDG 14 Life blow water

out projects alone. Big projects in the Arctic need joint mobilization of resources. This is also because institutions in the Arctic compete for resources with more powerful centers outside the Arctic. Third, many actors seem to have recognized the Arctic issues and, to some degree, they have a common understanding of the importance of resolving any disputes by consensus building.

Another common feature in many fields of cooperation is bottom-up rather than top-down processes from initiation to implementation. This is facilitated by a paucity of limitations on initiating joint initiatives and projects and the possibility to form and use informal discussion arenas that can lead to consensus. However, this is no mere discussion club. In many cases, such informal arenas serve to elucidate the subject matter, leading frequently to some sort of formalization via the establishment of relevant sets of rules, norms, or even formal agreements. Thus, the role of the formal institutions with corresponding formal mechanisms is equally important for success, for example, because these formalizations legitimize the work taking place on less formal arenas. Therefore, we propose exploring a new approach to understanding Arctic cooperation – a "self-reinforcing" cooperation mechanism (see Figure 16.1).

Arctic colocation and the mutual dependence of actors, their shared aspirations to develop the Arctic from perspectives of non-transactional values and assets (e.g., climate, wilderness, biodiversity, culture) fueled by a lack of access to financial resources leads to the formation of bottom-up initiatives. These initiatives result in the establishment of informal discussion arenas facilitated by the power of persuasion and not by the power of regulation. Informal discussion arenas, in turn, create joint projects which, should they lead to positive results, can be formalized and supported by governments (e.g., by moving from soft to

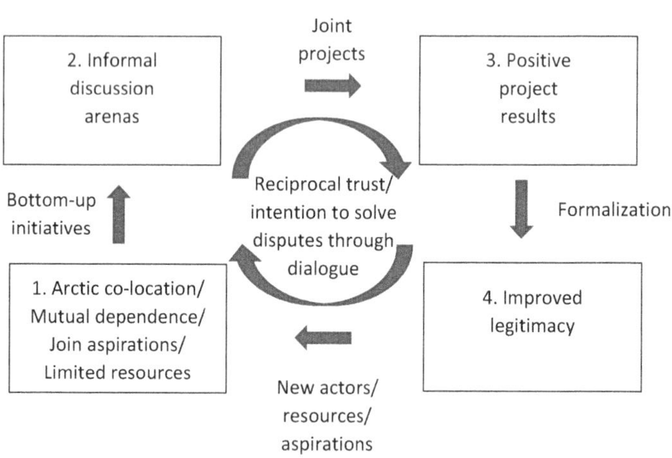

Figure 16.1. Arctic "self-reinforcing" cooperation mechanism.
Source: Authors.

binding frameworks), improving legitimacy of cooperation. Improved legitimacy can attract new actors, potentially more financial resources, and new aspirations. This circle is perpetuated by reciprocal trust and the intention to resolve any conflicts or disputes through dialog. Trust alleviates suspicion and tensions between actors. As history has shown, Arctic actors can cooperate on vital major issues despite fundamental political, economic, social, and technological differences.

The elements of the process described in Figure 16.1 have to some extent been studied before. For instance, the literature indicates that successful cooperation among many actors needs a first mover to lead to a "catalytic cooperation" (Hale, 2020). In this sense, local Arctic actors and the existence of informal forums can be seen as such important catalysators. Similarly, it is well known that reciprocal trust in relationships can be an important part of self-reinforcing processes in social systems (Korsgaard, 2018; Möllering and Sydow, 2019). Moreover, self-reinforcing processes have been shown to contribute to successful innovation ecosystems (do Vale, 2021; Wang, 2021). Likewise in political science, self-reinforcing feedback mechanisms have been shown to be key during policy implementation as they can affect the durability of those policies (Millar et al., 2021; Patashnik and Weaver, 2021). A self-reinforcing cooperation mechanism can explain the long-term economic and organizational successes of industrial parks (Lefdal and Eriksen, 2017); the success of all the companies in the industrial park was due to a shared geographical location and aspiration to achieve circular economics combined with effective logistics, legitimacy, and reciprocity. What is probably new in this research is that self-reinforcing cooperation processes can explain much of the successful cooperation in such a huge geographical region as the Arctic.

One way to test the proposition that "self-reinforcing" cooperation can explain successful cooperation results is to apply the model to the areas of Arctic development that were probably less successful (see Table 16.2). Table 16.2 shows that areas related to Arctic shipping via the Northern Sea Route (NSR), Norwegian-Russian gas cooperation, the creation of Arctic innovation ecosystems, and the development of smart cities in the Arctic were much less successful, also in terms of achieving SDGs. Could one reason be that those areas have not been subjected to self-reinforcing processes like the areas reported in Table 16.1?

If we analyze Table 16.2, we can see that international shipping and gas cooperation focused mostly on the development of technology and economic gains based on the reciprocities of actors trying to achieve economic synergies (e.g., somewhat reminiscent of a strategic alliance). The focus was also very much on transactional (economic) values with the purpose of utilizing Arctic resources (e.g., transportation routes, gas reserves) for commercial development. In this sense, cooperation based solely on economic synergies became a fragile structure because it became subject to its own (temporal) conditions on which the cooperation was initially built. When those conditions changed, the cooperation stalled. When it comes to Arctic innovation systems and smart city developments, the major challenge can be that common aspirations, bottom-up processes, and the creation of arenas for dialog are not yet in place.

Table 16.2 Characteristics of Arctic cooperation contributing less to SDGs

Area	*Who? Major Actors*	*What? Interests Related to the Arctic*	*Why? Reasons for Cooperation*	*How? Cooperation Mechanism*	*Why Not? Barriers*	*So What? Most Important SDGs*
Areas with Challenges for Cooperation and Outcomes						
International shipping and NSR	Russian and international shipping companies, Russian state	Saving time and costs	Sharing costs of expensive infrastructure; economy of scale through increased number of users	Inward-looking and less cooperative Russian policies for developing NSR; increasing monopolization	Perceived risks: environmental, punctuality, predictability	
Oil and gas (O&G)	O&G companies, states	Arctic O&G technology; access to Arctic O&G reserves	Beneficial exchange of know-how making the Arctic a conflict-free area	Building alliances subject to initial conditions	Changing political priorities and geopolitical climate interfere with and may even destroy business cooperation	Climate actions (SDG 13); improved health (SDG 3); access to clean energy (SDG 7)
Arctic innovations ecosystems	Researchers, commercial actors, companies, agencies, communities	Create technology-enabled innovative northern economies	Develop own circumpolar market and innovation potential	Failure to create an ecosystem for collective problem-solving innovations	Low level of commercial integration/small size of markets/ lack of capital	Sustainable economic growth (SDG 8); sustainable industrialization (SDG 9)
Smart City (SC) development dialogs	Local actors/ actors outside the Arctic regions	Cooperation in SC formation requires dialog for smart city development	Create SC solutions relevant to local stakeholders	Monologs led by actors outside the regions	Insufficient attention to local engagement mechanisms	Safe, resilient, and sustainable cities (SDG 11)

Arctic paradoxes

The emergence of the self-reinforcing cooperation model may also relate to the fact that the Arctic is a place of many interesting paradoxes. Having encountered the paradoxes they face, the actors initiate cooperation to address them. And the cooperation seems frequently to evolve in a self-reinforcing way. Below, we present six paradoxes identified from some of the chapters in this book:

- *The Arctic governance paradox*: Development in the Arctic is based on decisions regarding opportunities perceived by actors outside the Arctic and without considering the experience and aspirations of local residents.
- *The Arctic resource/climate change paradox*: Climate change improves access to natural resources, the original cause of warming of the Arctic.
- *The Arctic militarization paradox*: The more the Arctic is militarized with potential for loss of life in military conflicts, the more military capacity is becoming available for use in search and rescue operations through international cooperation in time of peace.
- *The Arctic NSR development paradox*: To develop NSR, Russia depends on the international markets that it cannot directly influence. But the attractivity of the NSR for these markets depends on that Russia succeed in developing policies that are legitime in these markets.
- *The Arctic security paradox*: Arctic states applaud and manifest the cooperative spirit of the Arctic while investing in their Arctic defense posture. A positive security dialog (hindering the spillover of tensions from other parts of the world) seems to need backing from military potential.
- *The Arctic technology innovation paradox*: In order to prosper, the Arctic needs technological innovations adapted to its own context, but it ends up adapting technology from outside the region that is not adapted to its context.

We are sure that more paradoxes can be found in the Arctic. For example, not addressed in this book but studied in detail by the project Business Index North, there is a socioeconomic paradox. It appears that most regions of Arctic Europe (including North Scandinavia and North West Russia) have greater economic growth than their national average. At the same time, the employment trend is different. Northern Scandinavia has shown a very slight growth in employment over the last ten years, far below the national averages. As for the Russian side, most of the regions experienced a decline in number of jobs (while the economic growth has been considerable). The economy of the north, mainly driven by natural resources and production industries physically located in the region, apparently needs only few people. We believe that this trend has to change.

Concluding reflections: how can the Arctic contribute to global sustainable development?

The main conclusion from this chapter is that the Arctic has probably developed a successful model of "self-reinforcing" cooperation. This needs further study, especially whether and how such a model could be adopted elsewhere in the world. Particularly, this can be an important contribution to achieving SDG 17 "Partnerships for the goals". Based on this chapter, we propose the following hypothesis: self-reinforcing cooperation is driven by paradoxes and associated challenges for the actors. Once big external economic motivations come into play (e.g., resource extractive industries), the cooperation pattern ceases to be self-reinforcing.

As we have seen, the self-reinforcing model has succeeded in many areas, notably in areas like climate change, cultural cooperation, and natural sciences. However, it has not yet been proven successful in social and economic dimensions, like community development, solving demographic problems, or increasing commercialization and innovations.[1] The reasons are likely that this cooperation model was not (yet) applied in these areas and, if ever applied, could have good potential. Even in those areas, some progress is discernible. For example, there are some indications of local involvement and bottom-up initiatives in the development of the regional supply industry for the oil and gas sector in the Russian North (e.g., Mineev and Bourmistrov, 2015). Another interesting example is the ecosystem of innovative companies within health care and IT in the city of Oulu in Northern Finland. More examples can be found throughout the Arctic.

For the rest of the world, this self-reinforcing model of cooperation can also potentially function as an aspiration to the adoption of collaborative approaches. As we have learned from the Arctic, promoting local-level umbrella cooperation mechanisms can have considerable effects on a larger scale, contributing to achieving the important balance between concerns of protection (e.g., of the natural environment, culture, values) and production (e.g., for the sake of local and global economic development). This can also facilitate a crucial problem-based approach – aspirations that identify the needs of society that drive science, and then leading governments and business to respond and accumulate knowledge. In this sense, the Arctic can become the region with the highest standards of sustainability, arousing interest from the rest of the world. The Arctic can become a "workshop" for sustainable industries, and in this sense, the Arctic seems to be well positioned to provide worldwide solutions.

The question of the transferability of the self-reinforcing model remains. Can such a self-reinforcing model be simply replicated in other parts of the world as a best practice? Or is it necessary to be exposed to a series of paradoxes as in the Arctic to be able to cooperate naturally in a self-reinforcing way? This issue deserves further attention. A more profound understanding of the Arctic paradoxes is needed and requires more research. If similar paradoxes can be found elsewhere, the application of the model proposed is indeed justified. When it comes to the further development of the Arctic, the results of paradox-oriented research have to be properly communicated to various actors. As Arctic social scientists, we can equip the actors with the model and raise their awareness of the paradoxes they are exposed to. But it should be their own decision whether to commit to cooperation and act upon.

December 2021

Ex-post reflections

The start of Russia's military offensive in Ukraine on February 24, 2022, divides history into before and after. Today, we know that the world will never be the same, but we do not know what is coming. The aim of this chapter, written before the outbreak of the Ukraine war, was to summarize the experience of Arctic cooperation. Therefore, it was based on the analysis of historical development. A notion of an Arctic "self-reinforcing" cooperation model introduced in the chapter was an aspiration to be useful to further international cooperation for sustainable development globally.

At present, the fact of war and its geopolitical character brings tensions to the Arctic. This dramatically challenges the very fundamental component in our model – "the reciprocal trust and intention to resolve disputes through dialogue" – the most important driving force that we think was unique to the Arctic. Although we hope and expect that the Arctic still will be a region of peace, future cooperation is likely to be different than what our model implies.

This does not, however, mean that the model of "self-reinforcing" international cooperation is inapplicable, as we think it can still be applied in some international contexts with strong collaborative fundaments. As we stated in the chapter, such fundamentals are associated with shared challenges and paradoxes. The model may even, in one way or another, be applicable to future cooperation in the Arctic, much depending on whether and eventually how reciprocal trust can be reestablished.

June 9, 2022

Note

1 Challenges for sustainable socioeconomic development in the European part of the Arctic are highlighted in detail in Business Index North reports available at www.businessindexnorth.com

References

Beerkens, E., and Derwende, M. (2007). The paradox in international cooperation: Institutionally embedded universities in a global environment. *Higher Education*, 53(1), pp. 61–79.

Bernauer, T., Mohrenberg, S., and Koubi, V. (2020). Do citizens evaluate international cooperation based on information about procedural and outcome quality? *The Review of International Organizations*, 15(2), pp. 505–529.

Curry, O. S., Mullins, D. A., and Whitehouse, H. (2019). Is it good to cooperate? Testing the theory of morality-as-cooperation in 60 societies. *Current Anthropology*, 60(1), pp. 47–69.

Dai, X., Snidal, D., and Sampson, M. (2010). International cooperation theory and international institutions. In *Oxford Research Encyclopedia of International Studies*.

do Vale, H. F. (2021). How to become innovative: The trajectory of social assistance programs in Brazil. *Latin American Policy*, 12(1), 97–115.

Håkansson, H., and Snehota, I. (1989). No business is an island: The network concept of business strategy. *Scandinavian Journal of Management*, 5(3), pp. 187–200.

Hale, T. (2020). Catalytic cooperation. *Global Environmental Politics*, *20*(4), pp. 73–98.

Korsgaard, M. A. (2018). Reciprocal trust: A self-reinforcing dynamic process. In *The Routledge companion to trust* (pp. 14–28). Routledge.

Lefdal, M. F and Eriksen, R. H. (2017). Balance between trust and control: A case study of management in long-term interorganizational cooperation. Master thesis, Nord University Business School (In Norwegian).

Millar, H., Bourgeois, E., Bernstein, S., and Hoffmann, M. (2021). Self-reinforcing and self-undermining feedback in subnational climate policy implementation. *Environmental Politics*, *30*(5), pp. 791–810.

Mineev, A., and Bourmistrov, A. (2015). Social capital at work: A case of adapting a Norwegian cooperation model in Russia. *Journal of East-West Business*, *21*(2), pp. 129–155, DOI: 10.1080/10669868.2015.1005326

Möllering, G., and Sydow, J. (2019). Trust trap? Self-reinforcing processes in the constitution of inter-organizational trust. In *Trust in contemporary society* (pp. 141–160). Brill.

Patashnik, E. M., and Weaver, R. K. (2021). Policy analysis and political sustainability. *Policy Studies Journal*, *49*(4), pp. 1110–1134.

Paulo, S. (2014). International cooperation and development: A conceptual overview. *German Development Institute/Deutsches Institut für Entwicklungspolitik Discussion Paper, 13.*

Peña, N. A., and de Arroyabe, J. C. F. (2002). Cooperation in game theory. In *Business cooperation* (pp. 57–73). Palgrave Macmillan, London.

Wang, X. (2021). Strategy on development of innovative ecosystem in customer values. *Business Process Management Journal*, *27*(7), pp. 2110–2122, DOI: https://doi.org/10.1108/BPMJ-12-2020-0584

Index

Note: **Bold** page numbers refer to tables, *italic* page numbers refer to figures and page numbers followed by "n" refer to end notes.